Classical Dynamics
and Its Quantum Analogues

David Park

Classical Dynamics and Its Quantum Analogues

Second Enlarged and Updated Edition

With 101 Figures

Springer-Verlag Berlin Heidelberg New York
London Paris Tokyo Hong Kong

Professor Dr. David Park

Department of Physics, Williams College, Williamstown, MA 01267, USA

The first edition appeared in the series
Lecture Notes in Physics, Vol. 110

ISBN 3-540-51398-1 2. Auflage   Springer-Verlag Berlin Heidelberg New York
ISBN 0-387-51398-1 2nd edition Springer-Verlag New York Berlin Heidelberg

ISBN 3-540-09565-9 1. Auflage Springer-Verlag Berlin Heidelberg New York
ISBN 0-387-09565-9 1st edition Springer-Verlag New York Berlin Heidelberg

Library of Congress Cataloging-in-Publication Data. Park, David Allen, 1919– . Classical dynamics and its quantum analogues / David Park. – 2nd enl. and updated ed. p. cm. Includes bibliographical references (p.) ISBN 0-387-51398-1 (U.S.) 1. Dynamics. 2. Mechanics. 3. Quantum theory. I. Title. QC133.P37   1989   531'.11– dc20   89-21889

Coverdesign: W. Eisenschink, D-6805 Heddesheim

2157/3150(3011)-543210 – Printed on acid-free paper

Preface

The short Heroic Age of physics that started in 1925 was one of the rare occasions when a deep consideration of the question: What does physics really say? was necessary in carrying out numerical calculations. In many parts of microphysics the calculations have now become relatively straightforward if not easy, but most physicists seem to agree that some questions of principle remain to be resolved, even if they do not think it is very important to do so. This situation has affected the way people think and write about quantum mechanics, a gingerly approach to fundamentals and a tendency to emphasize what fifty years ago was new in the new theory at the expense of continuity with what came before it. Nowadays those who look into the subject are more likely to be struck by unexpected similarities between quantum and classical mechanics than by dramatic contrasts they had been led to expect.

It is often said that the hardest part of understanding quantum mechanics is to understand that there is nothing to understand; all the same, to think quantum-mechanically it helps to have firm mental connections with classical physics and to know exactly what these connections do and do not imply. This book originated more than a decade ago as informal lecture notes [0][1], prepared for use in a course taught from time to time to advanced undergraduates at Williams College. The object was to present classical mechanics to students who know a little about classical and quantum mechanics but not too much of either, to teach them more of each, and to emphasize the continuity between them. Most undergraduates I know are more familiar with the hydrogen atom, for example, than with its planetary counterpart, and since I believe that historically and conceptually classical dynamics is the backbone of physics, I have treated planetary motion in some detail. Since the relation between wave optics and ray optics is much the same as that between wave mechanics and particle orbits, it has been possible to develop some ideas of optics at the same time.

In revising the notes I have added considerably to the sections on quantum mechanics, notably the sections on coherent states and field quantization, and in Chap. 10 I show how these combine to give a simple semiquantitative explanation of the Mössbauer effect. I have also written a number of new problems for this edition. A few, which I think are an order of magnitude more difficult than most, are marked with a *.

Classical mechanics, quantum mechanics, and optics – the scope is so vast that there has to be a principle of selection. I have mentioned those parts of elementary quantum mechanics that are of the most fundamental interest and utility,

---

[1] The bibliography is at the back of the book.

and explained those parts of classical mechanics that relate to them and illuminate them. The subjects in optics are pretty well determined by those choices. There are no insects of the kind that walk on a weightless spinning rod, no nonholonomic constraints, no nonconservative forces even. But the standard methods are here – Lagrangian, Hamiltonian and Hamilton-Jacobi together with their relations to quantum mechanics, and considerable discussion of methods of exact and approximate solution and the role of invariance. Believing that one learns a theory by seeing how it works, I have included several calculations in general relativity, including the deflection and retardation of light by the sun and the precession of planetary perihelia. Since the relativistic precession must in practice be disentangled from the precession produced by interaction with other planets, I have used that as an example of classical perturbation theory. Guided by the quantum analogues, most of the classical mechanics deals with one or two particles, but Chaps. 4, 9, and 10 refer to the motions of systems of particles, including solid bodies, and to the mechanics of continua, including fields. The quantum mechanics of rigid bodies is treated in analogy with the classical theory and contrasts are noted.

Inevitably, I have discussed parts of physics that appeal to me and made excuses for omitting parts that I find boring. Thus the selection is finally not a matter of logic so much as of personal taste. I trust there will be a few who share it.

The discussion of classical mechanics is fairly self-contained. Parts of elementary quantum mechanics not explained here are available in many texts. To avoid appearance of self-promotion, I have not usually given references to my own book [1], but all are to be found there.

This is intended to be a book about physics, more specifically, mathematical physics. I take it that the aim of mathematical physics is to calculate numerical answers, and therefore I have avoided the mathematical language of profound and abstract treatises like that of *Vladimir Arnold* [2]. Clarity is thereby sacrificed in the interest of techniques that can be generalized to serve other purposes.

In recent years universities in the United States have been fermenting over the question, What is the cultural tradition that a university should hold as its central concern and responsibility? One side says the current version of this tradition is simply the intellectual baggage that European intellectuals have accumulated over the centuries, and that though it may contain much that is of value it gives a one-sided and discriminatory picture of the world culture that surrounds us. Another side maintains that the hallowed classics of Western civilization are perhaps what anchor us most firmly to ideals of value and beauty that are being lost and might otherwise disappear. I catch echoes of this fundamental debate in the question, Why learn classical dynamics if it isn't right? I can only answer with the ultimate defense of the classicist after all the fine words about tradition have been spoken: Classics are classics because they are fun, because they give delight as they teach us about the world and about ourselves. Nothing remains a classic very long if people take no pleasure in it. I find classical dynamics full of surprises and clever thoughts, and of course that is the only reason I have written one more book about it.

Williamstown, January 1990   *David Park*

Contents

1. **Rays of Light** 1
   1.1   Waves, Rays, and Orbits 1
   1.2   Phase Velocity and Group Velocity 6
   1.3   Dynamics of a Wave Packet 10
   1.4   Fermat's Principle of Least Time 13
   1.5   Interlude on the Calculus of Variations 14
   1.6   Optics in a Gravitational Field 22

2. **Orbits of Particles** 29
   2.1   Ehrenfest's Theorems 29
   2.2   Oscillators and Pendulums 31
   2.3   Interlude on Elliptic Functions 35
   2.4   Driven Oscillators 38
   2.5   A Driven Anharmonic Oscillator 40
   2.6   Quantized Oscillators 46
   2.7   Coherent States 49

3. **Lagrangian Dynamics** 54
   3.1   Lagrange's Equations 54
   3.2   The Double Pendulum 61
   3.3   Planets and Atoms 64
   3.4   Orbital Oscillations and Stability 70
   3.5   Orbital Motion: Vectorial Integrals and Hyperbolic Orbits 72
   3.6   Other Forces 76
   3.7   Bohr Orbits and Quantum Mechanics: Degeneracy 77
   3.8   The Principle of Maupertuis and Its Practical Utility 79

4. **N-Particle Systems** 82
   4.1   Center-of-Mass Theorems 82
   4.2   Two-Particle Systems 89
   4.3   Vibrating Systems 90
   4.4   Coupled Oscillators 93
   4.5   The Virial Theorem 97
   4.6   Hydrodynamics 100

5. Hamiltonian Dynamics 103
   5.1 The Canonical Equations 103
   5.2 Magnetic Forces 107
   5.3 Canonical Transformations 114
   5.4 Infinitesimal Transformations 121
   5.5 Generating Finite Transformations from Infinitesimal Ones 129
   5.6 Deduction of New Integrals 132
   5.7 Commutators and Poisson Brackets 135
   5.8 Gauge Invariance 138

6. The Hamilton-Jacobi Theory 142
   6.1 The Hamilton-Jacobi Equation 142
   6.2 Step-by-Step Integration of the Hamilton-Jacobi Equation 148
   6.3 Interlude on Planetary Motion in General Relativity 151
   6.4 Jacobi's Generalization 157
   6.5 Orbits and Integrals 161
   6.6 "Chaos" 166
   6.7 Coordinate Systems 173
   6.8 Curvilinear Coordinates 175
   6.9 Interlude on Classical Optics 179

7. Action and Phase 184
   7.1 The Old Quantum Theory 184
   7.2 Hydrogen Atom in the Old Quantum Theory 190
   7.3 The Adiabatic Theorem 192
   7.4 Connections with Quantum Mechanics 197
   7.5 Heisenberg's Quantum Mechanics 200
   7.6 Matter Waves 207
   7.7 Schrödinger's "Derivation" 210
   7.8 Construction of a Wave Function 212
   7.9 Phase Shifts in Dynamics 220

8. Theory of Perturbations 225
   8.1 Secular and Periodic Perturbations 225
   8.2 Perturbations in Quantum Mechanics 228
   8.3 Adiabatic Perturbations 231
   8.4 Degenerate States 234
   8.5 Quantum Perturbation Theory for Positive-Energy States 238
   8.6 Action and Angle Variables 240
   8.7 Canonical Perturbation Theory 244
   8.8 Newtonian Precession 246
   Appendix.. 251

9. The Motion of a Rigid Body 253
   9.1 Angular Velocity and Momentum 253
   9.2 The Inertia Tensor 255

  9.3   Dynamics in a Rotating Coordinate System 257
  9.4   Euler's Equations 262
  9.5   The Precession of the Equinoxes 269
  9.6   Quantum Mechanics of a Rigid Body 274
  9.7   Spinors 279
  9.8   Particles with Spin 286

10. Continuous Systems 290
  10.1 Stretched Strings 290
  10.2 Four Modes of Description 291
  10.3 Example: A Plucked String 297
  10.4 Practical Use of Variation Principles 300
  10.5 More Than One Dimension 302
  10.6 Waves in Space 307
  10.7 The Matter Field 310
  10.8 Quantized Fields 313
  10.9 The Mössbauer Effect 317
  10.10 Classical and Quantum Descriptions of Nature 320

References 323

Notation 327

Subject Index 331

1. Rays of Light

1.1 Waves, Rays, and Orbits

Following the historical order of ideas is usually a good way of introducing people to physics, but the historical order is never the logical order. If it were, there would be no room for discovery, since each new development would merely follow deductively from the preceding ones. New discoveries usually involve radically new ideas not implicit in what went before, and deductive connections, when they can be made at all, must be made backwards. You cannot derive the formulas of relativistic mechanics from those of Newton. You can derive Newtonian formulas from relativistic ones by taking a low-velocity limit, but even this is not a derivation of Newtonian physics, since that physics takes place against a background of absolute space and time which Einstein abandoned. This example contains a lesson: We may be able to derive the equations of an earlier theory from those of a later one, but to recapture the theory itself requires imagination and scholarship.

In this book we are going to establish mathematical connections between classical and quantum mechanics, between ray optics and wave optics. The approach will be to consider classical mechanics as a limiting case of quantum mechanics, and ray optics as a limiting case of wave optics. The conceptual background will be discussed where necessary, but the reader ought to be fairly familiar with it anyhow. One result of this approach will be the revelation that classical and quantum theory are not so different conceptually as one thinks at first exposure. Quantum mechanics is probabilistic in content and classical mechanics is not: no amount of mathematical fiddling will cross that gap, but the gap is often more conceptual than practical.

The other transition, from waves to rays, is similar in many ways except that it has largely escaped the rhetoric of twentieth-century physics. Of course the question of quanta is introduced here if we claim that rays are the orbits of photons. We are then in the situation that light waves are classical and rays are quantum, while the orbits of particles are classical and their waves are quantum. Therefore building a bridge between waves and rays will be anti-historical either for light or for matter, whichever way we progress.

One of the useful by-products of this study will be to see whether there are any useful concepts that can be isolated and distinguished in the dichotomy classical-quantum. We shall discuss the question later in more detail, but for the moment it may be instructive to consider de Broglie's relation

$\lambda p = h$

connecting the wavelength and momentum of a beam of electrons. The "classical limit" of such a relation is the one in which h is replaced by zero. This does not mean that h is treated as a variable, but only that, in ordinary units, either λ or p is much larger than h. There are, in the limit, two possibilities. Either λ is large and p is effectively zero or p is large and λ is effectively zero. That is, either the theory contains the word wavelength but not the word momentum or the other way around. Both kinds of theory without h are perfectly familiar: examples are classical optics and classical mechanics. In the quantum theory one sometimes speaks of momentum and wavelength in the same sentence.

To begin the development, recall that from a historical point of view, at least, light waves and light rays both belong to the physics of the classical era; they can be discussed without the use of Planck's constant. Further, as will be seen in (1.5), the construction of Newtonian orbits from Schrödinger waves if formally just the same as the construction of light rays in a medium of varying index of refraction.

A necessary condition for any such construction is that the wave-length of the wave shall be "small"; that is, the beam itself and any lenses or holes it goes through are many wavelengths wide, and if the index of refraction n varies, it varies relatively little over one wavelength:

$$\lambdabar \left| \frac{\partial n}{\partial x} \right| \ll n \quad \lambdabar := \lambda / 2\pi \quad \text{or} \tag{1.1a}$$

$$\left| \frac{\partial n}{\partial x} \right| \ll kn \quad k =: 1/\lambdabar \quad . \tag{1.1b}$$

The condition on the width of the apertures makes it unnecessary to consider diffraction effects. Condition (1.1) will be built into our approximation later. It excludes a priori using this theory to discuss refraction at an air-glass interface, but we will find formulas that apply to this case also. We will consider waves of a single angular frequency ω. Other signals can be represented by Fourier superposition.

Waves and Rays. In explaining how light is affected by lenses and mirrors, one draws rays diagrams, lines which are straight except at the interfaces (Problem 1.1). If the medium is one in which the index of refraction varies continuously with position, the straight lines become curves and images are distorted — an example is that on clear evenings the sun's disc may appear flattened just before sunset. Our first task is to calculate the rays in such a case, starting from the differential equation of the wave.

A component of a monochromatic electromagnetic field satisfies d'Alembert's equation

$$\nabla^2 \varphi - \frac{n^2(\omega, \boldsymbol{r})}{c^2} \frac{\partial^2 \varphi}{\partial t^2} = 0 \tag{1.2l}$$

where c is the speed of light in a vacuum and where the wave velocity has been written as c/n. Schrödinger's equation uses potential energy to express what n expresses for light,

$$\frac{\hbar^2}{2m} \nabla^2 \varphi - V(\boldsymbol{r})\phi + i\hbar \frac{\partial \varphi}{\partial t} = 0 \quad . \tag{1.2m}$$

Now assume that $\varphi \sim e^{-i\omega t}$. We find for light

$$\nabla^2 \varphi + \frac{\omega^2}{c^2} n^2(\omega, \boldsymbol{r})\varphi = 0 \tag{1.31}$$

and for matter

$$\nabla^2 \varphi + \frac{2m}{\hbar^2}[E - V(\boldsymbol{r})]\varphi = 0 \quad , \quad E = \hbar\omega \quad . \tag{1.3m}$$

These are of the form

$$\nabla^2 \varphi + K^2(\omega, \boldsymbol{r})\varphi = 0 \quad \text{with} \tag{1.4}$$

$$K(\omega, \boldsymbol{r}) = \frac{\omega}{c} n(\omega, \boldsymbol{r}) \quad \text{or} \quad \sqrt{\frac{2m}{\hbar^2}[E(\omega) - V(\boldsymbol{r})]} \tag{1.5l, m}$$

in which (until V becomes larger than E) a varying potential energy produces the same effects as a varying index of refraction.

To solve (1.4) make the short-wavelength approximation that in quantum mechanics bears the name WKB (Wentzel-Kramers-Brillouin) but existed in the literature well before that [3]. We write the solution as

$$\varphi(\boldsymbol{r}) = a(\boldsymbol{r})e^{iw(\boldsymbol{r})} \tag{1.6}$$

where a, the amplitude, varies slowly and the term containing w, the phase, varies quickly (we will establish criteria for these words in a moment), and both are real. Putting (1.6) into (1.4) gives

$$\nabla^2 a + 2i\nabla a \cdot \nabla w + ia\nabla^2 w - a(\nabla w)^2 + K^2 a = 0$$

trading the complex quantity φ for two real ones. Separating real and imaginary parts gives

$$\nabla^2 a - a(\nabla w)^2 + K^2 a = 0 \tag{1.7r}$$

$$2\nabla a \cdot \nabla w + a\nabla^2 w = 0 \quad . \tag{1.7i}$$

In the first equation, if a varies much more slowly than w, the first term becomes small compared with the other two. We shall therefore drop it, postponing the mathematical justification until the end of this section.

The remaining terms of (1.7r) give

$$(\nabla w)^2 = K^2 \tag{1.8}$$

while equation (1.7i) is

$$\nabla \cdot (a^2 \nabla w) = 0 \quad . \tag{1.9}$$

Equation (1.8) is one of the central equations in this book, and Chap. 5 will show why: it leads, exactly, to the Newtonian dynamics of a particle. But suppose that in some small region of space the wave resembles a plane wave with a propagation constant $\boldsymbol{k}$ that is essentially constant. The wave here propagates like

$$\phi \simeq e^{i(\boldsymbol{k}\cdot\boldsymbol{r} - \omega t)} \tag{1.10}$$

3

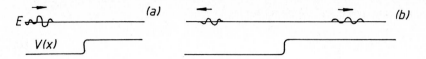

Fig. 1.1. Arriving at a sudden discontinuity of the medium, a wave packet splits. It is difficult to understand this in terms of the behavior of a particle

and w in (1.6) plays the role of $\mathbf{k} \cdot \mathbf{r}$. Taking the gradient, we have

$$\mathbf{k} \simeq \nabla w \quad , \quad k^2 \simeq (\nabla w)^2 = K^2 \tag{1.11}$$

so that ∇w gives the wave's wavelength and direction of propagation. The wavelength of any wave satisfying the equations is determined by (1.5), and its direction is determined not by any equation but by the initial conditions: where the wave comes from.

Many results flow from (1.11). The quanta of a wave of the form (1.10) have energy $\hbar\omega$ and momentum $\hbar k$. From (1.11) and (1.5) the momentum of a particle is

$$p = \hbar K = \sqrt{2m[E - V(\mathbf{r})]} \tag{1.12}$$

so that

$$\frac{p^2}{2m} + V(\mathbf{r}) = E \quad . \tag{1.13}$$

Of course, the last equation looks like Newtonian mechanics, but to see the real connection we must remember carefully the way it was derived. We assumed a wave that is approximately plane in a small region, and relations like the last three refer only to that region. The connection with the Newtonian mechanics of a particle is established only if the region is part of a larger wave packet, and if the potential $V(\mathbf{r})$ varies only slightly over the entire wave packet. (Otherwise, different parts of the wave packet would go off in different directions, see Fig. 1.1). Then p is the momentum associated with the wave packet and $V(\mathbf{r})$ is evaluated at the position of the wave packet. Equation (1.10) also has a meaning within quantum mechanics: if the system is in a state of frequency ω and energy $\hbar\omega$, a detector placed at $\mathbf{r}$ will detect particles having the given momentum. Note that Planck's constant is no longer present in these relations.[1]

The First Term of (1.7r). To end this section, return to the justification of dropping the first term of (1.7r). Since we are requiring that all parts of the wave packet be moving in the same direction, we can take the x axis parallel to that direction and calculate in one dimension. The relation to be approximated is

$$a\left(\frac{dw}{dx}\right)^2 = K^2 a + \frac{d^2 a}{dx^2}$$

---

[1] It is also natural to ask what is the momentum of a photon passing through a medium of index of refraction n. But since the field is continually in interaction with the medium, only the total momentum of field and medium is significant, and the attribution of part of it to the light is to some extent arbitrary. See *J.A. Arnaud* [4].

and the approximation consists in assuming that

$$\left|\frac{d^2a}{dx^2}\right| \ll \left|a\left(\frac{dw}{dx}\right)^2\right| \quad . \tag{1.14}$$

Let us see how this can be justified.

In writing down (1.6) we specified that a varies more slowly than the term containing w. This is expressed in terms of the fractional change of the functions:

$$\left|\frac{1}{a}\frac{da}{dx}\right| \ll \left|e^{-iw}\frac{d}{dx}e^{iw}\right| \quad \text{or}$$

$$\left|\frac{da}{dx}\right| \ll \left|a\frac{dw}{dx}\right| \quad . \tag{1.15}$$

Differentiating again gives

$$\left|\frac{d^2a}{dx^2}\right| \ll \left|\frac{da}{dx}\frac{dw}{dx} + a\frac{d^2w}{dx^2}\right| \quad .$$

In this one-dimensional case, (1.7i) is

$$2\frac{da}{dx}\frac{dw}{dx} + a\frac{d^2w}{dx^2} = 0$$

so that

$$\left|\frac{d^2a}{dx^2}\right| \ll \left|\frac{da}{dx}\frac{dw}{dx}\right|$$

and using (1.15) again gives

$$\left|\frac{d^2a}{dx^2}\right| \ll \left|a\left(\frac{dw}{dx}\right)^2\right|$$

which is (1.14), and was to be shown.

Finally, return to (1.9). If we write $\phi(r)$ in the form (1.6) it is easy to see that

$$a^2\nabla w = \text{Im}\,(\phi^*\nabla\phi) \quad .$$

The expression on the right is essentially the current vector in quantum mechanics, and in the steady-state regime we are discussing, (1.17i) reduces simply to the conservation of particles. The significance in the optical case is, of course, similar.

Comments. Since K is proportional to $\hbar^{-1}$, the approximation of small λ and large K can be viewed, even though $\hbar$ is not really a variable, as an approximation of small $\hbar$. For any $\hbar$, the approximation fails close to a point at which $E-V = 0$ and K therefore vanishes. Such points are called turning points, because they correspond to those points in classical mechanics at which a particle reaches the end of its allowed trajectory and reverses its motion (see Sect. 2.2). Here one solves the equation by using an approximation for small K, and finds ways to join it to the large-K form in the intermediate region ([91] Chap. 7). The subject belongs to quantum mechanics, and we need not pursue it here.

Finally, note that what has been proved is that a wave packet, within this approximation, has certain properties in common with a particle as described by Newtonian mechanics, and we will see in the next section that the analogy can be pressed much further. But this should not be understood to mean that a Newtonian particle is the same thing as a small wave packet. Most students of elementary quantum mechanics go through a period of believing that it is, not understanding that the quantum-mechanical description is purely statistical and that Newtonian mechanics has nothing statistical about it. The two theories have entirely different domains of discourse, and all correspondences between them must be expressed very carefully. In fact, one finds in quantum mechanics that a uniform plane wave produces the familiar effects from which the presence of particles is inferred, and that wave packets and particles have nothing to do with each other.

Problem 1.1. For practice and review, draw the ray diagram that explains the formation of a real image by a converging lens.

Problem 1.2. Draw the ray diagram that explains the formation of a virtual image by a converging lens used as a magnifying glass.

Problem 1.3. Draw the ray diagram that explains the formation of a virtual image by a convex mirror.

Problems 1.4–6. For the three cases above, explain the image not by drawing rays but by drawing a spherical wavefront leaving the object, encountering the lens or mirror, and leaving it as another spherical wavefront having a different center.

Problem 1.7. Prove by elementary vector analysis that the propagation vector k defined in (1.11) is directed perpendicular to the wavefront, i.e. to a surface of constant w. (If the corresponding ray and wave diagrams requested in the preceding problems are compared, the relevance of this result will be apparent.)

1.2 Phase Velocity and Group Velocity

The last section showed that in the short-wavelength approximation a wave packet follows a certain path and that along this path, if it is a matter wave, the relation between momentum and energy is the same as in particle mechanics. The next section will show how to calculate the paths themselves of light rays and particles. This section concerns the rate at which wave packets move, and leads to considerations with which many readers will already be familiar.

In the region in which the plane-wave approximation (1.10) is valid, choose the x axis parallel to the local value of k. Then

$$\varphi \simeq e^{i(kx - \omega t)} = e^{i\phi} \quad , \quad \phi = kx - \omega t \tag{1.16}$$

in which, in general, k is a function of ω, that is, ω is a function of k. The phase ϕ is constant for values of x and t satisfying

$$kx = \omega t + \text{const} \quad \text{or}$$

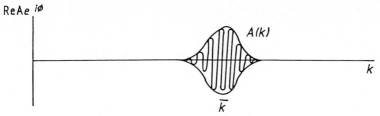

Fig. 1.2. A wave packet composed of momenta centered at $\bar{k}$

$$x = \frac{\omega}{k}t + \text{const} \quad .$$

The points of constant phase therefore move at a rate

$$v_\phi = \omega/k = \lambda\nu \quad . \qquad (1.17)$$

This is the *phase velocity*. If we try it for light we get

$$v_\phi = \frac{\omega}{k} = \frac{c}{n(\boldsymbol{r},\omega)} \qquad (1.18)$$

by (1.5), and this is the usual definition of the index of refraction. For matter, this is

$$v_\phi = \frac{\hbar\omega}{\hbar k} = \frac{E}{p} \quad . \qquad (1.19)$$

For a free particle, for which $E = p^2/2m$, $v_\phi = p/2m$, which is not the same as the particle velocity. The reason for this is well-known: since one cannot follow the crests and troughs of a de Broglie wave, v_ϕ is unobservable, and we must follow instead a modulation of the wave – a pulse or something like that. In fact, the old ways of measuring c also modulated the wave, and Michelson, making the first precise measurements, had to be reminded that he was measuring not v_ϕ but the group velocity[2] v_g (see Problem 1.11).

To evaluate v_g, use a construction that will be useful later. Make a wave packet out of components of roughly the same wave number k,

$$\varphi(x,t) = \int A(k)e^{i\phi}dk \quad , \qquad \phi = kx - \omega(k)t \qquad (1.20)$$

where $A(k)$ is as in Fig. 1.2. In general the rapid oscillations of the integrand will cause the integral to have a very small value, but not where ϕ varies slowly with k. In a matter wave, for example

$$\phi = kx - \frac{\hbar k^2}{2m}t \quad .$$

This is as in Fig. 1.3a, and $e^{i\phi}$ varies as in Fig. 1.3b. If k_1, where the phase varies most slowly, coincides with the value $\bar{k}$ where $A(k)$ is large, the integral will give an appreciable value for the amplitude of the wave packet. This occurs where

---

[2] The idea of group velocity goes back to W.R. *Hamilton* [Ref. 5, pp. 573, 578].

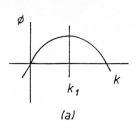
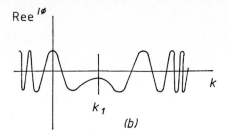

(a)

(b)

Fig. 1.3. The phase varies most slowly with k at the point k_1, determined by $d\phi/dk = 0$: $x - \hbar k_1 t/m = 0$

$$x = \frac{\hbar \overline{k}}{m} t \qquad (1.21)$$

and the velocity v_g of the center of the group now has the right dependence on the average momentum of the group, $v_g = \hbar \overline{k}/m$. More generally, the point of stationary phase occurs where

$$\frac{d\phi}{dk} = x - \frac{d\omega}{dk} t = 0 \quad \text{whence}$$

$$v_g = \frac{d\omega}{dk} \quad . \qquad (1.22)$$

This can also be written as

$$v_g = \frac{\hbar d\omega}{\hbar dk} = \frac{dE}{dp} \quad . \qquad (1.23)$$

If there is a conservative field of force the energy can be written as a function of coordinate and momentum; then it is called the Hamiltonian function $H(p, x)$ (see Chap. 5), and (1.23) becomes

$$v_g = \frac{\partial H}{\partial p} \qquad (1.24)$$

an equation we will encounter again. If $H = p^2/2m + V(x)$, this gives

$$v_g = p/m \qquad (1.24a)$$

which is the particle velocity in Newtonian physics.

The index of refraction of a medium is usually given (on physical grounds) as a function of the frequency. To find the group velocity write $\omega = ck/n(\omega)$, so that

$$d\omega = \frac{c}{n} dk - \frac{ckn' d\omega}{n^2} \quad \left(n' = \frac{dn}{d\omega}\right) \quad \text{and}$$

$$v_g = \frac{d\omega}{dk} = \frac{c/n}{1 + ckn'/n^2} = \frac{v_\phi}{1 + (\omega/n)dn/d\omega} \quad . \qquad (1.25)$$

Example 1.1. The relation between ω and k for waves in water of depth h [6] is

$$\omega = k\sqrt{\frac{g}{k} \tanh kh} \qquad \text{(g is the gravitational acceleration).}$$

Thus the phase velocity is

$$v_\phi = \sqrt{\frac{g}{k} \tanh kh} \quad .$$

In shallow water, $kh \ll 1$, and

$$\omega = k\sqrt{\frac{g}{k}\left(kh - \frac{1}{3}k^3h^3 + \ldots\right)} = \sqrt{gh}\left(k - \frac{1}{6}k^3h^2 + \ldots\right) \quad \text{so that}$$

$$v_\phi = \sqrt{gh}(1 - \tfrac{1}{6}k^2h^2 + \ldots) \quad , \quad v_g = \sqrt{gh}(1 - \tfrac{1}{2}k^2h^2 + \ldots) \quad \text{or}$$

$$v_\phi = \sqrt{gh}\left(1 - \frac{\omega^2 h}{6g} + \ldots\right) \quad , \quad v_g = \sqrt{gh}\left(1 - \frac{\omega^2 h}{2g} + \ldots\right)$$

and waves move more slowly as they reach the shore. In deep water, $\tanh kh \simeq 1$ and $\omega \simeq \sqrt{gk}$. Thus when $kh \gg 1$,

$$v_\phi \simeq \sqrt{g/k} \quad , \quad v_g \simeq \tfrac{1}{2}\sqrt{g/k} \quad .$$

Long waves move faster, and since in water both phase waves and wave groups are visible, one can produce a splash by dropping a stone into a pool and watch the phase waves passing through the group.

Problem 1.8. For waves of a given frequency, show qualitatively how the wavelength depends on the depth.

Problem 1.9. When sound passes down a pipe of square cross section or electromagnetic waves pass through a square waveguide of side a, the propagation in the lowest mode is governed by the relation

$$\omega^2 = c^2\left(k^2 + \frac{2\pi^2}{a^2}\right) \quad .$$

Find v_ϕ and v_g and note the simple relation between them. Is there any conflict with the results of relativity theory?

Problem 1.10. The relation between energy and momentum of a free particle in special relativity is

$$E^2 = c^2p^2 + (mc^2)^2$$

where m is the rest mass. With $E = \hbar\omega$ and $p = \hbar k$, same questions as in the preceding problem.

Problem 1.11. Careful measurement using an interrupted beam of light of wavelength $\lambda = 0.589\ \mu$m in air yields a speed equal to $299707.2 \pm 0.8\ \text{km s}^{-1}$. The index of refraction of air under standard conditions is

$$n = 1 + (27259.9 + 153.50\lambda^{-2} + 1.317\lambda^{-4}) \times 10^{-8}$$

when λ is given in μm. What correction must be applied to the measured value in order to find c? What correction is obtained if one forgets that group velocity was measured and corrects only for phase velocity?

Problem 1.12. A slow electron moves with velocity $100\,\mathrm{ms}^{-1}$ through the region enclosed by a metallic shell at a potential of 5×10^6 V. With K as in (1.5), find v_ϕ and v_g.

1.3 Dynamics of a Wave Packet

At this point in the discussion, someone who knew nothing of particle physics would have enough information to begin reconstructing it. In studying elementary quantum mechanics one learns to do this through Ehrenfest's theorems, but they involve the probabilistic interpretation of the wave function and it is perhaps more illuminating to see, using very simple mathematics, what is the relation between a wave and a ray. In one dimension, apply (1.13) to a wave packet centered at x, and ask how p changes with time. Since E is constant,

$$\frac{dp}{dt} = -\frac{m}{p}\frac{dV}{dx}\frac{dx}{dt} = -\frac{m}{p}\frac{dV}{dx}v_g = -\frac{dV}{dx}$$

because $p = mv_g$. In more interesting cases, when the force is not parallel to the momentum, one has to think a little harder but the argument is instructive and illustrates the importance of phase velocity.

Consider a beam moving from left to right in Fig. 1.4 which is deflected by some lateral inhomogeneity in $K(\boldsymbol{r})$. For the surfaces of constant s (see (1.6)) to stay in phase, the outside of the beam must move further than the inside,

$$\frac{v_\phi(\varrho)dt}{\varrho} = \frac{v_\phi(\varrho + d\varrho)dt}{\varrho + d\varrho} = \frac{v_\phi(\varrho) + d\varrho\,\partial v_\phi/\partial\varrho}{\varrho + d\varrho}dt$$

where ϱ is the radius of curvature of the path. Solving this for $1/\varrho$ gives

$$\frac{1}{\varrho} = \frac{1}{v_\phi}\frac{\partial v_\phi}{\partial\varrho} = \frac{\partial \ln v_\phi}{\partial\varrho} \quad.$$

Since by (1.18) and (1.11) $v_\phi = \omega/K$ with ω constant, this is

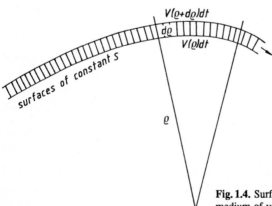

$V(\varrho+d\varrho)dt$

$d\varrho$

$V(\varrho)dt$

surfaces of constant s

ϱ

Fig. 1.4. Surfaces of constant s in a beam traversing a medium of varying index of refraction

$$\frac{1}{\varrho} = -\frac{\partial \ln K}{\partial \varrho} \quad . \tag{1.26}$$

Now we can compare the behavior of light and matter:

Light

$$K = \frac{\omega}{c} n(\mathbf{r})$$

$$\ln K = \ln n(\mathbf{r}) + \text{const}$$

$$\frac{1}{\varrho} = -\frac{\partial \ln n(\mathbf{r})}{\partial \varrho} \tag{1.27c}$$

Matter

$$K = \frac{2m}{\hbar^2}\sqrt{E - V(\mathbf{r})} \tag{1.27a, b}$$

$$\ln K = \tfrac{1}{2} \ln\left[E - V(\mathbf{r})\right] + \text{const}$$

$$\frac{1}{\varrho} = \frac{1}{2}\frac{\partial V/\partial \varrho}{E - V} = \frac{1}{2}\frac{\partial V/\partial \varrho}{mv_g^2/2}$$

so that

$$F_\varrho := -\frac{\partial V}{\partial \varrho} = -\frac{mv_g^2}{\varrho} \tag{1.28}$$

The first is, as we shall see, a generalization of Snell's law; the second is Huygens's celebrated formula for centripetal force (1659). Note that $-\partial V/\partial \varrho$, the rate of increase of V in the direction ϱ, is exactly the force transverse to the beam. Note also how Planck's constant got lost in taking the logarithmic derivative.

Example 1.2. Derive Snell's law for a horizontally layered medium.

Solution. Figure 1.5 shows the situation when n depends only on z. Since $d\vartheta = dl/\varrho$, we have

$$\frac{1}{\varrho} = \frac{d\vartheta}{dl} = -\frac{\partial \ln n}{\partial \varrho} = -\frac{d \ln n}{dz}\frac{\partial z}{\partial \varrho} \quad \text{or}$$

$$d \ln n + \frac{dz/dl}{\partial z/\partial \varrho} d\vartheta = 0 \quad .$$

From the figure, $dz/dl = \cos \vartheta$, $\partial z/\partial \varrho = \sin \vartheta$ and so

$$d\ln n + \cot \vartheta \, d\vartheta = d\ln n + d\ln \, \sin \vartheta = 0 \quad \text{or}$$

$$n \sin \vartheta = \text{const} \quad . \tag{1.29}$$

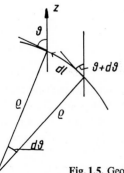

Fig. 1.5. Geometry of a curved ray

This is for a medium that varies continuously with z and not too fast with respect to λ. But λ has been lost sight of in this discussion; we may suppose it as small as we like, and so the change in n may take place very rapidly, though not, of course, discontinuously. In the limiting form (1.29) is Snell's law for the refraction of light at a plane interface, derived empirically in 1621.[3] Johann Bernoulli in 1696 seems to have been the first to recognize that (1.29) holds in the more general case of a continuously varying medium. We will return to his argument in Sect. 1.5. For the still more general case in which n varies in more than one direction, we will have to develop other methods below.

Problem 1.13. By how large an angle is the center of the sun's disc displaced by refraction as it sets? The index of refraction of air is about 1.000293 at ground level. Assume that the earth is flat and that sunset occurs at the moment the bottom of the sun's disc appears to touch the horizon. The calculation will be done for a curved earth in Sect. 6.9.

Problem 1.14. Assuming that the sun's angular diameter is $30'$ and defining the instant of sunset as in the preceding problem, find the change in the visual shape of the sun's disc as it sets.

Problem 1.15. Adapt the preceding argument so as to show directly that (1.29) holds at a sharp discontinuity in n such as occurs at an air-glass interface.

Problem 1.16. What theorem in mechanics corresponds to the optical law (1.29)?

Expression of Classical Mechanics in Terms of Phase. To conclude this section we forge another link with classical mechanics. The total time-dependent phase of the wave whose spatial part is (1.6) can be written as

$$\phi = w(r) - \omega t = \hbar^{-1} S(r, t) \quad \text{where} \tag{1.30}$$

$$S(r, t) := \hbar w(r) - Et \quad .$$

From (1.8) and (1.5) we have

$$(\nabla S)^2 = \hbar^2 K^2 = 2m[E - V(r)]$$

and by the definition of S this is

$$\frac{\partial S}{\partial t} + \frac{1}{2m}(\nabla S)^2 + V(r) = 0 \quad . \tag{1.31}$$

Further, (1.12) gives

$$p = \nabla S \quad . \tag{1.32}$$

Although derived from Schrödinger's equation these relations are once more independent of $\hbar$, and from their derivation they are some approximation to Newtonian mechanics. We might expect the relation to be subtle, since $S/\hbar$ is the phase of a quantum mechanical wave so that we are dealing with a classical formula that has a quantum interpretation. But equation (1.31) is *exactly* equivalent to Newton's laws

---

[3] Snell's discovery was anticipated by Thomas Hariot, probably before 1618 [7].

for the orbit of a particle. It was first derived in 1834 by *William Rowan Hamilton,* [Ref. 6, pp. 103, 162] and the function S, which Hamilton called the principal function, can be given a definition entirely within the classical system of ideas. We will do this in a leisurely way culminating in Chap. 5. For the moment note only that it is easier to cancel a significant number from an algebraic equation than to introduce it out of nowhere. We have derived (1.31) relatively painlessly from quantum mechanics by cancellation of $\hbar$, but to reconstruct quantum mechanics by introducing $\hbar$ through (1.30) and following the considerations backwards to derive Schrödinger's equation is somewhat harder. This will be done in Sect. 6.5. Before that, we shall take several chapters to study particle orbits and beams of light in more detail.

1.4 Fermat's Principle of Least Time

In about 60 AD, Hero of Alexandria noticed that there is a neat way to describe the reflection from a plane mirror: the reflected ray between two points takes the path that gets it from one to the other as quickly as possible. In 1657 Pierre de Fermat, the great amateur mathematician and founder of modern number theory, showed that one can derive Snell's law for refraction at a plane surface from the same principle. It is now called Fermat's *principle of least time,* but the phrase leads to confusing questions: Is it a phase wave or a signal that gets there in the least time? (It is a phase wave.) And since the speed is great, whose clock is used to measure the time? (The laboratory clock.) This second answer, if one is anthropomorphically inclined, leads to a further question: how does the ray know how to minimize its time with respect to somebody else's clock? Finally, you may be able to think of some simple situations involving curved mirrors in which the time is a maximum, not a minimum. Fermat would have saved trouble and confusion if he had known that light is a wave and spoken of a *principle of stationary phase.*

Problem 1.17. Prove that Hero's principle is equivalent to the law of equal angles. Is it true for a concave or convex mirror?

Problem 1.18. Show that Fermat's principle when properly stated is equivalent to Snell's law for plane optical surfaces.

It is easy to see that Fermat's principle implies the truth of (1.8) in the form

$$(\nabla w)^2 = \left(\frac{\omega n}{c}\right)^2 \ . \tag{1.33}$$

Let P_1 in Fig. 1.6 be a point source of monochromatic light in a medium of varying

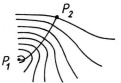

Fig. 1.6. Light ray traversing wave fronts from P_1 to P_2

index of refraction. The concentric curves represent, at a certain moment, the crests of waves which have issued from P_1. Now remove the point source and substitute a laser of the same frequency, and let P_2 be a point on the laser beam. We assert first that the ray from P_1 to P_2 intersects the wave surfaces orthogonally. This follows from Fermat's principle. The ray must cross all the wave fronts in the figure which lie between P_1 and P_2 in the shortest time, and since n does not vary appreciably from one wave crest to the next, the shortest path from each wave crest to the next is the orthogonal one.

If the ray is orthogonal to the surfaces of constant w, it is parallel or antiparallel to the vector ∇w, and we have for any displacement Δr along the ray

$$|\Delta w| = |\nabla w||\Delta r| \quad .$$

But if the wavelength locally is λ, then $|\Delta w| = 2\pi|\Delta r|/\lambda$ and from $2\pi/\lambda = \omega n/c$ we find finally

$$|\Delta w| = \frac{\omega n}{c}|\Delta r| = |\nabla w||\Delta r|$$

from which follows (1.33). (To be precise, an exception should be mentioned: this does not hold if, as in some crystals, the speed of light depends on the direction of the ray, but that case will not interest us here.)

Equation (1.33) is called the eikonal equation (from $\varepsilon i\kappa\acute{\omega}\nu$ = image, cf. *icon*). We have derived it first from wave optics and now from ray optics. The next problem is to solve it, that is, to construct the path of a ray through an optical medium. The necessary information is contained in (1.33) and we will see in Sect. 6.8 a general method for doing the calculation, but it is more direct and more historical to start with Fermat's principle. The necessary mathematical tool, needed also for the corresponding formulation of particle dynamics (Sect. 3.1) is the calculus of variations, and we must now develop the necessary formulas.

1.5 Interlude on the Calculus of Variations

Figure 1.7a shows a light ray in a layered medium with index of refraction $n(y)$, passing from a fixed point P_0 to another fixed point P_1. Let us find the formula $y(x)$ [or $x(y)$] that describes the path of the ray. If ds is an element of arc along the ray, the time for the phase wave to traverse it is $dt = c^{-1}n(y)ds$, and the total time from P_0 to P_1 is given by

$$cT = \int n(y)ds = \int n(y)\sqrt{1 + (dy/dx)^2}dx \quad . \tag{1.34a}$$

In this relation, n is a known function of y and y is an unknown function of x. Alternatively, this can be written as

$$cT = \int n(y)\sqrt{1 + (dx/dy)^2}dy \tag{1.34b}$$

where x is now an unknown function of y. Fermat's principle specifies the curve of the ray as that function $y(x)$ [or $x(y)$] that makes the integral for T a minimum.

14

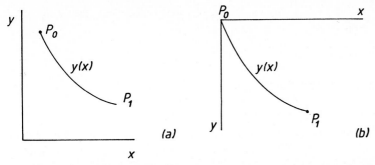

Fig. 1.7. Curve between two points, with coordinates suitable for (a) a ray of light, (b) the brachistochrone

The following problem in dynamics is classical: it goes back at least to Galileo. Let the curve $y(x)$ in Fig. 1.7b represent a wire running from P_0 to P_1, and let a bead slide down the wire without friction. Find the curve $y(x)$ that allows the bead to slide from P_0 to P_1 in the shortest time. The curve so defined is called the brachistochrone ($\beta\rho\acute{\alpha}\chi\iota\sigma\tau os$ = shortest $+\chi\rho\acute{o}\nu os$ = time). The bead's linear velocity is $\sqrt{2gy}$, so that

$$dt = ds(2gy)^{-1/2}$$

and the time of descent is given by

$$\sqrt{2g}\,T = \int_{P_0}^{P_1} y^{-1/2}\,ds = \int_{P_0}^{P_1} y^{-1/2}\sqrt{1 + (dy/dx)^2}\,dx \quad . \tag{1.35}$$

Evidently this is the same problem as (1.34a) with a particular form for $n(y)$. Working before the invention of calculus, Galileo had no techniques for setting up this problem, much less solving it, and he wrongly asserted that the desired $y(x)$ is an arc of a circle. Fifty-eight years later the calculus was in hand and Johann Bernoulli stated the problem again as a challenge to the mathematicians of Europe. Only he had the solution, and so has the reader of these lines, since Snell's law, generalized in (1.29) to continuously varying n, gives the path of a ray of light and therefore the form of the wire. The brachistochrone turns out to be a cycloid. (Example 1.4 below. A cycloid is the path of a point on the rim of a rolling wheel, in this case, the wheel is being rolled along the ceiling.) By one of the most brilliant pieces of analogical reasoning in the history of mathematics Bernoulli has obtained the answer without doing the work. Newton quickly solved the problem when it was presented to him and sent the solution to Bernoulli in a few anonymous lines which were, however, enough to allow him to identify the sender. The lion, Bernoulli murmured, is known from the mark of his claw. Now we have to do the work.

We generalize the problem as follows: Let $f[y(x), y'(x), x]$ be some given function of y, y' and x (higher derivatives of y are rarely useful) in which $y(x)$ is an unknown functional relation. Let

$$I = \int_{P_0}^{P_1} f[y(x), y'(x), x]\,dx \tag{1.36}$$

15

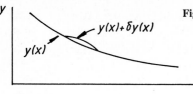

Fig. 1.8. A curve $y(x)$ and an infinitesimal variation of it

where P_0 and P_1 are given points (x_0, y_0) and (x_1, y_1). We are to find the function $y(x)$ that gives I a stationary value, maximum or minimum, subject to the boundary condition that the curve must pass through the given points P_0 and P_1. Modifications of this formulation are possible; we shall see some below.

The approach is the usual one in calculus: first select a good notation and then consider the effect of infinitesimal variations away from the correct answer. Let $y(x)$ be the desired curve, Fig. 1.8, and $y(x) + \delta y(x)$ be one that is slightly different. Here $\delta y(x)$ is a differentiable function which is arbitrary except that it is "small" in the sense that its square can be neglected and that by hypothesis, since the fixed points P_0 and P_1 are at (x_0, y_0) and (x_1, y_1), we have

$$\delta y(x_0) = \delta y(x_1) = 0 \quad . \tag{1.37}$$

In the neighborhood of the correct $y(x)$, I is stationary, meaning that to the first order in small quantities I does not vary when we vary y:

$$\delta I = \delta \int_{x_0}^{x_1} f[y(x), y'(x), x] dx = 0 \quad . \tag{1.38}$$

The variation in the integral is

$$\delta I = \int_{x_0}^{x_1} \left[\frac{\partial f}{\partial y} \delta y + \frac{\partial f}{\partial y'} \delta y' \right] dx = 0 \quad . \tag{1.39}$$

Now

$$\delta y' = y'_{\text{varied}} - y'_{\text{correct}} = \frac{d}{dx}(y_{\text{varied}} - y_{\text{correct}}) = (\delta y)'$$

so that (1.39) is

$$\int_{x_0}^{x_1} \left[\frac{\partial f}{\partial y} \delta y + \frac{\partial f}{\partial y'} (\delta y)' \right] dx = 0 \quad .$$

Integrate the second term by parts:

$$\int_{x_0}^{x_1} \left[\frac{\partial f}{\partial y} \delta y + \frac{d}{dx} \left(\frac{\partial f}{\partial y'} \delta y \right) - \frac{d}{dx} \left(\frac{\partial f}{\partial y'} \right) \delta y \right] dx = 0 \quad \text{or}$$

$$\int_{x_0}^{x_1} \left[\frac{\partial f}{\partial y} - \frac{d}{dx} \frac{\partial f}{\partial y'} \right] \delta y \, dx + \left[\frac{\partial f}{\partial y'} \delta y \right]_{x_0}^{x_1} = 0 \quad .$$

Because of (1.37) the integrated part is zero. And in the remaining integral, which is zero, the integrand contains a small but arbitrary function $\delta y(x)$. A little thought shows that the coefficient of δy must vanish for all values of x. We express this by writing

$$\frac{\delta f}{\delta y} = 0 \tag{1.40a}$$

where the notation is

$$\frac{\delta f}{\delta y} := \frac{\partial f}{\partial y} - \frac{d}{dx}\frac{\partial f}{\partial y'} \quad . \tag{1.40b}$$

This new kind of derivative is called a functional derivative. Note it does *not* denote a small variation in f divided by a small variation in y. The term implies that what is being varied is not the numerical value of the function, the usual df, but its functional form. Equations (1.40) were derived by Leonhard Euler in 1744 by rather complicated geometrical reasoning. Eleven years later he was surprised by a letter from a 19-year-old student in Italy named Ludovico de la Grange Tournier who derived the same results in a few lines of analysis, essentially as was done above. The equations themselves are called sometimes after Euler, sometimes Lagrange, sometimes both.

There is an assumption made in deriving (1.40a) that is not at first obvious. If f contains a power of y' higher than the first, the second term in (1.40b) will involve y''. Suppose that y' has a discontinuity in the range (x_0, x_1). Then y'' does not exist at that point and the integration by parts has no meaning. Therefore we must assume that the curve that makes the integral stationary has a continuous tangent.[4]

Returning to the problem of the ray of light, we see that f in (1.34a) is

$$f(y, y') = n(y)\sqrt{1 + y'^2} \quad . \tag{1.41}$$

We can if we wish calculate $\delta f/\delta y$, arriving at a second-order differential equation for $y(x)$, but it happens that there is a simplification we can use here and in many other cases. In the situation defined by (1.34a), f does not involve the independent variable x explicitly, and an integral of the differential equation (1.40) can be found at once.

Consider the function

$$F(y, y') := \frac{\partial f(y, y')}{\partial y'} y' - f(y, y') \quad .$$

Its derivative with respect to x is

$$\frac{dF}{dx} = \frac{d}{dx}\frac{\partial f}{\partial y'} y' + \frac{\partial f}{\partial y'} y'' - \frac{\partial f}{\partial y} y' - \frac{\partial f}{\partial y'} y'' = -\frac{\delta f}{\delta y} y' = 0$$

if f makes the integral I stationary, by (1.40). Thus a first integral of the equation

---

[4] For a famous example in which the curve giving the extremum has a discontinuous slope and is therefore inaccessible to the foregoing analysis see Newton's *Principia*, Book II, Prop. 34. This is the first use in history of the calculus of variations and is among the most puzzling. See *Goldstine* [8], Sect. 1.2.

in this case has the form

$$F = \frac{\partial f}{\partial y'} y' - f = \text{const} \quad . \tag{1.42}$$

Example 1.3. Snell's Law. With (1.41), (1.42) gives

$$\frac{n(y)y'^2}{\sqrt{1+y'^2}} - n(y)\sqrt{1+y'^2} = -C \text{ (const)} \quad \text{or}$$

$$\frac{n(y)}{\sqrt{1+y'^2}} = C \tag{1.43}$$

Since $y' = -\cot \vartheta$ (Fig. 1.5), this is

$$n(y) \sin \vartheta = C$$

which is Snell's law (1.29) again.

We can easily find an equation for the ray: Evaluate the constant at some point in the medium: $C = n_0 \sin \vartheta_0$. Then by solving (1.43) for y' we find x as a function of y,

$$x = \int dy \left[\left(\frac{n(y)}{n_0 \sin \vartheta_0} \right)^2 - 1 \right]^{-1/2} \quad . \tag{1.44}$$

Example 1.4. The Brachistochrone. From (1.35)

$$f(y, y') = y^{-1/2} \sqrt{1+y'^2}$$

and (1.42) gives

$$\frac{y'^2}{\sqrt{y(1+y'^2)}} - \frac{\sqrt{1+y'^2}}{\sqrt{y}} = \text{const} \cdot C \quad .$$

This quickly reduces to

$$y'^2 = \frac{1}{C^2 y} - 1 \tag{1.45}$$

which can be solved easily enough but the answer is uninformative. To recognize it as the differential equation for a cycloid, convince yourself that if ϑ is the angle through which the wheel generating the cycloid is rolled, the x and y coordinates of a point on the rim are given by

$$x = a(\vartheta - \sin \vartheta) \quad y = a(1 - \cos \vartheta)$$

where a is the radius of the wheel. Then note that

$$y' = \frac{dy}{dx} = \frac{a \sin \vartheta \, d\vartheta}{a(1 - \cos \vartheta) d\vartheta} = \frac{\sin \vartheta}{1 - \cos \vartheta}$$

and that equation (1.45) is satisfied provided that $aC^2 = \frac{1}{2}$.

Example 1.5. The Brachistochrone From a Point to a Line. A wire extends from the point $(0,0)$ to a line extending vertically downward from point $(p,0)$, reaching it at the point $(p, -q)$. What is the shape of the wire and the value of q that make the time of descent a minimum? This example is intended to show why one sometimes wants to consider variations δy that do not satisfy the condition (1.37).

First, the point (p, q) is some definite point, so that our previous solution applies and the curve is a cycloid. The integrand $f(y, y')$ is as in Example 1.4 and $\delta y(p) = 0$ but now $\delta y(q) \neq 0$. Equation (1.40) becomes $[\partial f/\partial y' \; \delta y]_q = [\partial f/\partial y']_q = 0$, and this is seen at once to require $y'(q) = 0$. Thus the wire must be arranged so as to reach the vertical line when the cycloid is at the bottom of its curve.

Variation with Auxiliary Conditions. There is a class of problems in the calculus of variations whose elementary prototype is the question "What figure whose perimeter is given encloses the largest area?", and anyone can see that the answer is a circle.

It is a special case of problems of the type: What function $y(x)$ maximizes or minimizes the integral

$$S = \int_{x_0}^{x_1} f(y, y', x)dx \tag{1.46}$$

where f is given, the variations vanish at the end points, and another integral,

$$T = \int_{x_0}^{x_1} g(y, y', x)dx \tag{1.47}$$

is at the same time required to have a given numerical value?

In varying $y(x)$ we are requiring that $\delta S = 0$ because S is stationary, while $\delta T = 0$ because T is fixed. Let $S' = S + \lambda T$, where λ is some number called a Lagrangian multiplier and to be chosen later. Then

$$\delta S' = 0 \tag{1.48}$$

is a necessary condition on $y(x)$, and its solution yields a $y(x, \lambda)$ depending on λ. If now it is possible to find a value of λ such that T as calculated by (1.47) takes on its given value, all the conditions are satisfied and the problem is solved.

What the Lagrangian multiplier enables one to do is remarkable and should be carefully noted. Suppose, for example, that we are given a piece of rope of fixed length and are to arrange it so as to maximize the area enclosed. One might at first suppose that the variations in the rope's position would have to be performed so that its length remains constant. Thus if the rope is pulled out at one place it would have to be pulled in somewhere else and we could no longer claim that $\delta y(x)$ is arbitrary. The proof would then fail. Lagrange's trick enables us to treat δy as arbitrary and impose the condition *after* the variation has been performed.

Problem 1.19. Prove that (1.40) follows from the steps before it.

Problem 1.20. Find the shape of a soap film forming a surface of revolution bounded by two parallel circular loops of wire on a common axis. (The physical hypothesis is that surface tension causes the film to assume the shape with the least area.)

Problem 1.21. Suppose that y depends on several independent variables $x_1 \ldots x_n$ and that the integral in (1.36) is a multiple integral over $x_1 \ldots x_n$. Show that the Euler equation is

$$\frac{\delta f}{\delta y} := \frac{\partial f}{\partial y} = \sum_{i=1}^{n} \frac{\partial}{\partial x_i} \frac{\partial f}{\partial(\partial y/\partial x_i)} = 0 \quad . \tag{1.49}$$

Problem 1.22. Suppose there is one independent variable but that f is a function of several variables $y_1(x) \ldots y_k(x)$ and their first derivatives. Show that Euler's equations are now k in number,

$$\frac{\delta f}{\delta y_i} = \frac{\partial f}{\partial y_i} - \frac{d}{dx} \frac{\partial f}{\partial y_i'} = 0 \quad , \quad i = 1, \ldots, k \tag{1.50}$$

Problem 1.23. What if there are n independent and k dependent variables?

There is another type of variational problem with auxiliary condition. Suppose for definiteness that there are one independent and k dependent variables and that we are trying to make the integral S defined in (1.46) stationary as before but that this time there is a condition of the form

$$h(y_1, \ldots y_k, y_1', \ldots, y_k', x) = H(\text{const})$$

that must hold however the y's are varied. Let

$$\delta \int_{x_0}^{x_1} F(y, y', x)dx = 0 \quad , \quad F := f(y, y', x) + \mu(x)h(y, y', x)$$

where y stands for all the y's and μ is to be determined. If we solve the equations $\delta F/\delta y = 0$ the result will be y's that depend on the form of the function $\mu(x)$. If now it is possible to choose μ so that the condition $h = H$ is satisfied, all the conditions will have been fulfilled. This procedure still works when there are several equations of constraints. Both the techniques explained here will be used later in this book. The recipe for this process can be stated as follows: Put in the term with the multiplier; then treat $y_1, y_2, \ldots y_k$, and μ as $k+1$ variables that can be varied independently.

Problem 1.24. What happens if f depends on y, y', y'', and x? (This case is rare in practice but not unknown.) Assume that $\delta y'$ as well as δy vanishes at the ends of the path of integration.

Problem 1.25. Calculate the path of the light ray taking y as the independent variable as in (1.34b).

Problem 1.26. Find the path that makes the shortest distance between two points in Cartesian and then in polar coordinates.

Problem 1.27. What is the limiting form of (1.44) when n is very close to 1? The density of the earth's atmosphere decreases roughly exponentially with height. Writing $n(y) = 1 + \varepsilon e^{-\alpha y}$, find the path of sunlight that enters the atmosphere obliquely and hits the (flat) earth at $x = y = 0$ at an angle ϑ_0. (Approximation for small ε.)

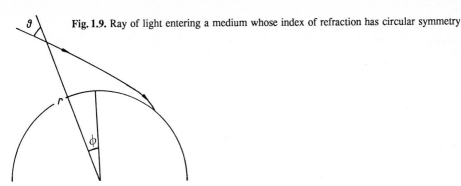

Fig. 1.9. Ray of light entering a medium whose index of refraction has circular symmetry

Problem 1.28. Consider the propagation of light in a material medium whose index of refraction $n(r)$ has spherical symmetry, Fig. 1.9. Introducing polar coordinates, show that

$$cdt = n(r)\sqrt{r^2 + r'^2}\,d\phi \quad , \quad r' = dr/d\phi \qquad (1.51a)$$

and that the equation of the ray is

$$\phi - \phi_0 = \int \frac{dr}{r\sqrt{(rn(r)/\alpha)^2 - 1}} \quad , \quad \alpha = \text{const} \quad . \qquad (1.51b)$$

[Compare (1.44)].

Problem 1.29. Show that the result of the last problem is equivalent to

$$rn(r)\sin\vartheta = \alpha \qquad (1.51c)$$

where ϑ is the angle between the ray at a point at a distance r from the center and a radial line drawn to that point.

Problem 1.30. When Queen Dido led a band of refugees to what is now Tunisia, she asked the local ruler for as much territory as could be covered by an ox-hide. "Well, sure." She then had the hide cut into a thin strip and using the straight shoreline as one boundary arranged the strip so as to enclose the maximum area of land (*Aeneid* I, 1. 367). What was the shape in which the strip was laid out? This is known as Dido's Problem, and tradition says she solved it correctly. Show that the equation of a circle with its point arbitrarily located satisfies the Euler-Lagrange equations.

Problem 1.31. A uniform chain of length l is suspended between the points $(-a, h)$ and (a, h). What is the equation for its curve? What role does the Lagrangian multiplier play in finding the answer?

Problem 1.32. Find $y_1(x)$ and $y_2(x)$ that make the integral

$$\int_0^1 \frac{1}{2}(y_1'^2 + y_2'^2 - py_2^2)\,dx$$

stationary when $y_2 - \sigma y_1 = 1$ (p, σ const).

Problem 1.33. Show that the prescription $y'(q) = 0$ derived in Example 1.5 uniquely defines the shape of the wire. What is q?

1.6 Optics in a Gravitational Field

According to the general theory of relativity a gravitational field is described not in terms of a potential (for particles) or an index of refraction (for light), but in terms of curvature of the underlying space-time. Optics and mechanics thus have a common basis. We shall see later that out of the geometrical description can be extracted a potential and an index of refraction that establish a conceptual link with other parts of physics, but here we will work directly from geometry. This is not the place to launch a discussion of the general theory of relativity, but I will make a few remarks on how to describe a curved surface (in any number of dimensions) in terms of the coordinate systems which can be drawn on it. The examples to follow will be in one and two dimensions; to go further one needs only a little pictorial imagination.

It is clear that coordinate systems are conditioned by the surfaces on which they are drawn. There is no way to cover a sphere with the Cartesian coordinates appropriate to a plane, or a plane with the coordinates of latitude and longitude that are used on a sphere. But there are many kinds of coordinates that can be used on a plane, and many on a sphere. Therefore coordinate systems are partly a matter of convention; the rest is imposed by the surfaces.

In the flat Minkowski space-time of special relativity x, y, z, and t correspond directly to measured intervals, but in a curved manifold the correspondence is less direct. The situation is the same as one encounters in geography: reading a flat map is easy if you neglect the earth's curvature, but in mapping a large region you must establish a convention for representing points on a curved surface in terms of the coordinates of a plane, and it is not immediately obvious how to measure the distance between two points. Several projections are used, in none of which, except in a small neighborhood, do distances measured on the map directly show distances measured along the ground. It is the element of arbitrariness that confuses students encountering geography or general relativity for the first time. They feel that there ought to be a "correct" representation, or at least one that is better than others, but there is not. Distances and angles on a map *represent* distances and angles on the earth. In Mercator's projection the angles are the same, but there is no constant scale of distance.

The spacetimes to be considered are not very different from flat ones, and we can take as a standard of comparison the invariant expression

$$c^2 ds^2 = c^2 dt^2 - dx^2 - dy^2 - dz^2 \tag{1.52}$$

which defines an element of proper time in special relativity. Consider the one-dimensional map projections of Fig. 1.10. The "surface" to be mapped is an arc of a

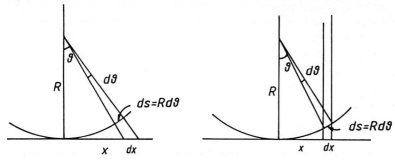

Fig. 1.10. Radial and vertical projections of a circular arc

semicircle, and among the various projections available to map it onto a flat surface two are shown. In the projections the actual distances $c\,ds = R\,d\vartheta$ along the curve are mapped differently onto coordinate distances dx measured on the map. In the radial projection,

$$x = R\tan\vartheta \quad \text{and} \quad dx = R\sec^2\vartheta\,d\vartheta \quad .$$

Thus

$$dx = R\,d\vartheta(1 + x^2/R^2) \quad \text{or}$$

$$c\,ds = \frac{dx}{1 + x^2/R^2} \quad . \tag{1.53a}$$

In the vertical projection, on the other hand,

$$x = R\sin\vartheta \quad , \quad dx = R\,d\vartheta\sqrt{1 - x^2/R^2} \quad \text{and}$$

$$c\,ds = \frac{dx}{\sqrt{1 - x^2/R^2}} \quad . \tag{1.53b}$$

These examples emphasize the conventional nature of coordinate systems, and one of the great problems of differential geometry, associated especially with the names of Gauss and Riemann, is to reconstruct the curvature of the underlying space, and ultimately its topological properties as well, when only a coordinate mapping is given.

The same diagrams can be used to show how the formulas look in two dimensions. Suppose that the arcs in Fig. 1.10 are the end views of segments of a cylinder, with y as the distance measured along its axis. Corresponding to given dx, dy on the map the formulas for actual distance measured along the curved surface of the cylinder are

$$c^2ds^2 = \frac{dx^2}{(1 + x^2/R^2)^2} + dy^2 \quad \text{(radial projection)}$$

and

$$c^2ds^2 = \frac{dx^2}{1 - x^2/R^2} + dy^2 \quad \text{(vertical projection)} \quad .$$

In the notation usual in relativity theory, x and y are called x^1 and x^2 and the coefficients of $(dx^1)^2$ and $(dx^2)^2$ are called g_{11} and g_{22}. Thus both formulas for distance can be written as

$$ds^2 = \sum_{i=1}^{2} g_{ii}(dx^i)^2$$

with appropriate g's.

In the spacetime of special relativity the coordinates t, x, y, z are again somewhat arbitrary. The numerical values of dt, dx, dy, and dz corresponding to the interval between a pair of events depend on the directions in which the observer is facing and moving, but ds^2 defined by (1.52) is postulated to be the same for all observers. In our notation this formula becomes

$$ds^2 = \sum_{i=0}^{3} g_{ii}(dx^i)^2 \qquad (1.54)$$

with $dx^0 = dt$, $g_{00} = 1$, $g_{11} = g_{22} = g_{33} = -1/c^2$. In the curved spacetime of general relativity the g's become functions of the coordinates, and in certain situations, which we shall not encounter here, the coordinates are no longer orthogonal and (1.54) is supplemented by terms of the form $g_{ij}dx^idx^j$ with $i \neq j$.

This will be explained further when it is needed. Here we will look at two optical effects, the deflection and the retardation of a light signal passing the sun. Both are capable of accurate quantitative measurement and are among the few quantitative tests of the theory. Another test, the precession of planetary orbits, is discussed in Sect. 6.3.

For dealing with the plane orbit of light passing the sun we use plane polar coordinates, in which (1.52) becomes

$$c^2 ds^2 = c^2 dt^2 - dr^2 - r^2 d\vartheta^2 \quad . \qquad (1.55)$$

In special relativity dt is an interval measured on a clock, and r, dr, and $d\vartheta$ are measurements made on scales at rest with respect to the clock. In the gravitational field of a spherical mass M, a convenient projection (known by the name of Schwarzschild) takes the form

$$c^2 ds^2 = c^2 g(r)dt^2 - g(r)^{-1}dr^2 - r^2 d\vartheta^2 \quad . \qquad (1.56a)$$

The function $g(r)$ is related to the gravitational potential of Newton's theory, and is derived by integrating Einstein's field equations just as the Coulomb potential comes from integrating Maxwell's equations. In these notes we shall take it, like Newton's and Coulomb's laws, as given us by Nature:

$$g(r) = 1 - 2\lambda/r \quad , \quad \lambda := GM/c^2 \qquad (1.56b)$$

where G is the Newtonian constant. For the sun,

$$\lambda_\odot = 1.477 \text{ km}$$

and the second term of g is an extremely small correction.

24

Problem 1.34. Starting from the Lorentz transformations of x, y, z, and t for motion along the x axis, give those for dx, dy, dz, and dt, and verify that ds^2 in (1.52) is an invariant.

Problem 1.35. Derive the formula for Mercator's projection in which any small region (i.e., a small lake or island) is mapped correctly as to shape but not as to size.

Problem 1.36. Derive a formula analogous to the spatial part of (1.56) for some mapping of a part of the earth's surface onto a plane. The quantity ds^2 should represent an actual terrestrial distance, while r, dr, and $d\vartheta$ represent distances and angles measured on the paper.

The Propagation of Light. In special relativity all observers describe the speed of a pulse of light in the same mathematical way: the invariant ds^2 is set equal to zero. The same feature carries over into general relativity. One again sets $ds^2 = 0$; this specifies the speed but not the path of the light ray. One needs an equation of motion in curved space-time. What comes out is simplicity itself: Fermat's principle, which will be derived in Sect. 6.3. The following example will show how the principle is applied.

Gravitational Deflection of Light. Figure 1.11 shows a ray of light originating in a distant star and passing close to the sun on its way to the earth. The label r represents a distance on the diagram but not as it would be measured in space. (I invite you to ponder on the difficulties of specifying operationally what is meant by distance in a strong gravitational field. Remember that the motion of light is affected by the field and that there can be no such thing as a perfectly rigid body.)

Setting $ds^2 = 0$ in (1.56a), we require that $\delta t = 0$ with

$$ct = \int \sqrt{g^{-1}r^2 + g^{-2}r'^2}\, d\vartheta \quad , \quad r' := dr/d\vartheta \quad . \tag{1.57}$$

Since the term on the right does not contain ϑ explicitly we can use (1.42):

$$\frac{g^{-2}r'^2}{\sqrt{g^{-1}r^2 + g^{-2}r'^2}} - \sqrt{g^{-1}r^2 + g^{-2}r'^2} = -\alpha\,(\text{const})$$

and solving this for r'^2 gives

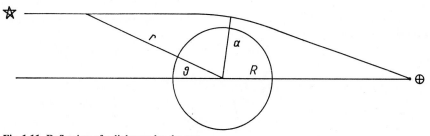

Fig. 1.11. Deflection of a light ray by the sun

25

$$r'^2 = \alpha^{-2}r^4 - gr^2 \quad . \tag{1.58}$$

The meaning of α is easy to see here. Where the ray passes closest to the sun, r is a minimum as a function of ϑ, and $r' = 0$. Thus

$$\alpha = g^{-1/2}r_{\min} \simeq \left(1 + \frac{\lambda}{r_{\min}}\right)r_{\min} = r_{\min} + \lambda \simeq r_{\min} \tag{1.59}$$

so that α is very nearly the beam's distance of closest approach.

Integration of (1.58) gives

$$\vartheta - \vartheta_0 = \int \frac{dr}{r\sqrt{r^2/\alpha^2 - 1 + 2\lambda/r}} \quad . \tag{1.60}$$

Comparing (1.57) with (1.51a) we see that r^2 and r'^2 are now multiplied by different quantities: it is like propagation through an anisotropic medium in which the radial and tangential velocities are different. Comparing (1.60) with (1.51b) we find an effective index of refraction

$$n = \sqrt{1 + \frac{2\alpha^2 GM}{c^2 r^3}} \quad . \tag{1.61}$$

This averages out the anisotropy but its value depends on the orbital constant α of the particular ray and so it is not a real index of refraction.

Because n in (1.61) is so nearly equal to 1, a ray of light passes the sun in nearly a straight line and its slight deflection is easy to find. Suppose a light pulse moves a distance ds along a path whose curvature ϱ^{-1} is given by (1.26), Fig. 1.12. The change in its direction is $d\delta = \varrho^{-1}ds$, with

$$\varrho^{-1} = -\frac{\partial \ln n}{\partial \varrho} \approx \frac{\partial \ln n}{\partial y} = -\frac{d \ln n}{dr}\frac{\partial r}{\partial y}$$

$$= -\frac{y}{r}\frac{d \ln n}{dr} \approx -\frac{\alpha}{r}\frac{d \ln n}{dr} = \frac{3\alpha^3 GM}{c^2 r^5} \tag{1.62}$$

so that, with $ds \simeq dx$ and $r^2 \simeq x^2 + \alpha^2$,

$$\delta = \int_{-\infty}^{\infty} \frac{dx}{\varrho} = \frac{3\alpha^3 GM}{c^2} \int_{-\infty}^{\infty} \frac{dx}{(x^2 + \alpha^2)^{5/2}} \approx \frac{4GM}{\alpha c^2} \quad . \tag{1.63}$$

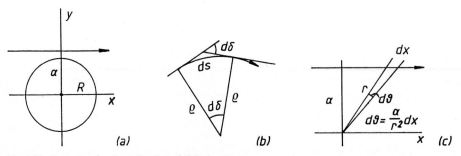

Fig. 1.12. Geometry for the deflection of light by the sun

At grazing incidence $\alpha \simeq R$, so that for the sun,

$$\delta \simeq \frac{4GM}{Rc^2} = \frac{4\lambda_\odot}{R} = 1.75'' \quad . \tag{1.64}$$

This formula was given by *Einstein* in his first paper on general relativity [9], translated in [10]. The best available verification of it is from measurements made with the Very Long Baseline Interferometer (VLBI) [11] which are so sensitive that they can detect the sun's pull on the light from a distant star essentially anywhere in the sky. The agreement is to a fraction of a percent. There are also confirmatory tests by radio astronomy [12].

Problem 1.37. Carry out the approximations in (1.62).

Problem 1.38. In 1801, *Johann Soldner* [13] gave a naive prerelativistic calculation of the deflection of light which is amusing historically and worth looking at here because it emphasizes the distinction between group and phase velocity. Assume that light of mass m travels at speed c in regions far away from the sun. Then by the conservation of energy it speeds up as it approaches the sun,

$$E = \tfrac{1}{2}mc^2 = \tfrac{1}{2}mv^2 + V(r) \quad . \tag{1.65}$$

In (1.26),

$$\begin{aligned}
\ln K &= \frac{1}{2}\ln\left[E - V(r)\right] + \text{const} \\
&= \frac{1}{2}\ln\left(\frac{1}{2}mc^2 + \frac{GmM}{r}\right) + \text{const} \quad \text{or}
\end{aligned}$$

$$\ln K = \frac{1}{2}\ln\left(1 + \frac{2GM}{rc^2}\right) + \text{const} \quad . \tag{1.66}$$

From this the calculation can be completed.

On the other hand, we can use the optical approach. From (1.65)

$$v = \frac{2}{m}\sqrt{E - V(r)} \quad \text{and}$$

$$n = \frac{c}{v} = \left[\frac{2}{mc^2}(E - V(r))\right]^{-1/2}$$

so that in (1.27a),

$$\ln K = \ln n + \text{const} = -\frac{1}{2}\ln\left(1 + \frac{2GM}{rc^2}\right) + \text{const} \quad . \tag{1.67}$$

The results (1.66) and (1.67) disagree. Resolve the discrepancy, calculate the deflection, and compare it with Einstein's value.

Problem 1.39. The special-relativistic version of (1.65) is

$$[E - V(r)]^2 = c^2p^2 + (mc^2)^2 \quad . \tag{1.68}$$

Calculate the deflection of light in this theory and compare it with the result of general relativity.

Delay of a Radar Signal Passing the Sun. With the advent of interplanetary radar techniques in the 1960s it became possible to test general relativity in another way by measuring directly the retardation of a radar signal reflected from a planet on the other side of the sun from ourselves.

We divide up the path into four sections and evaluate the time required for light to pass from the sun ($\vartheta = \frac{\pi}{2}$ in Fig. 1.11) to a planet at r_1, ϑ_1. The formulas are ready to hand. Putting (1.58) into (1.57) gives

$$ct_1 = \frac{1}{\alpha} \int_{\pi/2}^{\vartheta_1} g^{-1} r^2 \, d\vartheta$$

$$\simeq \frac{1}{\alpha} \int_{\pi/2}^{\vartheta_1} \left(1 + \frac{2\lambda}{r} \right) r^2 \, d\vartheta \quad .$$

To simplify this we remember that by Fermat's principle, the time is stationary under a small change of the path. We can therefore replace the curved path by a straight one with negligible error. By Fig. 1.12c,

$$r^2 \, d\vartheta = \alpha \, dx \quad \text{and}$$

$$ct_1 = \int_0^{x_1} \left(1 + \frac{2\lambda}{r} \right) dx \quad , \quad r = \sqrt{x^2 + \alpha^2} \tag{1.69}$$

which with $x_1 \gg \alpha$ is close to

$$ct_1 = x_1 + 2\lambda \ln \frac{2x_1}{\alpha} \quad . \tag{1.70}$$

The second term is the delay. For a round trip from the earth ($\oplus$) to Mars (σ) with the ray grazing the sun ($\alpha \approx R$) the delay is

$$\Delta t = \frac{4\lambda_\odot}{c} \ln \frac{4r_\oplus r_\sigma}{R^2} \quad . \tag{1.71}$$

This amounts to about $250 \, \mu s$, in a roundtrip that takes about $40 \, \text{min.}$, with respect to the time calculated from the known orbit of Mars without taking the delay into account. Because the sun is a very weak radio emitter, radio waves are better than light for testing this result. Recent measurements [14] agree with the theory to about 1 percent. More accurate measurements are difficult because of the planet's surface irregularities.

Problem 1.40. Design a test of (1.70) using as a signal source a distant pulsar that lies in the earth's orbital plane. Calculate how the pulsar's apparent frequency should vary throughout the year. Is the test worth trying? Could the method be used to determine the distance to the pulsar?

Problem 1.41. Ocean waves approach a straight, uniformly shelving beach, the line of the waves making an angle ϑ with the beach. In which direction do the waves curve as they approach the beach? Assuming that the waves break when the water reaches a depth k^{-1}, use the method of this section to find out how much their direction has changed. The answer is $7.80 \tan \vartheta$ degrees.

2. Orbits of Particles

If a system has only a very few degrees of freedom its equations of motion are often obvious. One writes down Newton's laws and sets about solving them. If they are not easy to solve it is because some equations are difficult. In complicated systems, even formulating the equations presents some problems, and most of this book is devoted to more advanced and general ways of writing down such equations and constructing their exact and approximate solutions. This chapter exhibits some relatively simple systems chosen so as to illustrate various forms of dynamical behavior. The first few examples need only Newton's laws and a little mathematics, but for the later ones it will be convenient to introduce new variables and write equations of motion in terms of generalized coordinates. This will be done using Lagrange's technique, which leads not only to general and useful equations of motion but also to the powerful methods of solution that follow.

2.1 Ehrenfest's Theorems

We have seen in Sect. 1.3 how to get Newton's laws out of Schrödinger's waves, but of course there is a nicer way to do it that dispenses with most of the talk. *Ehrenfest*'s theorems [15] are so easy to state and prove that I hope some readers, at least, will find the proofs not very explanatory and will learn more from the primitive discussion of Chap. 1.

The theorems relate not to the dynamical variables of quantum mechanics themselves but to their expectation values. The general equation for the time-dependence of the expectation value of a dynamical variable A in quantum mechanics is

$$\frac{d\langle \hat{A} \rangle}{dt} = \left\langle \frac{i}{\hbar}[\hat{H}, \hat{A}] + \frac{\partial \hat{A}}{\partial t} \right\rangle \tag{2.1}$$

where $\hat{A}$ is the operator representing A, $\hat{H}$ is the Hamiltonian operator, $[\hat{H}, \hat{A}]$ is the commutator $\hat{H}\hat{A} - \hat{A}\hat{H}$, and $\partial \hat{A}/\partial t$ is included in case the operator $\hat{A}$ depends explicitly on time. Postponing the introduction of magnetic fields till Sect. 5.2, we write the Hamiltonian for a particle in a field of force given by a potential function $V(r)$ as

$$\hat{H} = \frac{\hat{p}^2}{2m} + V(r) \quad . \tag{2.2}$$

We are in the Cartesian coordinate representation, in which positional coordinates are represented by numbers and momenta by operators according to

$$\hat{\boldsymbol{p}} = -i\hbar\nabla \quad . \tag{2.3}$$

The basic commutation relation is

$$[\hat{p}_j, x_k] = -i\hbar\delta_{jk} \quad . \tag{2.4}$$

It is trivial that

$$[\hat{A}, \hat{B}] = -[\hat{B}, \hat{A}] \tag{2.5a}$$

$$[\hat{A}, \hat{B} + \hat{C}] = [\hat{A}, \hat{B}] + [\hat{A}, \hat{C}] \quad \text{and} \tag{2.5b}$$

$$[\hat{A}, \hat{B}\hat{C}] = [\hat{A}, \hat{B}]\hat{C} + \hat{B}[\hat{A}, \hat{C}] \quad . \tag{2.5c}$$

For any component of position,

$$\begin{aligned}
\frac{d\langle x_j\rangle}{dt} &= \frac{i}{\hbar}\left\langle \left[\frac{\hat{\boldsymbol{p}}^2}{2m} + V(\boldsymbol{r}), x_j\right]\right\rangle \\
&= \frac{i}{2m\hbar}\langle[\hat{p}_x^2 + \hat{p}_y^2 + \hat{p}_z^2, x_j]\rangle = \frac{i}{2m\hbar}\langle[\hat{p}_j^2, x_j]\rangle \\
&= \frac{i}{2m\hbar}\langle\hat{p}_j[\hat{p}_j, x_j] + [\hat{p}_j, x_j]\hat{p}_j\rangle \\
&= \frac{1}{m}\langle\hat{p}_j\rangle \quad \text{or}^1
\end{aligned}$$

$$\langle \boldsymbol{p}\rangle = m\frac{d\langle\boldsymbol{r}\rangle}{dt} \quad . \tag{2.6}$$

Thus the relation between expectation values of momentum and velocity is the same as in classical physics. Now we find how $\langle\boldsymbol{p}\rangle$ varies in time.

$$\begin{aligned}
\frac{d\langle p_j\rangle}{dt} &= \frac{i}{\hbar}\left\langle\left[\frac{\hat{\boldsymbol{p}}^2}{2m} + V(\boldsymbol{r}), \hat{p}_j\right]\right\rangle \\
&= -\left\langle\frac{\partial V}{\partial x_j}\right\rangle \quad \text{or}
\end{aligned}$$

$$\frac{d\langle\boldsymbol{p}\rangle}{dt} = -\langle\nabla V\rangle \quad . \tag{2.7}$$

These identities are extremely general, for they refer to expectation values in any state at all. Connections with classical physics can be made only if the state resembles those of classical physics, in which we assume that it is possible to specify dynamical variables without uncertainty. This is an idealization even in classical physics; in quantum mechanics it is not even that, since the indeterminacy relations preclude simultaneous specification of certain pairs of variables. But indeterminacy rarely affects macroscopic parameters, and in all but a few very special situations we may assume wave packets which are, and remain, so narrow in all dynamical

---

[1] $\langle A\rangle$ is the same number as $\langle\hat{A}\rangle$. The notations suggest a slight difference in meaning: $\langle A\rangle$ is the expectation value of the dynamical variable A, while $\langle\hat{A}\rangle$ is the expectation value of the operator $\hat{A}$ that represents A.

variables that the quantum indeterminacy would be swamped by the errors in any physically possible program of measurement.

Problem 2.1. Derive (2.7) step by step from the basic commutation relations.

Problem 2.2. A 1-g point mass is dropped onto a target as accurately as possible from a height of 1 km. Neglect air pressure, the earth's rotation, etc., but assume that a strategy has been derived for minimizing the quantum error in the aim. Show that the ideally attainable accuracy is about 1 fm. Note: The initial Δx and Δp_x each contribute to the error. Use the fact that if a and b vary at random with variances $(\Delta a)^2$ and $(\Delta b)^2$, the variance in $a + b$ is the sum of the individual variances.

With the understanding that we are speaking of macroscopic parameters of macroscopic systems prepared in appropriate states, (2.6) and (2.7) reduce to the basic equations of Newtonian mechanics. Remember, though, that quantum mechanics is a theory of probabilities whereas Newtonian mechanics is a theory of definite dynamical quantities. It is not possible without further assumptions to derive one from the other because the two theories are talking about different subjects.

Quantum mechanics is the better and more general theory, and we now see (neglecting relativistic effects which can be similarly treated) that the equations of Newtonian dynamics, for the systems to which it was intended to apply, agree exactly with it.

2.2 Oscillators and Pendulums

It is not my intention to repeat the mechanics section of a course in elementary physics or to linger over the analogous parts of quantum mechanics, but a few remarks here on simple oscillating systems will be of use later.

A simple harmonic oscillator is a mass m bound to a fixed point by a spring of stiffness k. For motion in one dimension, the equation is

$$m\ddot{x} + kx = 0 \tag{2.8}$$

with the solution

$$x = A \cos (\omega t - \phi_0) \quad , \quad \omega^2 = k/m \quad . \tag{2.9}$$

To analyze sinusoidally oscillating systems like this the complex exponential is the method of choice, but it does not generalize readily to nonlinear oscillations like the pendulum to be discussed below. For this reason I will not use it in this section.

Although nothing could be simpler than the equation just integrated, it is a prototype of other equations of the second order that are appreciably harder, and we should note here that there is a way to integrate such equations that drops it from the second order to the first order. Multiply both sides of (2.8) by $\dot{x}$:

$$m\dot{x}\ddot{x} + kx\dot{x} = 0 \quad .$$

This can be integrated at once to give

$$\tfrac{1}{2}m\dot{x}^2 + \tfrac{1}{2}kx^2 = E(\text{const}) \tag{2.10}$$

31

in which the constant E is, of course, interpreted as the particle's energy, and the equation now becomes

$$\frac{dx}{dt} = \omega\sqrt{\frac{2E}{m\omega^2} - x^2} \tag{2.11}$$

which is integrated as

$$\int \frac{dx}{\sqrt{A^2 - x^2}} = \omega \int dt = \omega(t - t_0) \quad , \quad A = \sqrt{\frac{2E}{m\omega^2}} \quad \text{or} \tag{2.12a}$$

$$\sin^{-1} x/A = \omega(t - t_0) \tag{2.12b}$$

$$x = A \sin \omega(t - t_0) \quad . \tag{2.12c}$$

Several points in this elementary example should be noted. Generally, if mechanical energy is conserved we do not need to prove it, and so we could have started with (2.10). The general solution of the second-order equation contains two arbitrary constants, A and ϕ_0 or E and t_0. In the second solution we lowered the order of the equation from 2 to 1 and picked up a constant of integration, then integrated again and picked up t_0. For a system not subject to time-dependent influences from outside, t_0 is always one of the constants of integration and it is the least interesting, since it refers only to the arbitrary setting of clocks. In this simple case it is no great thing to have dropped the order of the equation from 2 to 1 before solving it, but more complex systems have more complex equations, and lowering their order with the inclusion of constants of integration in the remaining equation of lower order will be a useful device.

The characteristic feature of these oscillations is that the frequency is independent of the amplitude; that is what "harmonic" means, and it comes from the linearity of the equation of motion. In Chap. 4 we will discuss more complicated vibrating systems that are still harmonic, finding that linearity is a great help in their solution. Our next example will be one with a nonlinear equation.

Note on Arc-Sines. In (2.12c), t may be as large as one likes. Therefore this must also be true in (2.12a), and how it can be so may be puzzling if one looks at the obvious boundedness of the integral. To understand what is happening we must go back to (2.11) and note that it is not really correct, since $\dot{x}^2$ has two square roots, and $\dot{x}$ can be either positive or negative. For the quantity $\sqrt{A^2 - x^2}$ to become negative it must go through zero; therefore it changes sign as x goes through $\pm A$; we knew that, so the double sign is not written explicitly in (2.12a). The points where the sign changes are called turning points. The integral in (2.12a) is to be understood as going back and forth along the line from $-A$ to $+A$, the square root changing signs so as to be positive when dx is positive (forth) and negative when it is negative (back). In this way the value of the integral continually increases.

Problem 2.3. For how long, subject to quantum limitations, can a pencil be balanced on its point? Do Brownian fluctuations in air pressure affect your figure?

Problem 2.4. A particle moves in the potential $V(x) = a/x^2 + bx^2 - (a/b)^{1/4}$. (The constant changes nothing; why was it put in?) If its total energy is E, what is its frequency of oscillation? If you notice anything about your result that is peculiar, comment on it. The formula

$$\int_a^b \frac{dx}{\sqrt{(x-a)(b-x)}} = \pi$$

may be useful.

Problem 2.5. Use the integral formula (2.12a) to find the period and amplitude of the harmonic oscillator; then do the same for an oscillator with potential energy $\mu x^4/2$.

Problem 2.6. Show that the harmonicity of an oscillation comes from the linearity of the equation of motion, and that a magic oscillator obeying $d^4x/dt^4 = -Kx$ is also harmonic.

Problem 2.7. Find the motion of an oscillator if it has a resistive force proportional to the velocity. (Note that there is no energy integral here.)

Problem 2.8. A hole is drilled through the earth along a diameter. Assuming no friction, no rotation, and uniform density, find the motion of an object allowed to fall from the earth's surface. How soon does it arrive at the antipodes? How does this compare with the time between the same two points for a satellite in an orbit at treetop height?

The Simple Pendulum. This pendulum is a point mass m on the end of a weightless string of length l. If the mass is distributed one speaks of a physical pendulum. We shall do so in Chap. 9.

The mass moves in a plane whose coordinates may be taken as x and y, but one variable is enough to tell where it is: the angle ϑ with the vertical. It is a system in two dimensions but with one degree of freedom. To set up Newton's second law in terms of torques is easy but unnecessary, for it is easy to write down the energy:

$$E = \tfrac{1}{2}ml^2\dot\vartheta^2 - mgl\cos\vartheta \qquad (2.13)$$

where we have taken the zero level of potential energy at the point of suspension. (This makes E negative for amplitudes less than $\frac{\pi}{2}$.) The differential equation is solved by the integral

$$\int \frac{d\vartheta}{\sqrt{E/mgl + \cos\vartheta}} = \sqrt{\frac{2g}{l}} \int dt \quad . \qquad (2.14)$$

The integration cannot be carried out in terms of elementary functions unless we are willing to assume that the energy is so small that ϑ never gets very large and use the small-angle approximation

$$\cos\vartheta \simeq 1 - \tfrac{1}{2}\vartheta^2 \quad .$$

Then we have

$$\int \frac{d\vartheta}{\sqrt{E/mgl + 1 - \frac{1}{2}\vartheta^2}} = \sqrt{\frac{2g}{l}} \int dt \qquad (2.15)$$

which is formally the same as (2.12a).

Problem 2.9. Complete the solution in the small-angle approximation.

Problem 2.10. Starting from (2.13), find the second-order equation of motion in ϑ.

To reduce the exact equation to a standard form introduce the angular amplitude of swing, ϑ_0. From (2.13) with $\dot{\vartheta} = 0$, $E = -mgl \cos \vartheta_0$, so that (2.14) requires the integral

$$\int \frac{d\vartheta}{\sqrt{\cos \vartheta - \cos \vartheta_0}} \quad .$$

We have already seen that $t(\vartheta)$ reduces to something familiar (an arc-sine) when ϑ_0 is small. To introduce a parameter that is small when ϑ_0 is small, write

$$\cos \vartheta - \cos \vartheta_0 = 2(\sin^2 (\vartheta_0/2) - \sin^2 (\vartheta/2))$$

and then, to measure sin ϑ against the amplitude: $\sin (\vartheta/2) = \sin (\vartheta_0/2) \sin \phi$. With this,

$$\tfrac{1}{2} \cos (\vartheta/2)d\vartheta = k \cos \phi \, d\phi \quad , \quad k = \sin (\vartheta_0/2)$$

and

$$d\vartheta = \frac{2k \cos \phi \, d\phi}{\sqrt{1 - k^2 \sin^2 \phi}}$$

so that finally,

$$t = \sqrt{\frac{l}{2g}} \int_0^\phi \frac{2k \cos \phi \, d\phi}{\sqrt{2k^2(1 - \sin^2 \phi)(1 - k^2 \sin^2 \phi)}} \quad \text{or}$$

$$t = \sqrt{\frac{l}{g}} \int_0^\phi \frac{d\phi}{\sqrt{1 - k^2 \sin^2 \phi}} \qquad (2.16)$$

where we have chosen the limit so that the clock starts as the pendulum goes through its lowest point.

The amplitude determines k, and for ordinary pendulums k^2 is pretty small. To find the period T, note that in a quarter-period ϕ goes from 0 to $\frac{\pi}{2}$, so that

$$T = 4\sqrt{\frac{l}{g}} \int_0^{\pi/2} \frac{d\phi}{\sqrt{1 - k^2 \sin^2 \phi}}$$

$$= 4\sqrt{\frac{l}{g}} \int_0^{\pi/2} \left(1 + \frac{1}{2}k^2 \sin^2 \phi + \frac{3}{8}k^4 \sin^4 \phi + \dots \right) d\phi \qquad (2.17)$$

or

$$T = 2\pi \sqrt{\frac{l}{g} \left(1 + \frac{k^2}{4} + \frac{9k^4}{64} + \ldots \right)} \quad . \tag{2.18}$$

As Galileo noted many years ago, the pendulum's period does not depend much on the amplitude. Even for $\vartheta_0 = \frac{\pi}{2}$, with $k^2 = \frac{1}{2}$, the period is increased by only 17 percent.

Problem 2.11. This pendulum may also be regarded as a little ball that rolls in a trough of circular curvature. Suppose now that the curvature is that of a cycloid. How does the period depend on the amplitude? It will be convenient, using the equations for a cycloid below (1.45), to introduce a new variable u defined by $u^2 = 1 - \cos \vartheta$. The result you are about to obtain was embodied by *Huygens* in the manufacture of clocks in 1657, when Isaac Newton was 15 years old [16].

2.3 Interlude on Elliptic Functions

The integral in (2.17) is called a complete elliptic integral of the first kind, [17] and the indefinite integral in (2.16) defines an elliptic function. Since we will meet them again and they are important in many branches of physics, I will briefly survey them here.

In (2.16) make the substitution $x = \sin \phi$. Omitting the irrelevant constant the integral takes the form

$$t = \int_0^s \frac{dx}{\sqrt{(1 - x^2)(1 - k^2 x^2)}} \quad . \tag{2.19}$$

This is another standard form of the elliptic integral. Look at it first in the limit $k = 0$, where everything is familiar. Evaluation gives

$$t(s) = \sin^{-1} s \quad . \tag{2.20a}$$

The notation suggests a more convenient form for the answer,

$$s(t) = \sin t \tag{2.20b}$$

since whereas (2.20a) defines a multi-valued function of s, (2.20b) defines a single-valued function of t.

If one knew nothing about the sine function, one could find out about it as follows. Go back to the integral and write down its differential equation,

$$\frac{dt}{ds} = \frac{1}{\sqrt{1 - s^2}} \quad , \quad k = 0 \quad .$$

Since we are interested in s as a function of t, invert it,

$$\frac{ds}{dt} = \sqrt{1 - s^2} \tag{2.21a}$$

and find the differential equation for s by differentiating again:

$$\frac{d^2 s}{dt^2} = -\frac{s}{\sqrt{1-s^2}} \frac{ds}{dt} = -s \quad . \tag{2.21b}$$

It follows from (2.19) that $s(0) = 0$ and from (2.21a) that $\dot{s}(0) = 1$. The sine is defined by (2.21b) and these boundary conditions.

Let us go through the same steps with the elliptic integral (2.19). We find

$$\frac{dt}{ds} = \frac{1}{\sqrt{(1-s^2)(1-k^2 s^2)}} \tag{2.22a}$$

and by the same maneuver as before,

$$\frac{d^2 s}{dt^2} = -(1+k^2)s + 2k^2 s^3 \quad . \tag{2.22b}$$

This nonlinear equation and the same boundary conditions $s(0) = 0$, $\dot{s}(0) = 1$ define a periodic function of t that was named

$$s(t) = \operatorname{sn} t \tag{2.23}$$

(pronounced *ess-en tee*) by Jacobi in 1827. Equation (2.19) can now be rewritten so as to make the definition explicit:

$$t = \int_0^{\operatorname{sn} t} \frac{dx}{\sqrt{(1-x^2)(1-k^2 x^2)}} \quad . \tag{2.22c}$$

From our previous work, cf. (2.17), the period of $\operatorname{sn} t$ is $4K(k)$, where

$$K(k) := \int_0^1 \frac{dx}{\sqrt{(1-x^2)(1-k^2 x^2)}} \quad . \tag{2.24}$$

(It has another period with respect to the imaginary part of t, but we will not need that.) For small k, $\operatorname{sn} t$ resembles the sine. Figure 2.1 shows $\operatorname{sn} t$ for two larger values of k.

If sn is analogous to the sine, what is analogous to the cosine? We define

$$\operatorname{cn} t := \sqrt{1 - \operatorname{sn}^2 t} \tag{2.25}$$

with the same sign conventions in taking the square root as for the cosine. Then by differentiating (2.25)

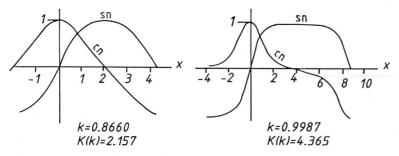

$$k=0.8660$$
$$K(k)=2.157$$

$$k=0.9987$$
$$K(k)=4.365$$

Fig. 2.1. Jacobian elliptic functions $\operatorname{sn} x$ and $\operatorname{cn} x$ for two values of k

$$\frac{d\,\mathrm{cn}\,t}{dt} = -\frac{\mathrm{sn}\,t}{\mathrm{cn}\,t}\frac{d\,\mathrm{sn}\,t}{dt} \quad .$$

If we invert (2.22a) and write it as

$$\frac{d\,\mathrm{sn}\,t}{dt} = \mathrm{cn}\,t\,\mathrm{dn}\,t \quad , \quad \mathrm{dn}\,t := \sqrt{1 - k^2\,\mathrm{sn}^2\,t} \tag{2.26a}$$

we have

$$\frac{d\,\mathrm{cn}\,t}{dt} = -\mathrm{sn}\,t\,\mathrm{dn}\,t \quad . \tag{2.26b}$$

These are easy to remember: write the relations you would expect by analogy with sines and cosines and multiply by $\mathrm{dn}\,t$. Differentiating (2.26b) shows that $\mathrm{cn}\,t$ is a solution of

$$\frac{d^2c}{dt^2} = -(1 - 2k^2)c - 2k^2c^3 \quad . \tag{2.27}$$

Note that $\mathrm{cn}\,t$ does not satisfy the same equation as $\mathrm{sn}\,t$. From (2.26b);

$$dt = \frac{d\,\mathrm{cn}\,t}{\mathrm{dn}\,t\,\mathrm{sn}\,t} = -\frac{d\,\mathrm{cn}\,t}{\sqrt{(1 - \mathrm{cn}^2\,t)(1 - k^2 + k^2\,\mathrm{cn}^2\,t)}}$$

and since $\mathrm{cn}\,(0) = 1$, this is

$$t = \int_{\mathrm{cn}\,t}^{1} \frac{dx}{\sqrt{(1 - x^2)(1 - k^2 + k^2x^2)}} \tag{2.28}$$

to be compared with (2.22c).

Problem 2.12. Find the differential equation and definite integral for $\mathrm{dn}\,t$. Remember that $\sqrt{1 - k^2} \le \mathrm{dn}\,t \le 1$.

Problem 2.13. By expanding the integrands in (2.19) and (2.28), express $\mathrm{sn}\,t$ and $\mathrm{cn}\,t$ in terms of the sine and cosine to the first order in k^2.

Solution for $\mathrm{sn}\,t$. Expanding (2.19) and integrating the first two terms gives

$$t = \sin^{-1} s + \tfrac{1}{4}k^2[\sin^{-1} s - s\sqrt{1 - s^2}] \quad \text{or}$$

$$(1 + \tfrac{1}{4}k^2)\sin^{-1} s = t + \tfrac{1}{4}k^2 s\sqrt{1 - s^2} \quad .$$

We reduce this, always omitting higher powers of k^2:

$$s = \sin\left[(1 - \tfrac{1}{4}k^2)t + \tfrac{1}{4}k^2 s\sqrt{1 - s^2}\right]$$

$$= \sin\,(1 - \tfrac{1}{4}k^2)t + \tfrac{1}{4}k^2 s\sqrt{1 - s^2}\cos\,(1 - \tfrac{1}{4}k^2)t \quad .$$

Let $\omega = 1 - k^2/4$, and in the small second term let $s = \sin\,\omega t$,

$$s = \sin\,\omega t + \tfrac{1}{4}k^2\,\sin\,\omega t\,\cos^2\,\omega t$$

so that finally

$$s = \mathrm{sn}\,t = (1 + \tfrac{1}{16}k^2)\sin\,\omega t + \tfrac{1}{16}k^2\,\sin\,3\omega t + O(k^4) \quad . \tag{2.29}$$

Thus, in lowest order, k^2 changes the frequency and introduces an anharmonicity, though the function still oscillates between -1 and $+1$. The change in frequency corresponds to the change in the period of a pendulum found earlier. Now follow the same procedure for cn t and check your results by verifying that $\operatorname{sn}^2 t + \operatorname{cn}^2 t = 1$.

Problem 2.14. Show that sn t is an odd function and cn t is an even function.

Problem 2.15. What is the limiting form of sn t when $k = 1$?

Problem 2.16. The general solution of a second-order equation contains two arbitrary constants. Show that the general solution of (2.8) can be written either as the sum of two elementary solutions ($x = A \cos \omega t + B \sin \omega t$) or as a solution with displaced origin ($x = C \sin \omega(t - t_0)$). Remembering that sn t and cn t satisfy different, nonlinear equations, construct the general solution of (2.22b).

Problem 2.17. An oscillator is made with a spring having a nonlinear restoring force, $F = -ax - bx^3$. It is started $t = 0$ with $\dot{x} = 0$, $x = A$. Show that its subsequent motion is given by $x = A\operatorname{cn} \omega t$ with a suitable ω and k. How will the oscillator's frequency depend on A for small A?

Problem 2.18. Show that the perimeter of an ellipse with semi-axes a and b is

$$P = 4a \int\limits_0^{\pi/2} \sqrt{1 - k^2 \sin^2 \phi}\, d\phi$$

where k is the eccentricity, $k^2 = 1 - b^2/a^2$. The integral defines a complete elliptic integral of the second kind, and shows how elliptic integrals got their name.

Problem 2.19. Starting from the equation derived in Problem 2.10 and aided by (2.13), derive (2.22b) directly by change of variable. [Remember that the constant factor $\sqrt{g/l}$ was omitted in writing (2.19).]

Problem 2.20. The amplitude, am, of the argument of an elliptic function is defined by

$$\operatorname{sn} t = \sin \operatorname{am} t \quad , \quad \operatorname{cn} t = \cos \operatorname{am} t \quad .$$

Show that omitting the constant factor, am t is the ϕ defined by (2.16) and that

$$\operatorname{am} t = t + \tfrac{1}{8} k^2 \sin 2\omega t + \dots \quad , \quad \omega = 1 - \tfrac{1}{4} k^2 + \dots$$

$$\frac{d}{dt} \operatorname{am} t = \operatorname{dn} t \quad . \tag{2.30}$$

2.4 Driven Oscillators

When an oscillator is driven by an outside force, phenomena of resonance occur. The case familiar from elementary physics is that in which the driving force $f(t)$ is sinusoidal and applied for a long time. We will discuss the harmonic oscillator in the more general case of arbitrary $f(t)$ and then give an elementary treatment of resonance to an applied sinusoidal force in an anharmonic oscillator, in which some entirely new phenomena occur.

Driven Harmonic Oscillator. Assume that there is a damping force proportional to the velocity and that the force per unit mass is an arbitrary function $f(t)$. The equation to be solved is

$$\ddot{x} + 2r\dot{x} + \omega_0^2 x = f(t) \tag{2.31}$$

where r represents the resistance due to friction. The general solution for the motion without an applied force is

$$x(t) = (a \cos \omega_1 t + b \sin \omega_1 t)e^{-rt} \tag{2.32}$$

where a and b are constants,

$$\omega_1 = \sqrt{\omega_0^2 - r^2} \tag{2.33}$$

and the cosine and sine are replaced by hyperbolic functions if r is large enough to make ω_1 imaginary. We are going to solve (2.31) by assuming that the oscillating particle's present position at time t is produced by the entire past history of the force $f(t)$, and there is an influence function Φ which tells how the value of f at a particular past moment t' influences the present value of $x(t)$. The obvious formula that embodies these ideals is

$$x(t) = \int_{-\infty}^{t} \Phi(t - t')f(t')dt' \tag{2.34}$$

where the lower limit is taken as $-\infty$ for convenience only, and may be any fixed value.

Differentiating (2.34) gives

$$\dot{x}(t) = \Phi(0)f(t) + \int \dot{\Phi}(t - t')f(t')dt'$$

$$\ddot{x}(t) = \Phi(0)\dot{f}(t) + \dot{\Phi}(0)f(t) + \int \ddot{\Phi}(t - t')f(t')dt' \quad .$$

Substitution into the original equation (2.31) then gives

$$\ddot{x} + 2r\dot{x} + \omega_0^2 x = \Phi(0)\dot{f}(t) + [\dot{\Phi}(0) + 2r\Phi(0)]f(t) + \int [\ddot{\Phi}(t - t')$$

$$+ 2r\dot{\Phi}(t - t') + \omega_0^2\Phi(t - t')]f(t')dt' = f(t)$$

and this is satisfied provided that Φ satisfies (2.31) with $f = 0$, together with the boundary conditions

$$\Phi(0) = 0 \quad , \quad \dot{\Phi}(0) = 1 \quad . \tag{2.35}$$

Given the general solution (2.32), this is easy to arrange:

$$\Phi(\tau) = \omega_1^{-1}e^{-r\tau} \sin \omega_1\tau \quad , \quad \tau = t - t' \tag{2.36}$$

and the general solution of (2.31), assuming that the force f accounts for all the oscillator's present motion, is

$$x(t) = \int_{-\infty}^{t} e^{-r(t-t')} \frac{\sin \omega_1(t - t')}{\omega_1} f(t')dt' \quad . \tag{2.37}$$

Note the role of the resistance r here: if the interval $t - t'$ is long, the exponential is small, and the corresponding values of $f(t')$ have little influence on the present $x(t)$. That is, the oscillator forgets its ancient history. The idealized case with $r = 0$ has the peculiar feature that such an oscillator forgets nothing.

Problem 2.21. Show that (2.32) is the general solution of (2.31) when $f = 0$.

Problem 2.22. A hammer blow at time t_0 started the oscillator moving. The force of the blow was such that the hammer, of mass m travelling with velocity v, was brought to rest. Where is the oscillator now?

Problem 2.23. A sinusoidally driven anharmonic oscillator has an additional cube-law restoring force $-\varepsilon x^3$. Write the equation of motion as

$$\ddot{x} + \omega_0^2 x = -\varepsilon x^3 + f_0 \cos \omega t \quad , \quad f_0 = \text{const}$$

and find the first-order perturbed solution $x = x_0 + x_1$ by finding a solution x_0 without ε and then, for a suitable Φ, letting

$$x_1(t) = \int_{-\infty}^{t} \Phi(t - t')[-\varepsilon x^3(t')]dt' \quad . \tag{2.38}$$

We will construct a nonperturbative solution below.

Problem 2.24. Let $r = 0$ and let the applied force be a slow push of the form

$$f(t) = ct_1^{-1} e^{-(t-t_0)^2/t_1^2}$$

centered at t_0, and let $t - t_0 \gg t_1$ so that the push was completed long ago and the upper limit of the integral may be replaced by ∞. Discuss in the light of common experience the difference in behavior according as $\omega_0 t_1$ is small or large compared to unity.

Problem 2.25. Discuss the response of the harmonic oscillator to a sinusoidal driving force $f_0 \cos \omega t$ (f_0 const), paying attention to the possibility that r^2 may exceed ω_0^2 in (2.33).

Problem 2.26. Use the example of a driven harmonic oscillator to discuss what life would be like in a laboratory in which mechanical oscillators with negative r could be built.

Problem 2.27. Discuss the response of a resistanceless oscillator to a sinusoidal force $f_0 \cos \omega t$ that was turned on at time t_0.

2.5 A Driven Anharmonic Oscillator

If an anharmonic (i.e., nonlinear) oscillator is driven by a periodic force there are resonance phenomena, but they are considerably more interesting than those of the linear oscillator. For convenience I illustrate with a simple case.

Let the oscillator's anharmonicity be produced by a cubic restoring force, so that the equation of motion (Duffing's equation) is of the form

$$\ddot{x} + \omega_0^2 x + \varepsilon x^3 = f_0 \cos \omega t \qquad (2.39)$$

where the parameter ε will not be considered small. To include damping would only complicate the picture. Equation (2.29) suggests we look at a solution of the form

$$x = a \cos \omega t + b \cos 3\omega t + \ldots \quad .$$

(Note that as with the harmonic oscillator this substitution will not give the general solution.) Substitute this into (2.39) and reduce the nonlinear term by trigonometric identities, keeping only terms of frequency ω and 3ω:

$$-(\omega^2 - \omega_0^2)a \cos \omega t - (9\omega^2 - \omega_0^2)b \cos 3\omega t$$

$$+\frac{3}{4}\varepsilon \left[(a^3 + a^2 b + 2ab^2) \cos \omega t + \left(\frac{a^3}{3} + 2a^2 b + b^3 \right) \cos 3\omega t \right]$$

$$+\ldots = f_0 \cos \omega t \qquad (2.40)$$

where the omitted terms are of frequency 5ω, 7ω, and 9ω. Evidently there is coupling between the various modes, and driving the oscillator with frequency ω will produce resonant responses at all the higher frequencies. Another way to put this is to say that the system has several resonant frequencies and can be made to oscillate in them by driving it at lower frequencies called subharmonics. The multiple resonances can be clearly heard in the ear, which for protective purposes is made highly nonlinear. Listen to a sine-wave tone (monitor it with an oscilloscope to make sure that the amplifier stays in the linear range!) and turn up the volume. A shrillness is heard corresponding to the excitation of higher frequencies in the resonating structures of the inner ear.

The analysis of an equation like (2.40) is wearisome; let us study only the fundamental resonance. To do this set $b = 0$ and ignore the higher frequency:

$$(-\Delta + \tfrac{3}{4}\varepsilon a^2)a = f_0 \quad , \qquad \Delta = \omega^2 - \omega_0^2 \quad \text{or}$$

$$\Delta = \frac{3}{4}\varepsilon a^2 - \frac{f_0}{a} \quad . \qquad (2.41)$$

This function is sketched in Fig. 2.2 for $\varepsilon > 0$ and $f_0 > 0$, and a few easily-identified points are labelled. A given value of ω corresponds to a horizontal line at $\Delta = \omega^2 - \omega_0^2$. If one varies ω^2, starting at negative values of Δ, there will be a steady-state motion, but with a remaining finite as long as ε is not zero, along the right-hand branch of the curve. If one starts at large positive values of Δ, there are three values of a possible, but the part of the curve marked as unstable does not represent an observable motion, since we shall see below that the slightest disturbance will throw the system onto one of the stable branches of the curve. The two stable branches correspond to different modes; if $a > 0$ the motion is in phase with the driving force while in the other case it is out of phase. If the system is started at positive Δ in the out-of-phase mode and ω is decreased below the minimum value allowed for that mode, the oscillator will make an abrupt transition to in-phase motion.

Problem 2.28. Derive Eq. (2.40).

Problem 2.29. Use (2.40) to discuss resonance at frequencies in the neighborhood of $\omega_0/3$. (Questions of stability of these resonances are difficult and can be ignored.)

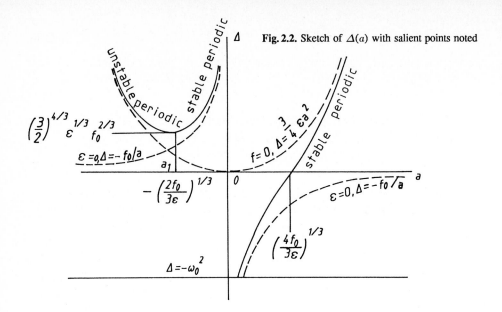

Fig. 2.2. Sketch of $\Delta(a)$ with salient points noted

Problem 2.30. What happens when the anharmonic force is $-\varepsilon x^2$ instead of $-\varepsilon x^3$?

Problem 2.31. Analyze the nonlinear system by quantum mechanics, taking both the anharmonicity and the driving force as perturbations. (The way to do this is to let both perturbations be turned on adiabatically from $t = -\infty$ and use time-dependent perturbation theory.)

The Question of Stability. Imagine that some part of a moving dynamical system is tapped gently with a hammer. Three kinds of behavior are possible: the displacement from the unperturbed path may become smaller and smaller, so that the effect of the perturbation ultimately disappears. Or the system may adopt a new motion that fluctuates around the unperturbed path but never goes far from it. Or finally, the system may wander away in a new path which, in time, takes it as far as one likes from the old one. In this case the motion is said to be unstable; the other two represent different kinds of stability. The criteria for different kinds of stability have been variously stated and studied by some of the best mathematicians of the last hundred years (Poincaré, Kolmogorov, Lyapounov), and it would be folly to go into generalities here, but mathematical facts known to most readers of this book are enough to treat the stability of the anharmonic oscillator discussed above, [19] and in Sect. 9.4 we shall briefly note a simpler case involving the free precession of a solid body.

Suppose that the motion of an oscillator satisfying (2.39) has been slightly perturbed, so that $x(t)$ becomes $x(t) + \xi(t)$, with ξ initially small. After the perturbation the system still satisfies (2.39):

$$\ddot{x} + \ddot{\xi} + \omega_0^2(x + \xi) + \varepsilon(x + \xi)^3 = f_0 \cos \omega t \quad . \tag{2.42}$$

If we keep only the first two terms in $(x + \xi)^3$ and subtract (2.39), we find that ξ, as long as it is small, satisfies the linear equation

$$\ddot{\xi} + (\omega_0^2 + 3\varepsilon x^2)\xi = 0 \quad .$$

With $x \simeq a \cos \omega t$, this is

$$\ddot{\xi} + (\omega_0^2 + \tfrac{3}{2}\varepsilon a^2 + \tfrac{3}{2}\varepsilon a^2 \cos 2\omega t)\xi = 0 \quad . \tag{2.43}$$

The prototype of this equation is known as Mathieu's equation [17], [20], often written in the standard form

$$\frac{d^2\xi}{dx^2} + (\alpha + \beta \cos 2x)\xi = 0 \quad , \quad x = \omega t \quad . \tag{2.44}$$

It is a somewhat advanced topic in the theory of special functions, but the solutions have a qualitative feature that you may have encountered in another context. Write it as

$$\frac{d^2\xi}{dx^2} + \frac{2m}{\hbar^2}[E - V(x)]\xi = 0 \quad \text{where} \tag{2.45}$$

$$E = \frac{\hbar^2\alpha}{2m} \quad , \quad V = -\frac{\hbar^2\beta}{2m}\cos 2x \quad . \tag{2.46}$$

It is now the Schrödinger equation for a particle in a sinusoidally periodic lattice, and we know that the eigenvalues E for which the equation has solutions that remain finite in x lie in bands.

In the present context, x represents the time, and we see that, for a given β in (2.44), there are only certain ranges of α for which the solution is stable for large values of t. To derive these limits of stability is not especially simple, but they can be expressed as series of which a few terms are shown in Fig. 2.3.

The question before us is to see under what circumstances the solutions of (2.43) are stable. Setting $x = \omega t$ in (2.43) and comparing with (2.44) gives

$$\omega^2\alpha = \omega_0^2 + \tfrac{3}{2}\varepsilon a^2 \quad , \quad \omega^2\beta = \tfrac{3}{2}\varepsilon a^2 \quad . \tag{2.47}$$

Since $\alpha > 0$, the condition that α lie in the lowest stable band can be read off the curves in Fig. 2.3,

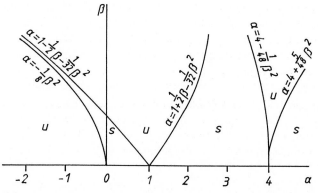

Fig. 2.3. Stable and unstable regions of the $\alpha\beta$ plane together with approximate expressions for their boundaries

43

$$\omega^2 \alpha < \omega^2 - \tfrac{1}{2}\omega^2 \beta - \dots$$

or by (2.47)

$$\omega_0^2 + \tfrac{3}{2}\varepsilon a^2 < \omega^2 - \tfrac{3}{4}\varepsilon a^2 + \dots \quad .$$

By (2.41), and neglecting the higher terms, this is

$$\frac{3}{2}\varepsilon a^2 < -\frac{f_0}{a} \quad .$$

We can express f_0 in terms of the amplitude a_1 in Fig. 2.2 at which the left-hand branch of the resonance curve has its minimum, given by

$$f_0 = -\tfrac{3}{2}\varepsilon a_1^3 \quad (a_1 < 0) \quad .$$

Thus

$$a^2 < \frac{a_1^3}{a} = -\frac{|a_1|^3}{a} \quad .$$

This implies that $a_1 < a < 0$. A similar calculation, using the other curve that meets the axis at $\alpha = 1$, shows that the whole right-hand branch is stable, and so the criterion for stability becomes

$$a_1 < a \quad . \tag{2.48}$$

If the oscillator is started with parameters that place it exactly on the curve of unstable periodic motions in Fig. 2.2, it will stay there in the sense that a pencil might in principle be exactly balanced on its point, but eventually it will wander away from the curve and move aperiodically.[2]

Instabilities of this kind have been much studied in recent years as it has become obvious that they occur in almost all branches of physics and engineering. We have just seen that in the unstable region two trajectories that are initially close together will move apart, and a detailed study [18, Chap. 6] shows that the divergence begins exponentially. Problem 2.36 will show how fast it occurs in a numerical calculation. Consider what this means if one intends to use the equations of Newtonian mechanics to predict the position of an actual nonlinear oscillator at some future moment when the motion is in the unstable regime. One must supply the initial data, but the slightest error will grow until the calculated position has no relation to what actually happens. We will return to this matter in Sect. 6.6 to calculate the rate of divergence in a toy model, and it will turn out that for any practical purpose one cannot predict what is going to happen. This will lead to some thoughts as to whether one has a right to call Newtonian mechanics a deterministic theory.

Is there any obvious difference in behavior between oscillators in the stable and unstable regimes? Suppose we start two nonlinear oscillators of this kind, A and B, so that A is on a stable periodic curve and B is off the curve but close to A. Then A will oscillate at the same frequency as the driving force, while B will move nearly in phase with it. On the unstable branch this will not happen, since B soon moves

---

[2] There are graphical ways to represent the behavior of linear systems that open an intuitive understanding of them [21, of which Volume 1 illustrates Duffing's equation].

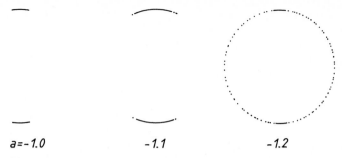

a=-1.0   -1.1   -1.2

Fig. 2.4. Synchrony between oscillator motion and driving force, as described in the text. In these plots $a_1 = -1.1006$, Δ is set close to the critical value, and a is set at -1.0, -1.1, and -1.2

entirely differently from A and therefore out of synchrony with the driving force. A simple machine calculation (Problem 2.32) illustrates this behavior. Represent the periodic driving force by a point that moves in a circle at frequency ω, and put a dot on this circle when the position of the oscillator goes through zero. This will show the relation between the position of the oscillator and the phase of the driving force. Figure 2.4 shows typical results. In these plots the transition between the stable and unstable regimes is not sharp; the motion just becomes less coordinated with the driving force. Far from the curves of periodicity the instability is still determined by (2.48), but there is no obvious difference in behavior between the stable and unstable regimes.

Problem 2.31. What differences occur in the foregoing discussions if $\varepsilon < 0$?

Problem 2.32. Write the program described in the preceding paragraph and use it to study the transition between the stable and unstable regimes.

Problem 2.33. Using a computer, study the effect of a damping force proportional to $\dot{x}$ in the oscillator considered above [6, 18, 26].

Problem 2.34. Use a computer to study the motion of the Duffing oscillator by plotting orbits in phase space, the coordinates of which are x horizontal, $\dot{x}$ vertical.[3] Start with $f_0 = 10$, $\omega_0 = 1$, $\varepsilon = 5$, and see what happens when $\omega = 1.55$, $x_0 = 1.5$, and $\dot{x}_0 = 0$. Can you explain what is happening? Try some other orbits. It will be helpful to look at the results by plotting x against t.

Problem 2.35. Disconnect the driving force by letting $f_0 = 0$ and study some phase-space orbits of free motion. Can you explain qualitatively the changes seen as the oscillator's energy is raised?

Problem 2.36. Study the stability of periodic orbits as follows: Program a computer to plot pairs of phase-space orbits starting at the same value of $\dot{x}$ but slightly different values of x; then let one of the orbits be a periodic one. Is (2.48) verified?

Problem 2.37. Use the program of the previous problem to study pairs of initially neighboring trajectories neither of which is periodic.

---

[3] The representation of the state of a dynamical system by a point in phase space will be important in Chap. 5.

2.6 Quantized Oscillators

We have analyzed how some oscillators move. What is the quantum analogue of such motions? To answer this one might start by quantizing an oscillator and exhibiting its stationary states, but the result, exactly because the states are stationary, doesn't resemble the motion as described in classical physics or as observed in nature. The states we ordinarily observe are not stationary, but what are they? The following sections will discuss the question under several topics: the quantization of the simple harmonic oscillator (briefly, because everybody should know how to do it), the Heisenberg picture of the oscillator, the description of it in terms of coherent states, and finally the calculation of its behavior when subjected to an external force.

Quantized Oscillator. An oscillating particle of mass m and natural frequency ω is described in the Schrödinger picture by the equation of motion

$$\hat{H}|n\rangle = E_n|n\rangle \quad , \quad \hat{H} = \frac{1}{2}\left(\frac{\hat{p}^2}{m} + m\omega^2\hat{x}^2\right)$$

where ^ is the sign of an operator and the operators $\hat{p}$ and $\hat{x}$ obey the algebraic rule

$$[\hat{p}, \hat{x}] = -i\hbar \quad . \tag{2.49}$$

In addition to $\hat{p}$ and $\hat{x}$, introduce two new *nonhermitian* operators obtained by factoring $\hat{H}$ in an obvious way,

$$\hat{a} := \sqrt{\frac{m\omega}{2\hbar}}\left(\hat{x} + \frac{i}{m\omega}\hat{p}\right) \quad , \quad \hat{a}^\dagger := \sqrt{\frac{m\omega}{2\hbar}}\left(\hat{x} - \frac{i}{m\omega}\hat{p}\right) \quad . \tag{2.50}$$

The commutation relation between these operators follows from (2.49):

$$[\hat{a}, \hat{a}^\dagger] = 1 \tag{2.51}$$

and the entire calculation is carried out with these operators.

It is a simple matter to show that the Hamiltonian becomes

$$\hat{H} = \frac{1}{2}\hbar\omega(\hat{a}^\dagger\hat{a} + \hat{a}\hat{a}^\dagger) = \hbar\omega(\hat{N} + \frac{1}{2}) \tag{2.52}$$

where $\hat{N}$, the number operator, is defined as $\hat{a}^\dagger\hat{a}$. The calculation will be done by finding the eigenvalues and eigenfunctions of $\hat{N}$.

First we can show (Problem 2.38) that $\hat{a}^\dagger$ and $\hat{a}$ are the creation and annihilation operators for $\hat{N}$ in the sense that if $|n\rangle$ is the normalized eigenfunction of $\hat{N}$ belonging to the eigenvalue n, then

$$\hat{a}^\dagger|n\rangle \propto |n+1\rangle \quad , \quad \hat{a}|n\rangle \propto |n-1\rangle \quad . \tag{2.53}$$

It is clear from the definition of $\hat{H}$ that all the eigenvalues of $\hat{H}$ are positive; therefore those of $\hat{N}$ are all greater than $-\frac{1}{2}$. But it would appear from (2.53) that by successive applications of $\hat{a}$ we could reduce the eigenvalue as far as we like. Therefore there is a point where the process fails, and we define an eigenfunction $|0\rangle$ as that for which

$$\hat{a}|0\rangle = 0 \quad . \tag{2.54}$$

Clearly, $\hat{N}|0\rangle = 0$ also, and therefore zero is the lowest eigenvalue of $\hat{N}$. The eigenvalues of $\hat{H}$ are read off from (2.52) and (2.53),

$$E_n = (n + \tfrac{1}{2})\hbar\omega \quad , \quad n = 0, 1, 2, \ldots \tag{2.55}$$

so that the eigenvalues of $\hat{N}$ tell which state the operator is in.

Higher states of the oscillator are generated by operating with $\hat{a}^\dagger$ on $|0\rangle$, and a detailed study using the completeness relations [1, App. 2] shows that these are in fact all the states. Next we must pause for a moment to see how they are normalized. Write the nth normalized eigenfunction of $\hat{N}$ as $C_n\hat{a}^{\dagger n}|0\rangle$, where C_n is a normalizing constant to be determined:

$$|n\rangle = C_n\hat{a}^{\dagger n}|0\rangle = C_n\hat{a}^\dagger\hat{a}^{\dagger n-1}|0\rangle = \frac{C_n}{C_{n-1}}\hat{a}^\dagger|n-1\rangle \quad .$$

Then

$$\langle n|n\rangle = \left|\frac{C_n}{C_{n-1}}\right|^2 \langle n-1|\hat{a}\hat{a}^\dagger|n-1\rangle \tag{2.56}$$

where we have used the hermiticity of $\hat{p}$ and $\hat{x}$. If all the eigenfunctions are normalized, then by (2.51) this is

$$1 = \left|\frac{C_n}{C_{n-1}}\right|^2 n \quad \text{so that}$$

$$C_n = \frac{1}{\sqrt{n}}C_{n-1} = \ldots = \frac{1}{\sqrt{n!}}C_0 = \frac{1}{\sqrt{n!}}$$

since evidently $C_0 = 1$. Thus finally

$$|n\rangle = \frac{1}{\sqrt{n!}}\hat{a}^{\dagger n}|0\rangle \quad . \tag{2.57}$$

To find $|0\rangle$ we use (2.50) to calculate the normalized solution to (2.54), and in the coordinate representation it comes out as

$$|0\rangle = \left(\frac{m\omega}{\pi\hbar}\right)^{1/4}\exp\left(-\frac{m\omega}{2\hbar}x^2\right) \quad . \tag{2.58}$$

From this all the stationary states can be constructed.

The Heisenberg Picture. The foregoing derivation is neat and leads to useful formulations of more advanced problems but it has little to do with one's intuitive idea of an oscillator. The obscurity of the connection between the two theories originates in the simple fact that classical mechanics exhibits dynamical variables like p as functions of time, whereas in the Schrödinger form of quantum mechanics the operator $\hat{p}$ is independent of time; all time dependence of this description is expressed by the wave function, which has no counterpart in classical theory. The original form of quantum mechanics as invented by Heisenberg had no wave function and all the time dependence resided in the operators representing dynamical variables, so that it was conceptually somewhat closer to classical mechanics than the theory as later developed by Schrödinger. The connection between the two can be developed as follows.

The observable quantities of a quantum system are calculated from the Schrödinger wave function $\psi_S(x, t)$ (itself not an observable) by finding the expectation value of some dynamical variable A according to the formula

$$\langle A \rangle(t) = \int \psi_S(x, t)^* \hat{A} \psi_S(x, t)(dx) \tag{2.59}$$

where $\hat{A}$ is the operator representing the variable A. The time dependence is prescribed by Schrödinger's equation

$$i\hbar \frac{\partial}{\partial t} \psi_S(x, t) = \hat{H}(x) \psi_S(x, t)$$

which can be integrated formally to give

$$\psi_S(x, t) = e^{-i\hbar^{-1}\hat{H}t} \psi_S(x, 0) \quad.$$

Put this into (2.59) and find

$$\langle A \rangle(t) = \int \psi_H(x)^* \hat{A}_H(x, t) \psi_H(x)(dx) \quad \text{where}$$

$$\psi_H(x) = \psi_S(x, 0) \quad \text{and} \quad \hat{A}_H(x, t) := e^{i\hbar^{-1}\hat{H}t} \hat{A} e^{-i\hbar^{-1}\hat{H}t} \quad. \tag{2.60}$$

H stands for Heisenberg and the time dependence has been transferred to the operator representing the dynamical variable. The equation of motion for $\hat{A}_H$ is found by differentiation,

$$\frac{d}{dt} \hat{A}_H = \frac{i}{\hbar} [\hat{H}, \hat{A}_H] \quad. \tag{2.61}$$

Note that $\hat{H}$ does not need a subscript since it commutes with itself and is therefore independent of time. Of course the expression for $\hat{A}_H$ in (2.60) solves this equation, but except in special cases it is hard to evaluate.

To see how this works let us see it in a familiar context, the motion of a particle. We will need the commutator

$$\begin{aligned}
[\hat{p}_H, \hat{x}_H] &= [e^{i\hbar^{-1}\hat{H}t} \hat{p} e^{-i\hbar^{-1}\hat{H}t}, e^{i\hbar^{-1}\hat{H}t} \hat{x} e^{-i\hbar^{-1}\hat{H}t}] \\
&= e^{i\hbar^{-1}\hat{H}t} [\hat{p}, \hat{x}] e^{-i\hbar^{-1}\hat{H}t}
\end{aligned}$$

so that as in the Schrödinger representation,

$$[\hat{p}_H, \hat{x}_H] = -i\hbar \quad. \tag{2.62}$$

Problem 2.39 will show that if we start from the usual Hamiltonian the Heisenberg operators satisfy (in 3 dimensions)

$$\frac{d\hat{x}_H}{dt} = \frac{\hat{p}_H}{m} \quad, \quad \frac{d\hat{p}_H}{dt} = -\nabla V(\hat{x}_H, t) \quad. \tag{2.63}$$

These are of the familiar Newtonian form except that because they involve noncommuting quantities they are normally much harder to solve.

There are a few simple cases. For a free particle, for example, we can integrate directly to get

$$\hat{p}_\text{H}(t) = \text{const} \quad , \quad \hat{x}_\text{H}(t) = \hat{x}_\text{H}(0) + \frac{\hat{p}_\text{H}}{m} t \quad .$$

Suppose that at time 0 the wave function ψ_H represents a wave packet centered at $\langle \hat{x} \rangle(0)$. Then, using always the same initial wave packet, the particle's later position will be given by

$$\langle \hat{x}_\text{H} \rangle(t) = \langle \hat{x}_\text{H} \rangle(0) + \langle \hat{p}_\text{H} \rangle \frac{t}{m} \quad .$$

The simple harmonic oscillator is also easy to solve; it is left for Problem 2.41.

The commutation relation (2.62) and the equation of motion (2.61) suffice in principle for the calculation of Heisenberg operators and therefore, given also the theory's probabilistic machinery, for all calculations in the domain we are discussing. Of course, the scope of the theory extends further; dynamical variables like spin and isospin require further postulates and also we are not confined to Cartesian coordinates. Curvilinear coordinates can be introduced but they will not be needed here.

Problem 2.38. Show that the operators $\hat{a}^\dagger$ and $\hat{a}$ raise and lower the eigenvalues of $\hat{N}$ as in (2.53).

Problem 2.39. Show that $\hat{x}_\text{H}$ and $\hat{p}_\text{H}$ satisfy the Newtonian equations (2.63).

Problem 2.40. Find constants that replace the proportionalities in (2.53) by equalities.

Problem 2.41. Solve the Heisenberg equations of motion for the simple harmonic oscillator, with and without a driving force. (To introduce a driving force $f(t)$, add a term $-\hat{x} f(t)$ to the Hamiltonian operator.)

Problem 2.42. A free particle starts moving in a Gaussian wave packet. Show, in the Heisenberg picture, how the packet broadens as it moves forward.

2.7 Coherent States

The Heisenberg picture helps to bridge the gap between quantum and classical theories; at least the dynamical variables depend on time in the expected way. But the gap remains, since these dynamical variables are operators, not numbers, and in addition, we can no longer, as in the Schrödinger picture, visualize the particle's probability distribution in space. What we need is a picture of the particle's motion in terms of states that represent a moving distribution. Such states were introduced long ago by *Schrödinger* [22, 23], and have been studied since the 1960s; they are called coherent states.

Equation (2.54) defines the ground state $|0\rangle$ as the eigenstate of $\hat{a}$ belonging to the eigenvalue 0. Coherent states are represented by this and other eigenfunctions of $\hat{a}$. Since $\hat{a}$ is not hermitian, the eigenvalues, which we shall call α, are in general complex numbers. Introducing a convenient constant factor and ignoring time dependence for the moment, write

$$\hat{a}|\alpha\rangle = \sqrt{\frac{m\omega}{2\hbar}}\alpha|\alpha\rangle \quad . \tag{2.64}$$

By the definition of $\hat{a}$ this is

$$\left(x + \frac{\hbar}{m\omega}\frac{d}{dx}\right)|\alpha\rangle = \alpha|\alpha\rangle$$

of which the normalized solution is

$$|\alpha\rangle = \left(\frac{m\omega}{\pi\hbar}\right)^{1/4}\exp\left[-\frac{m\omega}{2\hbar}(x-\alpha)^2\right] \quad . \tag{2.65}$$

(It will be understood in this discussion that all ket-vectors are expressed in the coordinate representation.) These are gaussian wave functions the same as (2.58) but displaced through a (complex) distance α.

Now let us see how these states depend on time. To do this it will be convenient to express them in terms of the eigenstates $|n\rangle$ of $\hat{N}$, which have a simple time dependence. Let

$$|\alpha\rangle = \sum_{n=0}^{\infty}|n\rangle\langle n|\alpha\rangle$$

where the expansion coefficients are given by

$$\langle n|\alpha\rangle = \left\langle 0\left|\frac{\hat{a}^n}{\sqrt{n!}}\right|\alpha\right\rangle = \left(\frac{m\omega}{2\hbar}\right)^{n/2}\frac{\alpha^n}{\sqrt{n!}}\langle 0|\alpha\rangle \tag{2.66}$$

so that

$$|\alpha\rangle = \sum_{n=0}^{\infty}\frac{1}{\sqrt{n!}}\left(\frac{m\omega}{2\hbar}\right)^{n/2}\alpha^n|n\rangle\langle 0|\alpha\rangle \tag{2.67}$$

(see Problem 2.43). All these quantities except for $|n\rangle$ are independent of time. Now put in the time dependence of $|n\rangle$, proportional to $e^{-in\omega t}$:

$$|\alpha\rangle(t) = \sum_{n=0}^{\infty}\frac{1}{\sqrt{n!}}\left(\frac{m\omega}{2\hbar}\right)^{n/2}(\alpha e^{-i\omega t})^n|n\rangle\langle 0|\alpha\rangle \quad .$$

This is the same as (2.67) except that α has acquired a time dependence, and (2.65) shows at once how to write it:

$$|\alpha\rangle(t) = \left(\frac{m\omega}{\pi\hbar}\right)^{1/4}\exp\left[-\frac{m\omega}{2\hbar}(x-\alpha e^{-i\omega t})^2\right] \quad . \tag{2.68}$$

This is the coherent state $|\alpha\rangle$ in the Schrödinger representation. If the time dependence were sinusoidal and real, the interpretation would be simple: the wave packet just swings back and forth, but it is complex and we must look at it more closely.

In (2.68), assume that α is real (its phase can be included additively in t):

$$(x-\alpha e^{-i\omega t})^2 = (x-\alpha\cos\omega t)^2 + 2i\alpha\sin\omega t(x-\alpha\cos\omega t) - \alpha^2\sin^2\omega t \quad .$$

The last term is independent of x and must be absorbed in the normalization. The middle term contains phase information, but if we ask for the particle's probability distribution only the first term remains:

$$|\,|\alpha\rangle(t)|^2 = \sqrt{\frac{m\omega}{\pi\hbar}}\exp\left[-\frac{m\omega}{\hbar}(x - \alpha\cos\omega t)^2\right] \quad . \tag{2.69}$$

The wave packet oscillates as one would expect; what is perhaps surprising is that it does not spread out as time goes on. This peculiar feature of harmonic oscillators was first noticed by Schrödinger.

Finally, if an oscillator is known to be in the state $|\alpha\rangle$, what is the probability, call it P_n, that it would be found in the nth stationary state? For this we need the probability amplitude (2.66), and this gives

$$P_n = |\langle n|\alpha\rangle|^2 = \frac{1}{n!}\left(\frac{m\omega|\alpha|^2}{2\hbar}\right)^n|\langle 0|\alpha\rangle|^2 \quad .$$

To find $|\langle 0|\alpha\rangle|^2$,

$$\sum_{n=0}^{\infty} P_n = \exp\left[\frac{m\omega|\alpha|^2}{2\hbar}\right]|\langle 0|\alpha\rangle|^2 = 1$$

so that P_n becomes

$$P_n = \frac{1}{n!}\left(\frac{m\omega|\alpha|^2}{2\hbar}\right)^n\exp\left[-\frac{m\omega|\alpha|^2}{2\hbar}\right] \tag{2.70}$$

which is a Poisson distribution: if the average number of rare and uncorrelated events occurring in a certain time interval is $\bar{n}$, the probability that an observation during an equal time interval will yield exactly n events is

$$P_n = \frac{\bar{n}^n}{n!}e^{-\bar{n}} \quad . \tag{2.71}$$

Comparison with (2.70) shows that

$$\bar{n} = \frac{m\omega|\alpha|^2}{2\hbar} \quad . \tag{2.72}$$

From (2.68) we see that the half-width of the distribution in the oscillator's ground state is $d = \sqrt{\hbar/m\omega}$ and that $|\alpha|$ is the amplitude of the oscillations of the peak of the coherent state; therefore $\bar{n}$ can be interpreted as $|\alpha/d|^2/2$.

The Forced Oscillator. There is another way to look at (2.67): associate the factor $e^{-i\omega t}$ with the operator $\hat{a}$ from the beginning. That is, put $\hat{a}$ into the Heisenberg picture. This can be done by using (2.50):

$$\frac{d}{dt}\hat{a}_{\mathrm{H}} = i\omega\left[\hat{a}_{\mathrm{H}}^{\dagger}\hat{a}_{\mathrm{H}}, \hat{a}_{\mathrm{H}}\right] = -i\omega\hat{a}_{\mathrm{H}}$$

and similarly for $\hat{a}_{\mathrm{H}}^{\dagger}$, so that

$$\hat{a}_{\mathrm{H}} = \hat{a}e^{-i\omega t} \quad , \quad \hat{a}_{\mathrm{H}}^{\dagger} = \hat{a}^{\dagger}e^{i\omega t} \quad .$$

Define the time-dependent coherent states as those generated by $\hat{a}_{\text{H}}$. It is now easy to solve the forced oscillator. Write the Hamiltonian as (2.52) with an added term $-f(t)x$. With (2.50) this is

$$\hat{H} = \hbar\omega\left(\hat{a}_{\text{H}}^{\dagger}\hat{a}_{\text{H}} + \frac{1}{2}\right) - \sqrt{\frac{\hbar}{2m\omega}}\, f(t)(\hat{a}_{\text{H}} + \hat{a}_{\text{H}}^{\dagger}) \tag{2.73}$$

and (2.61) gives

$$\frac{d}{dt}\hat{a}_{\text{H}} = -i\omega\hat{a}_{\text{H}} + \frac{i}{\hbar}\sqrt{\frac{\hbar}{2m\omega}}\, f(t) \quad . \tag{2.74}$$

This can be solved by the substitution (2.34), but this time (Problem 2.47) $\Phi(\tau) = e^{-i\omega\tau}$ and

$$\hat{a}_{\text{H}}(t) = \hat{a}e^{-i\omega t} + i(2\hbar m\omega)^{-1/2}\int_{-\infty}^{t} e^{i\omega(t'-t)} f(t')dt' \tag{2.75}$$

where $\hat{a}$ is arbitrary. Suppose for simplicity that the oscillator is initially at rest, in the state $|\alpha = 0\rangle$, so that $\hat{a} = 0$; then α in (2.68) is replaced by

$$\sqrt{\frac{2\hbar}{m\omega}}(2\hbar m\omega)^{-1/2}i\int = \frac{i}{m\omega}\int \tag{2.76}$$

where $\int$ represents the integral in (2.75). Separating its real and imaginary parts as $\mathcal{R} + i\mathcal{I}$ and comparing with (2.65) we see that

$$|\alpha\rangle \propto \exp\left(-\frac{m\omega}{2\hbar}\left(x - \frac{1}{m\omega}(i\mathcal{R} - \mathcal{I})\right)^{2}\right) \quad . \tag{2.77}$$

As in (2.69) the coherent state is centered at

$$\bar{x} = -\frac{1}{m\omega}\mathcal{I} = \frac{1}{m\omega}\int_{-\infty}^{t} f(t')\, \sin\omega(t - t')dt'$$

and this is the same displacement as given by (2.37) (where m was taken as 1). The conclusion is that an oscillator that starts in the state $|\alpha = 0\rangle$ ends with the wave packet no larger in size but oscillating according to the classical solution. Of course this result, though perhaps enlightening as to the connection between quantum and classical physics, is of no practical use if we have to miraculously create states with exactly the right spatial form. But this is not necessary, since it can be shown that arbitrary wave packets can be expressed as superpositions of coherent states with suitable (complex) α.

The preceding discussion suggests an intuitive picture of a coherent state and one way in which it might be prepared. Imagine an oscillator in its ground state. At time t_0 give it a sharp impulse with a force $f(t) = I\delta(t - t_0)$. Afterwards the oscillator is in a coherent state for which, by (2.76),

$$|\alpha|^2 = (I/m\omega)^2 \quad .$$

Energy eigenstates are stationary; nothing moves, but this state oscillates and $|\alpha|$ gives the amplitude of the oscillation. On the average, by (2.72), the oscillator's state of excitation will be

$$\bar{n} = \frac{I^2}{2\hbar m \omega} \quad .$$

Not all the phase information has been lost, but also not all information about the state of excitation. The coherent state preserves something of each.

The existence of coherent states does not depend on the parabolic shape of the oscillator potential or the equal spacing of its energy levels. Coherence is quite a general property and the states can be constructed by group-theoretical methods [24]. Also, not all oscillators are mechanical. As will be seen in Sect. 10.8, the states of a quantized field are represented by oscillators and the equal spacing of levels represents the fact that the field contains an integral number of quanta. Thus the theory so developed has applications where fields are quantized: in lasers, superconductivity, and superfluids [25]. An application to the Mössbauer effect will be mentioned in Sect. 10.8, but further discussion here would escape the boundaries of this book.

Problem 2.43. Derive (2.67) by using (2.57).

Problem 2.44. Use the Heisenberg picture to discuss what happens to the probability distribution of a simple harmonic oscillator that starts at $t = 0$ in a Gaussian wave packet.

Problem 2.45. Evaluate the Heisenberg product $\Delta x \, \Delta p$ for the coherent state defined by α.

Problem 2.46. Show that the creation operator $\hat{a}^\dagger$ has no eigenfunctions.

Problem 2.47. Derive the influence function $\Phi(\tau)$ used to solve (2.74).

Problem 2.48. Newton's law of motion is of second order in time and therefore requires two arbitrary constants, but (2.74) is of first order and requires only one. How is it possible to duplicate the classical solution that contains two constants?

3. Lagrangian Dynamics

The conscientious reader in the eighteenth century, trying to follow the proofs in Newton's *Principia* or John Bernoulli's derivation of elliptical orbits (see Box III, below) was left with the belief that yes, it was almost certain that God had designed the universe along mathematical lines and again yes, that he gave to a few of his children a tremendous mental power that could discern a little of that order, but that no, the reader was not one of those people and could only follow, not lead. The trouble was that every new problem demanded a fresh approach. The authors would start with a clever drawing. In Newton that led to long geometrical arguments; in Bernoulli and Euler, to intricate algebraic substitutions as well. It is comparatively easy to represent dx as a line on a diagram, but it takes a measure of genius to represent d^2x (try it). In 1788, which was 99 years after the *Principia, Joseph Louis Lagrange* published a short book [27] called *Méchanique analitique* explaining a general method applicable to a wide variety of calculations in dynamics. "One will find no figures in this work," he wrote in the Introduction, and indeed their lack makes the reading a little harder, but the claim must be taken as an affirmative statement. The method is analytical and rests on general arguments; figures are not needed.

The Lagrangian approach has proved to be remarkably fertile. Not only is it the beginning of the arguments that led to the theories of Hamilton and Jacobi (Chaps. 5 and 6, below), but also, as shown by Feynman in the 1940s, it provides the most direct and plausible route from classical theory to quantum theory. (I say plausible because the assumptions of quantum theory are different from those of classical theory and one does not follow mathematically from the other.) This chapter will apply Lagrangian theory to simple problems involving particles and Chaps. 9 and 10 will use it in the discussion of the motion of rigid bodies and continuous systems. The Feynman argument will be sketched in Chap. 7.

3.1 Lagrange's Equations

The preceding calculations have used two kinds of coordinates, linear and angular, and in the next one, planetary motion, both will be used at once, since the planet's position is to be specified in polar coordinates. We could perfectly well go on for a while, treating successively more complicated systems by ad-hoc devices, but eventually the returns would diminish, and it would be necessary to start using more general methods. We are going to start using them here. The first step will be a simple and general way of introducing different kinds of coordinates into the equations of motion. In subsequent chapters we shall start looking at ways to construct the solutions, not just the equations, with a minimum of labor.

Let us start with an N-particle system in which the forces can be derived from a potential function $V(x)$. We shall denote the x, y, and z coordinates of all the particles by the general letter x, giving it an index where necessary: x_i, $i = 1 \ldots 3N$. The system's kinetic energy can be written in the form

$$T = \frac{1}{2} \sum_{i=1}^{3N} m_i \dot{x}_i^2 \qquad (3.1)$$

and with this define the Lagrangian function

$$L(x, \dot{x}) := T(\dot{x}) - V(x) \qquad (3.2)$$

resembling the energy except for a change in sign. Newton's equations of motion can now be written as

$$\frac{\partial L}{\partial x_i} - \frac{d}{dt} \frac{\partial L}{\partial \dot{x}_i} = 0 \quad , \quad i = 1, \ldots 3N$$

which we abbreviate in our earlier notation (1.40b) as

$$\frac{\delta L}{\delta x_i} = 0 \quad , \quad i = 1, \ldots 3N \quad . \qquad (3.3)$$

Two facts flow from this formula. The first is that L can be used in a variational formulation of the laws of motion that will be closely analogous to Fermat's principle in optics. We have already seen the fundamental similarity of mechanics and optics starting from their description in terms of waves. This new analogy is historically older, and it will enable us to approach the same similarity starting from the principles formulated by Fermat and Newton in the 17th century.

The second important fact about (3.3) is that it can immediately be generalized to other kinds of coordinates. Suppose the system has f degrees of freedom; that is, suppose that it requires f numbers $q_1, \ldots q_f$ to specify where everything is. If the system is a gas, every atom moves separately and f will be about 10^{23}. If it is a rigid pendulum (a real one, not a point mass on a weightless cord) there may still be 10^{23} atoms but the position of every one is determined as soon as the pendulum's angular orientation is given. Thus, $f \leq 3N$, and in expressing the coordinates x_i in terms of $q_1, \ldots q_f$ we will usually introduce economies. This is obviously true for solid bodies, but even for a system like the solar system we will profit from being able to separate out the motion of the center of mass, since no aspect of the solar system that interests us depends on how the whole thing is moving through space.

The remarkable feature of Lagrange's equations now to be proved is that if we write

$$x_i = x_i(q_1, \ldots q_f) \quad , \quad i = 1, \ldots 3N$$

or $x(q)$ for short, the functional derivative of L with respect to q becomes

$$\frac{\delta L}{\delta q_n} = \sum_{i=1}^{3N} \frac{\delta L}{\delta x_i} \frac{\partial x_i}{\partial q_n} \quad , \quad n = 1, \ldots f \qquad (3.4)$$

so that from (3.3), the laws of motion take exactly the same form in generalized coordinates,

$$\frac{\delta L(q, \dot{q})}{\delta q_n} = 0 \quad , \quad n = 1, \ldots f \quad .$$ (3.5)

In writing down the proof I will use the summation convention: if in any expression a certain index occurs twice, the expression is to be summed over the range of that index: equation (3.4), for example, becomes

$$\frac{\delta L}{\delta q_n} = \frac{\delta L}{\delta x_i} \frac{\partial x_i}{\partial q_n} \quad , \quad n = 1, \ldots f \quad .$$

In these expressions i runs from 1 to $3N$; m and n from 1 to f. The summation is over i and not over n. To change variables,

$$\frac{\partial L}{\partial q_n} = \frac{\partial L}{\partial x_i} \frac{\partial x_i}{\partial q_n} + \frac{\partial L}{\partial \dot{x}_i} \frac{\partial \dot{x}_i}{\partial q_n} \quad .$$

In the last term interchange the order of differentiations.

$$\frac{\partial L}{\partial q_n} = \frac{\partial L}{\partial x_i} \frac{\partial x_i}{\partial q_n} + \frac{\partial L}{\partial \dot{x}_i} \frac{d}{dt} \frac{\partial x_i}{\partial q_n} \quad .$$

Since x_i does not depend on $\dot{q}$, we have next

$$\frac{\partial L}{\partial \dot{q}_n} = \frac{\partial L}{\partial \dot{x}_i} \frac{\partial \dot{x}_i}{\partial \dot{q}_n} \quad .$$

To simplify this note that

$$\dot{x}_i = \frac{\partial x_i}{\partial q_m} \dot{q}_m$$

so that differentiating with respect to one of the $\dot{q}$'s, say $\dot{q}_n$, gives

$$\frac{\partial \dot{x}_i}{\partial \dot{q}_n} = \frac{\partial x_i}{\partial q_n} \quad .$$

Now assemble the foregoing into

$$\frac{\partial L}{\partial q_n} - \frac{d}{dt} \frac{\partial L}{\partial \dot{q}_n} = \frac{\partial L}{\partial x_i} \frac{\partial x_i}{\partial q_n} + \frac{\partial L}{\partial \dot{x}_i} \frac{d}{dt} \frac{\partial x_i}{\partial q_n} - \frac{d}{dt} \left(\frac{\partial L}{\partial \dot{x}_i} \frac{\partial x_i}{\partial q_n} \right)$$

$$= \left(\frac{\partial L}{\partial x_i} - \frac{d}{dt} \frac{\partial L}{\partial \dot{x}_i} \right) \frac{\partial x_i}{\partial q_n} = 0$$

which is (3.4), as promised. When Lagrange wrote these equations in 1760 he actually said more than is said here, since he included the possibility of nonconservative forces. The more general form and its derivation are given in many books, e.g. [38].

Before applying Lagrange's equations we note by comparison with (1.40) the single variational principle by which all f equations of motion can be expressed:

$$\delta \int L[q(t), \dot{q}(t)] dt = 0$$ (3.6)

where the q's are not to be varied at the endpoints. In this form it was given by *Hamilton* in 1834 [5, p. 103]. It is called Hamilton's principle or the principle of least action; the former is more exact because the integral is not necessarily a minimum. As in optics, any small segment of the path minimizes the integral, but the whole path may make it a maximum or a point of inflection. "Stationary action" would be better, but it is rarely used.

Hamilton's principle has a formal simplicity about it that makes one feel that it says something very profound about nature. This is true. Almost every quantum or classical dynamical theory can be put into this form, but we must not forget that the equations of motion, guessed from simple observations and verified through endless examples, are logically prior to their expression as a variational principle. The question "Why must the variation vanish at the endpoints?" is answered "Because that leads to the equations of motion". Later, in Sect. 7.8, we will see that there is a different way to look at Hamilton's principle, suggested this time by quantum mechanics. But there again, it is the equations of motion that come first.

There is obviously a certain difficulty of physical interpretation if Hamilton's principle is taken as the fundamental law of mechanics: since both endpoints are fixed, a system has to know where it is going to end up before it starts, and the dynamical principle merely chooses the right way to get there. The French mathematician P.L.M. de Maupertuis, who made the first sketch of a principle of least action in 1744, was well aware of the implication and was proud to have shown how the science of mechanics explains the working of God's plan for the universe [28]: "These laws, so simple and so beautiful, are perhaps the only ones that the Creator needed to produce all the phenomena of the visible word." We will see in Sect. 7.8 that there is another way to explain the physical content of the action principle: a particle tries everything.

Suppose we add to L the time derivative of any function of q and t:

$$L(q, \dot{q}) \rightarrow L(q, \dot{q}) + \dot{F}(q, t) \quad .$$

Since the integral in (3.6) involves F evaluated at the ends where q is not varied, the addition does not change the content of the equations of motion in any way, though it changes their form and may, with suitable F, be used to simplify them. We will see how this is done in Sect. 5.3.

Finally, we note that since by hypothesis neither T nor V involves t, we can use the theorem of (1.42) to derive the existence of a constant of the motion,

$$\dot{q}_n \frac{\partial L}{\partial \dot{q}_n} - L = E \,(\text{const}) \quad . \tag{3.7}$$

This is the energy integral.

Problem 3.1. Verify that if the q's are taken to be the Cartesian coordinates x_i of an N-particle system, E is the system's total energy. Then taking q as the angle from the vertical, set up Lagrange's equations for the simple pendulum of Sect. 2.2 and show that E is again the energy.

Problem 3.2. A pendulum consisting of a massive dot and a comparatively massless rod 1 m long is oriented with the dot 1 Å (10^{-10} m) away from the point directly above the point of support and then released. How long will it take the pendulum to fall through its lowest point? What if the dot starts 1 Å away from the lowest point?

Box I

Lagrange's Equations in the *Méchanique analitique (1788)*

226 MÉCHANIQUE ANALITIQUE.

9. De cette maniere la formule générale du mouvement $\Gamma + \Delta = 0$ (art. 2) sera transformée en celle-ci,

$$\Xi\,\delta\xi + \Psi\,\delta\psi + \Phi\,\delta\varphi + \&c = 0,$$

dans laquelle on aura

$$\Xi = d.\ \frac{\delta T}{\delta d\xi} - \frac{\delta T}{\delta\xi} + \frac{\delta V}{\delta\xi}$$

$$\Psi = d.\ \frac{\delta T}{\delta d\psi} - \frac{\delta T}{\delta\psi} + \frac{\delta V}{\delta\psi}$$

$$\Phi = d.\ \frac{\delta T}{\delta d\varphi} - \frac{\delta T}{\delta\varphi} + \frac{\delta V}{\delta\varphi}$$

$$\&c,$$

en fuppofant

$$T = S\left(\frac{dx^2 + dy^2 + dz^2}{2\,dt^2}\right)m,\ V = S\,\Pi\,m,$$

$$\&\ d\Pi = P\,dp + Q\,dq + R\,dr + \&c.$$

Si donc dans le choix des nouvelles variables ξ, ψ, φ, &c, on a eu égard aux équations de condition données par la nature du fyftême propofé, enforte que ces variables foient maintenant tout-à-fait indépendantes les unes des autres, & que par conféquent leurs variations $\delta\xi, \delta\psi, \delta\varphi$, &c, demeurent abfolument indéterminées, on aura fur le champ les équations particulieres $\Xi = 0$, $\Psi = 0$, $\Phi = 0$, &c, lefquelles ferviront à déterminer le mouvement du fyftême; puifque ces équations font en même nombre que les variables ξ, ψ, φ, &c, d'où dépend la pofition du fyftême à chaque inftant.

Mais quoiqu'on puiffe toujours ramener la queftion à cet état, puifqu'il ne s'agit que d'éliminer par les équations de condition, autant de variables qu'elles permettent de le faire, & de prendre enfuite pour ξ, ψ, φ, &c, les variables

Courtesy of the Chapin Library of Williams College

Here $d.$ means d/dt, δ means ∂, and S means Σ. One of Lagrange's great innovations is potential energy. He defines Π (potential energy per unit mass) in terms of force $\times$ infinitesimal distance along a path, $d\Pi = P\,dp + \ldots$, but regards it as a function of the present positions of the particles, expressed in terms of the generalized coordinates $\xi, \psi, \varphi, \ldots$.

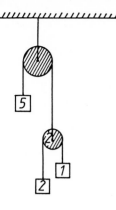

Fig. 3.1. To illustrate Problem 3.3

Problem 3.3. The device of Fig. 3.1, in which the numbers give the masses of the movable elements, is constructed and let go. Neglecting the rotational inertia of the pulleys, find the acceleration of the 5-unit mass.

There is another way in which constants of the motion can be found. Suppose that L does not explicitly involve one of the coordinates, say q_m. Then (2.53) becomes

$$\frac{d}{dt}\frac{\partial L}{\partial \dot{q}_m} = 0 \quad , \quad \frac{\partial L}{\partial \dot{q}_m} = p_m \text{ (const)} \quad .$$

A familiar example is the flight of a projectile, where the potential energy depends on the height but not the horizontal coordinate x. Thus

$$L = \tfrac{1}{2}m(\dot{x}^2 + \dot{y}^2) - V(y) \quad , \quad \partial L/\partial x = 0 \quad \text{and}$$

$$p_x = m\dot{x} = \text{const} \quad .$$

This is the momentum corresponding to x, and in its honor we call the general p_m a momentum also; it is the generalized momentum corresponding to (or *conjugate* to) the generalized coordinate q_m. The missing coordinate q_m is traditionally called an *ignorable coordinate* or a *cyclic coordinate*. We will look carefully for them in what follows, for conservation laws are very important in physics.

Motion Under Constraints. When Lagrange's equations are written in terms of the variational principle (3.6) it becomes easy to treat systems which operate under some kinds of constraints. Suppose for example that the coordinates q_m are not allowed to vary independently because there is some relation of the form $f(q, \dot{q}, t) = 0$ that connects them. This situation was already encountered in Sect. 1.5. There it was shown that if one adds to the integrand a term $\mu(t)f(q, \dot{q}, t)$, finds the variational equations, and then chooses $\mu(t)$ so as to make f equal to zero, all the conditions are satisfied. The following example is trivial but illustrates the method.

Example 3.1. A bead slides down a wire in the shape of a helix of radius a and pitch h. Find the z-component of its motion.

The auxiliary condition is $z = h\vartheta/2\pi$ and the Lagrangian including it is

$$L = \frac{1}{2}m(a^2\dot\vartheta^2 + \dot z^2) - mgz + \mu\left(z - \frac{h}{2\pi}\vartheta\right) \quad .$$

The equations of motion following from this are

$$ma^2\ddot\vartheta = \frac{h}{2\pi}\mu \quad , \qquad m\ddot z = -mg + \mu$$

and these are compatible with the auxiliary condition provided that

$$\mu = \frac{ma^2 g}{a^2 + (h/2\pi)^2} \quad .$$

(In this simple case, but not in general, it is constant.) With this the z-equation becomes

$$\ddot z = \frac{h^2 g}{(2\pi a)^2 + h^2}$$

as can easily be verified in other ways.

Note a curious fact: If you pretend that the system has $f+1$ degrees of freedom instead of f and treat μ as the extra variable, then no auxiliary condition need be stated, for $\delta L/\delta\mu = 0$ imposes it automatically. This device is sometimes seen in the literature and it is sometimes useful.

Problem 3.4. Find the period of a simple pendulum as an example of motion under constraint by writing the Lagrangian as

$$L = \tfrac{1}{2}(\dot x^2 + \dot y^2) - gy + \tfrac{1}{2}\mu(x^2 + y^2 - l^2) \quad .$$

Show that $\mu = (3gy - 2E)/l^2$ (where E is the energy) and complete the calculation to find $x(t)$ in the small-angle approximation.

Problem 3.5. A pendulum is made by sliding a frictionless bead on a wire in the form of an ellipse with horizontal and vertical semi-axes a_h and a_v. Find the period in the small-angle approximation.

The Meaning of Lagrangian Multipliers. The Lagrangian multiplier in an isoperimetric problem (Sect. 1.5) is a constant. Constants in dynamics represent conserved quantities and have a certain physical prominence, so it usually becomes clear what they mean. The following note is a digression intended to show what μ means when the auxiliary condition is of the form $f(q_1, \ldots q_n) = 0$ and μ is not a constant. (Generalization to constraints involving velocities is not difficult.) Following the prescription for such a case we add to the potential energy a term $\mu(t)f(q)$ (note the abbreviation q for $q_1, \ldots q_n$). When the resulting equations have been solved, the system moves along a path given by $q_0(t)$, defined as a set of coordinates which satisfy

$$f_0 := f(q_0) = 0 \tag{3.8}$$

and they also satisfy the equations of motion

$$\frac{\partial L_0}{\partial q_{0i}} - \mu(t)\frac{\partial f_0}{\partial q_{0i}} - \frac{d}{dt}\frac{\partial L_0}{\partial \dot{q}_{0i}} = 0 \quad i = 1, \ldots n \tag{3.9}$$

where L_0 is the original Lagrangian expressed in terms of coordinates q_0 and $\dot{q}_0$.

Now let us try to impose the auxiliary condition in a completely different way. Use coordinates q that do not exactly satisfy (3.8) and impose the auxiliary condition by adding to the potential a term

$$\frac{1}{2\varepsilon}f^2(q)$$

which for small ε represents a narrow valley in q-space rising steeply on each side of $q = q_0$. Values of q will never wander far outside this valley, so write

$$q_i(t) = q_{0i}(t) + q_{1i}(t) \quad , \quad q_{1i}(t) = \varepsilon\kappa_i(t) \quad .$$

The equations of motion with this added potential are

$$\frac{\partial L}{\partial q_i} - \frac{1}{\varepsilon}f\frac{\partial f}{\partial q_i} - \frac{d}{dt}\frac{\partial L}{\partial \dot{q}_i} = 0 \quad . \tag{3.10}$$

Functions of q become for example

$$f(q) = f_0(q_0) + \varepsilon\kappa_j\frac{\partial f_0}{\partial q_{0j}}$$

and similarly for derivatives. Expand the equation of motion (3.10), remember that $f_0 = 0$, and let ε approach zero. The result is

$$\frac{\partial L_0}{\partial q_{0i}} - \kappa_j\frac{\partial f_0}{\partial q_{0j}}\frac{\partial f_0}{\partial q_{0i}} - \frac{d}{dt}\frac{\partial L_0}{\partial \dot{q}_{0i}} = 0 \quad .$$

Comparing this equation with (3.9) shows that the two are the same provided that

$$\mu(t) = \frac{1}{\varepsilon}q_{1j}\frac{\partial f_0}{\partial q_{0j}} \quad .$$

Thus $\mu(t)$ measures the change in f produced by the infinitesimal deviations from the constrained path that would occur if the constraint were imposed by a potential and not by fiat [29].

3.2 The Double Pendulum

To show how Lagrange's equations are used, here is an example worked out in detail. Suppose that one simple pendulum is suspended from another and both are set swinging through small angles, in such a way that they swing periodically thereafter. (There are other conditions after which they do not swing periodically – see Sect. 4.3.) The periodic modes of motion are called *normal modes*. Figure 3.2(a) shows the coordinates. To simplify formulas, assume that $m_1 \gg m_2$, that $l_1 < l_2$, and that all angles are small. The solution to follow involves more algebra than one perhaps enjoys, but later we can see how to do it more neatly. If we neglect ϑ_1^3 and ϑ_2^3 the Cartesian coordinates of the two masses are

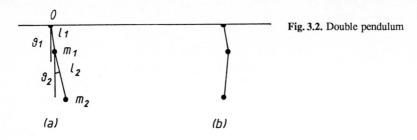

Fig. 3.2. Double pendulum

(a)   (b)

$$x_1 = l_1\vartheta_1 \quad , \quad y_1 = -l_1(1 - \vartheta_1^2/2)$$
$$x_2 = l_1\vartheta_1 + l_2\vartheta_2 \quad , \quad y_2 = -l_1(1 - \vartheta_1^2/2) - l_2(1 - \vartheta_2^2/2) \quad .$$

From these,

$$L \simeq \tfrac{1}{2}m_{12}l_1^2\dot\vartheta_1^2 + \tfrac{1}{2}m_2(l_2^2\dot\vartheta_2^2 + l_1l_2\dot\vartheta_1\dot\vartheta_2) - \tfrac{1}{2}m_{12}gl_1\vartheta_1^2 - \tfrac{1}{2}m_2gl_2\vartheta_2^2 \qquad (3.11)$$

plus a constant, where $m_{12} = m_1 + m_2$. Lagrange's equations of motion are

$$m_{12}l_1\ddot\vartheta_1 + \tfrac{1}{2}m_2l_2\ddot\vartheta_2 + m_{12}g\vartheta_1 = 0$$
$$m_2l_2\ddot\vartheta_2 + \tfrac{1}{2}m_2l_1\ddot\vartheta_1 + m_2g\vartheta_2 = 0 \quad . \qquad (3.12a)$$

Except for the small-angle approximation, these are exact.

Now we state that the desired solution is periodic in ω,

$$\vartheta_1 = a_1 \cos (\omega t + \phi) \quad , \quad \vartheta_2 = a_2 \cos (\omega t + \phi)$$

where ϕ is arbitrary and ω is to be determined. Substitution into (3.12a) gives

$$m_{12}(g - l_1\omega^2)a_1 - \tfrac{1}{2}m_2l_2\omega^2 a_2 = 0$$
$$-\tfrac{1}{2}m_2l_1\omega^2 a_1 + m_2(g - l_2\omega^2)a_2 = 0 \quad . \qquad (3.12b)$$

For these, considered as equations in a_1 and a_2, to have any solution at all, their determinant must vanish. This gives a quadratic equation in ω^2 whose solutions are

$$\omega^2 = \frac{g}{2(m_{12} - m_2/4)l_1l_2}\left\{m_{12}(l_1 + l_2) \pm \sqrt{m_{12}^2(l_2 - l_1)^2 + m_{12}m_2l_1l_2}\right\} \quad .$$

The simplifying assumptions allow us to write $m_2l_1l_2 \ll m_{12}(l_2 - l_1)^2$:

$$\omega^2 = \frac{g}{2m_{12}l_1l_2}\left(1 + \frac{1}{4}\frac{m_2}{m_{12}}\right)\left[m_{12}(l_1 + l_2)\right.$$
$$\left. \pm m_{12}(l_2 - l_1)\left(1 + \frac{m_2l_1l_2}{2m_{12}(l_2 - l_1)^2}\right)\right] + \dots \quad \text{or}$$

$$\omega^2 \simeq \begin{cases} \dfrac{g}{l_1}\left(1 + \dfrac{m_2}{4m_{12}}\dfrac{l_2}{l_2 - l_1} + \dots\right) \\[3mm] \dfrac{g}{l_2}\left(1 - \dfrac{m_2}{4m_{12}}\dfrac{l_1}{l_2 - l_1} + \dots\right) \end{cases} \quad .$$

There are thus two normal modes, with frequencies quite close to those of the individual pendulums. To see how the pendulums move, we set the first (larger) value of ω^2 into the first of (3.12b):

$$a_2 = -\frac{l_1}{2(l_2 - l_1)}a_1 \quad , \quad \omega_2 \simeq \frac{g}{l_1}$$

whereas for the lower frequency,

$$a_2 = \frac{2m_{12}}{m_2}\frac{l_2 - l_1}{l_2}a_1 \quad , \quad \omega^2 \simeq \frac{g}{l_2} \quad .$$

The first one looks like Fig. 3.2(b); the second looks like (a), and the two normal modes are clearly distinguished, for in one the pendulums are exactly in phase and in the other they are exactly out of phase.

Note. In the following problems potential energy curves with central instabilities may arise as in Fig. 3.3(a, b). In the first case we have an integral of the form

$$\frac{1}{4}T = \int_{x_1}^{x_2} \frac{dx}{\sqrt{(x^2 - x_1^2)(x_2^2 - x^2)}} = \frac{1}{x_2}\int_{1}^{k^{-1}} \frac{du}{\sqrt{(u^2 - 1)(1 - k^2 u^2)}} \tag{3.13a}$$

where $k = x_1/x_2$. To reduce this to standard form, make the substitution

$$u = \frac{1}{\sqrt{1 - k'^2 v^2}} \quad , \quad k'^2 = 1 - k^2 \quad . \tag{3.13b}$$

The second case gives rise to

$$\int_{0}^{1} \frac{du}{\sqrt{(1 - u^2)(k'^2 + k^2 u^2)}} \quad . \tag{3.13c}$$

In this, let

$$u = \sqrt{1 - v^2} \tag{3.13d}$$

and see what happens. The 24 possible substitutions of this type are given in Peirce's Tables.

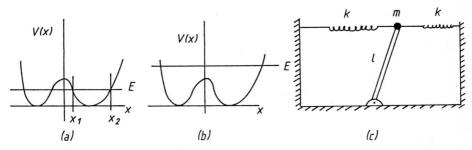

Fig. 3.3. (a, b) Potential energy curve with a central maximum, showing discontinuous behavior as E increases. **(c)** To illustrate Problem 3.6

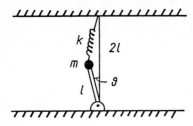

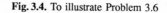 Fig. 3.4. To illustrate Problem 3.6

Problem 3.6. What have the formulas (3.13a) and (3.13c) to do with the situations illustrated in Fig. 3.3a,b?

Problem 3.7. Analyze the behavior of the system shown in Fig. 3.3(c) as completely as you can. Assume that the horizontal distances are long enough so that the springs remain effectively horizontal.

Problem 3.8. Figure 3.4 shows an inverted pendulum of length l supported by a stiff spring (to keep the angles small) whose unstressed length is l. Sketch the potential energy as a function of ϑ and find the (approximate) points at which the pendulum is in equilibrium. Find the frequency of small oscillations around the equilibrium.

3.3 Planets and Atoms

To find the orbits under a central force that varies as the inverse square of the radius is historically the most important problem in physics. In Newton's *Principia*, the agreement of the calculated and observed positions of the planets validated the entire system. In the history of quantum mechanics it was the derivation of Balmer's formula successively by Bohr using his semi-classical theory, Pauli and Dirac using quantum mechanics, and Schrödinger using wave mechanics that first convinced people of the importance of these theories.

In quantum mechanics the calculation of wave functions and energies is straight-forward. In classical mechanics one asks where the planet is at a given time. The answer hinges on the solution of an equation first given by Kepler in about 1609; it was hard to solve then and it is hard now. We shall derive an approximation. Finally, if the energy of the system is positive, we can derive Rutherford's formula for Coulomb scattering as a bonus.

Let us first assume that the sun, or the nucleus, is clamped. We shall see in Sect. 4.2 that nothing changes much if it is allowed to move.

We choose the origin at the center of force (Fig. 3.5), use polar coordinates, and note that the orbit all lies in the same plane (why?). The Lagrangian is

$$\tfrac{1}{2}m(\dot{r}^2 + r^2\dot{\vartheta}^2) - V(r) \tag{3.14}$$

where $V(r)$ is the gravitational or electrostatic potential, or it may be some other central force altogether. Lagrange's equations are second-order equations for r and ϑ; there are four time derivatives and there will be four constants of motion. We

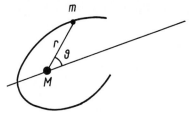

Fig. 3.5. Particle of mass m orbiting an attractive center

speak of the equations as a fourth-order system. Such a thing is liable to be quite complicated, and we search for ways to solve it stepwise.

The Angular Motion. Let us look first at the equation in ϑ. Since V does not depend on ϑ, ϑ is cyclic, and

$$\dot{p}_\vartheta = 0 \quad , \quad p_\vartheta = mr^2\dot{\vartheta} = L\,(\text{const}) \quad . \tag{3.15}$$

The quantity p_ϑ is the momentum conjugate to the angle ϑ; more briefly, the angular momentum. The use of L for the numerical value of the angular momentum as well as the Lagrangian should cause no confusion. Its conservation in this case has given us a constant of integration. The remaining problem is of the third order.

Problem 3.9. A particle of energy E approaches a center of force described by a potential $V(r) = -a/r - b/r^2$ (a and $b > 0$), travelling along a line that passes at a distance h from the center of force. What is the capture cross-section for the particle? (Note that as the particle approaches the center, $\dot{r} < 0$ and if it leaves, $\dot{r} > 0$. Under what conditions will $\dot{r}$ change sign?)

Problem 3.10. It is usual in physics to equip angular momentum with direction as well as a magnitude and make it a vector perpendicular to the orbital plane by writing

$$\boldsymbol{p}_\vartheta = m\boldsymbol{r} \times \dot{\boldsymbol{r}} \quad . \tag{3.16}$$

Show that the magnitude of p_ϑ thus defined is $mr^2\dot{\vartheta}$.

Problem 3.11. Show that if the force is directed towards or away from the point $r = 0$,

$$\frac{dp_\vartheta}{dt} = 0 \quad . \tag{3.17}$$

The angular momentum is a dynamical quantity but it has a geometrical meaning also. Suppose the planet moves from a to b (Fig. 3.6) in a time dt, so that its radius

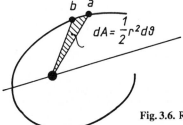

Fig. 3.6. Relation between area and angular momentum

sweeps out an area $r^2 d\vartheta/2$. The rate at which this area is swept out is

$$\frac{dA}{dt} = \frac{1}{2}r^2\dot\vartheta = \frac{L}{2m} \text{ (const)} \quad . \tag{3.18}$$

This is *Kepler's* second law [31, p. 165][1]. It is of more general derivation than the other two, for it requires only a central force; laws 1 and 3 require that the force be inverse-square.

To find r, we use another constant of the motion (such constants are usually called integrals of the motion). This one is the energy

$$E = \tfrac{1}{2}m(\dot r^2 + r^2\dot\vartheta^2) + V(r) \tag{3.19}$$

and incorporating the integral just found gives

$$E = \frac{m\dot r^2}{2} + \frac{L^2}{2mr^2} + V(r) \quad . \tag{3.20}$$

Comparing this with (2.10), we see that it is as if the motion were taking place in one dimension instead of two, in an effective potential

$$V_{\text{eff}} = \frac{L^2}{2mr^2} + V(r) \quad . \tag{3.21}$$

If $V(r)$ is an attractive potential of the form $-\gamma/r$, a graph of V_{eff} looks like Fig. 3.7. Suppose the energy E of the orbiting planet is negative. The situation resembles that of a harmonic oscillator: outside the region between r_1 and r_2 the kinetic energy would be negative. In classical mechanics this is impossible while in quantum mechanics the wave function decreases strongly. Trapped between r_1 and r_2, the planet oscillates back and forth. If $E \geq 0$ there is still a minimum r_1 but the planet is unbound.

How would one have to observe the planet for it to look like a linear oscillator? One would have to stand at the sun and turn around at a varying rate, always facing the planet. The planet would then appear to be subject to the force

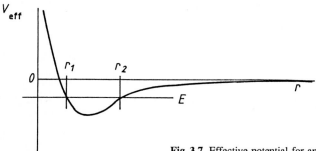

Fig. 3.7. Effective potential for an attractive inverse-square force

---

[1] Kepler's laws:

I) The planets move in ellipses having the sun at one focus.
II) The radius drawn from the sun to any planet sweeps out equal areas in equal times.
III) The ratio of (mean distance from the sun)^3 to (orbital period)^2 is the same for all planets of the Solar System.

Box II

Newton's Proof of the Law of Areas
(Principia, Prop. 1, Th. 1)

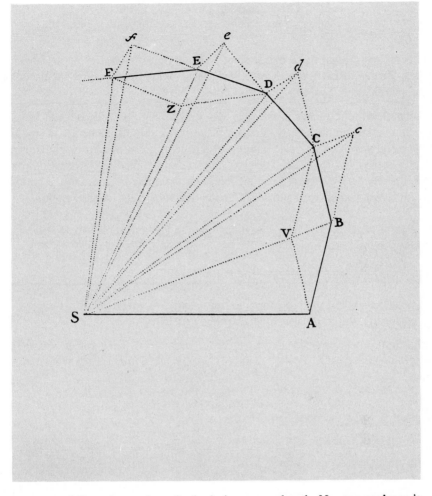

Unable to follow the motion of a body in a curved path, Newton replaces it with a polygon, [30, p. 40] and he replaces the continuously acting force of gravity with a succession of hammer blows (*impulsi*) of arbitrary strength, at equal intervals of time and directed toward the center, S. During the first interval the radius vector sweeps out the area ABS, and if the force did not act again it would proceed to c. Obviously $\triangle BcS = \triangle ABS$. The impulse at B, however, sends the body to C. Since $cC \parallel BS$, $\triangle BCS = \triangle BcS = \triangle ABS$, and so on along the path. "Now let the number of those triangles be augmented, and their breadth diminished *in infinitum*", and Newton argues that the limit is a continuous force producing a motion that continuously sweeps out equal areas in equal times. This ingenious and suggestive argument should not be mistaken for a proof.

$$F_{\text{eff}} = -\frac{dV_{\text{eff}}}{dr} = \frac{L^2}{mr^3} - \frac{dV}{dr} \quad . \tag{3.22}$$

In terms of angular velocity, the first part of F_{eff} is

$$\frac{L^2}{mr^3} = mr\dot{\vartheta}^2 \quad .$$

This is the centrifugal force that is sometimes an object of confusion in elementary studies. We see that it is a real force, with observable effects, which one encounters if one places oneself in a rotating coordinate system.

Problem 3.12. The planet in the potential well of Fig. 3.7 will oscillate back and forth at a certain frequency. If this happens to be exactly the same as the frequency of the angular motion, the planet will return to its original position with its original velocity at the end of a cycle. The orbit repeats and is called re-entrant. Considering orbits that vary only slightly from a circle, fit a harmonic potential to the bottom of the effective potential well and see if this is true for inverse-square motion.

The Radial Motion. The next step in solving the Kepler problem is to solve (3.20) for $\dot{r}$ and separate the variables:

$$\int \frac{dr}{\sqrt{2(E - V(r))/m - L^2/m^2r^2}} = t - t_0 \quad . \tag{3.23}$$

If the goal in solving the problem is to find r and ϑ as functions of t, there are now two reasonable programs that might be followed:

Program A. Integrate (3.23) to get $t(r)$. Invert to find $r(t)$. Then to find $\vartheta(t)$ integrate $\dot{\vartheta} = L/mr^2$:

$$\vartheta - \vartheta_0 = \frac{L}{m} \int \frac{dt}{r^2(t)} \quad . \tag{3.24}$$

Program B. Find the orbit first. Since

$$dr/dt = (L/mr^2)dr/d\vartheta \quad ,$$

(3.20) gives

$$\int \frac{dr}{\sqrt{2mr^4(E - V(r))/L^2 - r^2}} = \vartheta - \vartheta_0 \quad . \tag{3.25}$$

Integrate this to get $\vartheta(r)$, invert to get $r(\vartheta)$ and introduce t by writing $L = mr^2\dot{\vartheta}$:

$$t - t_0 = \frac{m}{L} \int r^2(\vartheta)d\vartheta \quad . \tag{3.26}$$

This gives $t(\vartheta)$. Solve for $\vartheta(t)$ and put into $r(\vartheta)$ to get $r(t)$.

Program A fails at the second step (Problem 3.12). We can find $t(r)$ but cannot solve it (at least, not in closed form) to find $r(t)$.

Program B can be carried a little further. Putting

$$V(r) = -\gamma/r \quad , \quad \gamma = \begin{cases} \pm e^2 \\ GMm \end{cases}$$

we find an integral that can be evaluated and inverted to give

$$r(\vartheta) = \frac{L^2/m\gamma}{1 - e\cos\vartheta} \tag{3.27}$$

where e represents not electric charge but eccentricity:

$$e = \sqrt{1 + \frac{2EL^2}{m\gamma^2}} \quad . \tag{3.28}$$

The orbit is a conic section: an ellipse for $E < 0$, $e < 1$ (see Problem 3.15), a parabola for $E = 0$, $e = 1$, and a hyperbola for $E > 0$, $e > 1$. For bound orbits we have derived *Kepler's* first law [31, p. 285].

For elliptical orbits ($E < 0$), the semi-axes are

$$a = \frac{\gamma}{-2E} \quad , \quad b = \frac{L}{\sqrt{-2mE}} \quad , \tag{3.29a, b}$$

From the first one, we see that $E = -\gamma/2a$; the energy depends only on the major axis and not at all on the eccentricity. (This is why the Bohr theory can derive all the energy levels of hydrogen by considering only the circular orbits.)

Problem 3.13. Verify the failure of Program A.

Problem 3.14. Derive (3.27–28).

Problem 3.15. Starting from $x^2/a^2 + y^2/b^2 = 1$ and the definition of eccentricity as $e = \sqrt{1 - b^2/a^2}$, show that (3.27) represents an ellipse with the origin at one focus. Find the expressions (3.29a, b) for the semi-axes.

Problem 3.16. Suppose that $V(r)$ varies as a power of r: $V(r) = \lambda r^n$. For what (positive and negative) values of n are the orbits re-entrant (i.e., closed)? This is the same as: for what values of n can the integral in (3.25) giving $\vartheta(r)$ [or $r(\vartheta)$] be expressed in terms of trigonometric functions? It will be convenient to change the variable in (3.25) to $u = r^{-1}$.

Problem 3.17. Is it possible to find a force law for which orbits close after several turns?

Having found the orbits, we must now find how they are traversed. The total time of revolution is easily determined from the area law (3.18):

$$T = \frac{2m}{L} \times \pi ab = \frac{\sqrt{2\pi\gamma m}}{(-2E)^{3/2}} \quad , \tag{3.30}$$

and substituting $-2E$ from (3.29) gives

$$T = 2\pi\sqrt{\frac{m}{\gamma}}a^{3/2} \quad . \tag{3.31}$$

The period depends only on the energy or only on the major axis. For planetary motion, $\gamma = GMm$ and

$$T = \frac{2\pi}{\sqrt{GM}}a^{3/2} \quad . \tag{3.32}$$

This is almost Kepler's third law. *Kepler* has the mean orbital radius instead of the semi-major axis [32, p. 189], but the planetary orbits are so nearly circular that it makes little difference.

To continue with Program B, we return to (3.26), which with $t_0 = 0$ is

$$t = \frac{T}{2\pi}(1 - e^2)^{3/2} \int \frac{d\vartheta}{(1 - e \cos \vartheta)^2} \quad . \tag{3.33}$$

This can be integrated, but the results cannot be solved for $\vartheta(t)$, so Program B comes to a halt unless we are willing to get our hands dirty.

Problem 3.18. Carry out the integration in (3.33) and expand the result to show that for small e,

$$\Omega t = \vartheta + 2e \sin \vartheta + \tfrac{3}{4}e^2 \sin 2\vartheta + \dots \tag{3.34}$$

where Ω is the mean orbital frequency

$$\Omega = 2\pi/T \quad . \tag{3.35}$$

Problem 3.19. Solve (3.34) to get

$$\vartheta = \Omega t - 2e \sin \Omega t + \tfrac{5}{4}e^2 \sin 2\Omega t + \dots \tag{3.36}$$

Problem 3.20. Complete Program B by evaluating

$$r = (1 - e^2)a(1 + e \cos \vartheta + e^2 \cos^2 \vartheta + \dots)$$
$$r = a\left(1 + \frac{e^2}{2} + e \cos \Omega t - \frac{e^2}{2} \cos 2\Omega t + \dots\right) \tag{3.37}$$

and verify that ϑ and r are correctly given at the two ends of the elliptical orbit, where they can be evaluated by elementary geometry.

At this point we run out of results that are easily accessible to pencil and paper. Astronomers of the last century carried the series developments much further and introduced computational methods that are of great mathematical interest, but today computers have taken over the work of calculation.

3.4 Orbital Oscillations and Stability

I mentioned in Problem 3.11 that a closed orbit can be regarded as one in which a particle carries out a whole number of radial oscillations while the angular variable runs from 0 to 2π. (More complicated closed orbits result if the angle is 4π, 6π, etc.) The situation is easily studied for nearly circular orbits, and we are led to interesting questions of stability as well.

We have seen in (3.22) that a particle in orbital motion under a central force is subject to a radial force

$$F_{\text{eff}} = \frac{L^2}{r^3} - V'(r) \quad (m = 1)$$

and the orbit will be a circle of radius a if F_{eff} is zero, tending neither to increase nor decrease the radius,

$$\frac{L^2}{a^3} - V'(a) = 0 \quad . \tag{3.38}$$

To study small variations from circular motion, let $r = a + \alpha$ ($\alpha \ll a$) and expand F_{eff} around $r = a$. With $m = 1$,

$$\ddot{r} = \ddot{\alpha} = \frac{L^2}{a^3} - \frac{3L^2}{a^4}\alpha - V'(a) - V''(a)\alpha$$

or by (3.38),

$$\ddot{\alpha} = -\left[V''(a) + \frac{3}{a}V'(a)\right]\alpha \quad . \tag{3.39}$$

If

$$\omega_{\text{osc}}^2 := V''(a) + \frac{3}{a}V'(a) > 0 \quad \text{(stable)} \tag{3.40}$$

α oscillates with a frequency ω_{osc}. If $\omega_{\text{osc}}^2 < 0$ the slightest departure from circular motion will grow exponentially and the particle will wander away from its circular orbit.

The orbital frequency is readily obtained from (3.38),

$$\omega_{\text{orb}}^2 = \frac{1}{a}V'(a)$$

and one finds at once that if $V(r) = -\gamma/r$ the orbital and oscillatory frequencies are the same.

Suppose now that the force is not exactly inverse-square but has a slight inverse-cube part,

$$V(r) = -\frac{\gamma}{r} - \frac{\varepsilon}{r^2} \quad , \quad \varepsilon > 0 \quad . \tag{3.41}$$

We find

$$\omega_{\text{osc}}^2 = \frac{\gamma}{a^3} \quad , \quad \omega_{\text{orb}}^2 = \frac{\gamma}{a^3} + \frac{2\varepsilon}{a^4} \quad .$$

The period of a radial oscillation is

$$T_{\text{osc}} = \frac{2\pi}{\omega_{\text{osc}}} = 2\pi\sqrt{\frac{a^3}{\gamma}}$$

and in this time the planet's angular position advances by

$$\omega_{\text{orb}}T_{\text{osc}} = 2\pi\sqrt{\frac{a^3}{\gamma}}\sqrt{\frac{\gamma}{a^3} + \frac{2\varepsilon}{a^4}} = 2\pi\sqrt{1 + \frac{2\varepsilon}{\gamma a}}$$

$$\simeq 2\pi + \frac{2\pi\varepsilon}{\gamma a} \quad .$$

Because $\omega_{\text{orb}}T_{\text{osc}} > 2\pi$, the orbit has precessed forward through an angle

$$\delta\vartheta = \frac{2\pi\varepsilon}{\gamma a}$$

when r resumes its initial value. In Sect. 6.3 it will be seen that general relativity introduces just such a perturbation into $V(r)$, with

$$\varepsilon = 3\gamma^2/c^2$$

if, as here, the particle has unit mass. This yields a precession

$$\delta\vartheta = \frac{6\pi GM}{ac^2} \qquad (3.42)$$

an estimate still not exact enough to compare with observation since it applies only to infinitesimal deviations from a circular orbit, and the orbit of Mercury, in which the effect is largest, is appreciably flattened. A more exact calculation will be performed in Sect. 6.3.

Problem 3.21. For what power-law central forces are all circular orbits stable?

Problem 3.22. Invent a potential function that gives bound orbits and, for a given value of the radius a, an unstable circular orbit. Use a computer to study the behavior of nearby orbits.

3.5 Orbital Motion: Vectorial Integrals and Hyperbolic Orbits

We close the discussion of orbital motion by showing that there is another integral independent of E and L which can be used to simplify the integration of the orbital equations even further. Start with the equation of motion in vectorial form,

$$m\ddot{\boldsymbol{r}} = -\gamma\frac{\boldsymbol{r}}{r^3}$$

and the formula (3.16) for the constant vectorial angular momentum,

$$m\boldsymbol{r} \times \dot{\boldsymbol{r}} = \boldsymbol{L} \quad .$$

Form

$$\ddot{\boldsymbol{r}} \times \boldsymbol{L} = m\ddot{\boldsymbol{r}} \times [\boldsymbol{r} \times \dot{\boldsymbol{r}}] = -\frac{\gamma}{r^3}\boldsymbol{r} \times [\boldsymbol{r} \times \dot{\boldsymbol{r}}] \quad .$$

By the vector identity

$$\boldsymbol{a} \times [\boldsymbol{b} \times \boldsymbol{c}] = (\boldsymbol{a} \cdot \boldsymbol{c})\boldsymbol{b} - (\boldsymbol{a} \cdot \boldsymbol{b})\boldsymbol{c} \qquad (3.43)$$

we have

$$\frac{1}{r^3}\boldsymbol{r} \times [\boldsymbol{r} \times \dot{\boldsymbol{r}}] = -\frac{1}{r^3}[r^2\dot{\boldsymbol{r}} - (\boldsymbol{r} \cdot \dot{\boldsymbol{r}})\boldsymbol{r}]$$

$$= -\left[\frac{\dot{\boldsymbol{r}}}{r} - \frac{\boldsymbol{r} \cdot \dot{\boldsymbol{r}}}{r^3}\boldsymbol{r}\right] = -\frac{d}{dt}\frac{\boldsymbol{r}}{r} \quad . \qquad (3.44)$$

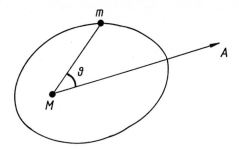

Fig. 3.8. The vectorial integral A

Therefore

$$\ddot{\boldsymbol{r}} \times \boldsymbol{L} = \frac{d}{dt}(\dot{\boldsymbol{r}} \times \boldsymbol{L}) = \gamma \frac{d}{dt}\frac{\boldsymbol{r}}{r}$$

and so the vector

$$\boldsymbol{A} := \boldsymbol{L} \times \dot{\boldsymbol{r}} + \gamma \frac{\boldsymbol{r}}{r} \tag{3.45}$$

is a constant vector in the plane of the orbit. To find the planetary orbit, take the dot product of this equation with $m\boldsymbol{r}$. Since

$$m\boldsymbol{r} \cdot \dot{\boldsymbol{r}} \times \boldsymbol{L} = m\boldsymbol{r} \times \dot{\boldsymbol{r}} \cdot \boldsymbol{L} = L^2$$

we have

$$-L^2 + m\gamma r = m\boldsymbol{A} \cdot \boldsymbol{r} = mAr \cos \vartheta$$

where ϑ is the angle between $\boldsymbol{A}$ and $\boldsymbol{r}$. Solving for r gives

$$r = \frac{L^2/m\gamma}{1 - \gamma^{-1}A\cos\vartheta} \ . \tag{3.46}$$

This is the same as (3.27) if

$$A = \gamma e \tag{3.47}$$

and we see also from (3.46) that $\boldsymbol{A}$ points from the origin of the ellipse in the direction of the point furthest from the "sun", the aphelion, Fig. 3.8. The foregoing analysis is due to *Laplace* [34, vol. 1, pp. 160–168] but the result is already latent in a brilliant analysis by Johann Bernoulli in 1710. Because the method used illustrates so well the virtuosity with which simple methods of integration were used during Newton's lifetime, I give in Box III Bernoulli's discussion in a letter commenting on an earlier attempt by Jakob Hermann and a translation into modern terms. Laplace's vector is known by every name but his: Hamilton, Runge, Lenz, Pauli, ...

What has been accomplished here by the discovery of a new integral of the equations of motion is that, if the values of the integrals A, L, and E are specified (subject to restrictions mentioned below), there is only one orbit compatible with these values. We will not often be so lucky, but in Sect. 6.7 it will become clear that even if the orbit is not uniquely determined, it is at least restricted within certain limits by the integrals that can be found.

Box III

From Bernoulli's letter to Hermann, 7 Oct. 1710 [32]

EXTRAIT DE LA REPONSE

De Monfieur B E R N O U L L I *à Monfieur* H E R M A N,
dattée de Basle *le* 7 *Octobre* 1710.

Dans vôtre équation differentio-differentielle — $a\,ddx =$
$(y\,dx — x\,dy) \times (x\,y\,dx — x\,x\,dy) : (x\,x + y\,y)^{\frac{3}{2}}$ je ne
mets pas feulement [comme vous] — $a\,dx$ pour l'intégrale de
— $a\,ddx$, mais — $a\,dx \pm$ une quantité conftante, c'eft-à-
dire, — $a\,dx \pm e(y\,dx — x\,dy)$; pour le refte je le fais
comme vous; de forte qu'en intégrant vôtre précédente équation differentio-differentielle, je trouve — $a\,dx \pm e\,(y\,dx$
— $x\,dy) = — y(y\,dx — x\,dy) : \sqrt{(x\,x + y\,y)} = (x\,y\,dy$
— $y\,y\,dx) : \sqrt{(x\,x + y\,y)}$, ou — $a\,b\,dx : x\,x \pm e\,b\,(y\,dx$
— $x\,dy) : x\,x = (b\,x\,y\,dy — b\,y\,y\,dx) : x\,x\,\sqrt{(x\,x + y\,y)}$,
dont l'intégrale eft $ab : x \pm e\,by : x \pm c = b\sqrt{(x\,x + y\,y)} : x$
c'eft-à-dire (en prenant $h = eb$, & en reduifant l'équation)
$ab \pm hy \pm cx = b\sqrt{(x\,x + y\,y)}$: laquelle équation, quoiqu'elle renferme hy, que la vôtre ne renfermoit pas, eft ce-
pendant [comme elle] aux trois Sections Coniques.

The equation in x is $\quad m\ddot{x} = -\gamma x/r^3$.

Write this, for reasons which will be clear in a moment, as

$$m\ddot{x} = -\frac{\gamma m^2}{l^2}\frac{x(y\dot{x} - x\dot{y})^2}{r^3} \quad (y\dot{x} - x\dot{y} = l/m = \text{const}) \qquad \text{or}$$

$$-a\ddot{x} = x(y\dot{x} - x\dot{y})^2/r^3 \quad a = l^2/m\gamma \quad .$$

This integrates directly to

$$-a\dot{x} = -y(y\dot{x} - x\dot{y})/r \qquad + \text{const} \quad .$$

The trick is to write the constant as $\mp e(y\dot{x} - x\dot{y})$ and introduce the integrating factor bx^{-2}:

$$-\frac{ab\dot{x}}{x^2} \pm eb\frac{y\dot{x} - x\dot{y}}{x^2} = -\frac{by(y\dot{x} - x\dot{y})}{x^2 r} \quad .$$

Once more, everything can be integrated. Let $eb = h$.

$$\frac{ab}{x} \mp h\frac{y}{x} = b\frac{r}{x} \pm \text{const } c \quad .$$

Thus $\quad br \pm hy \pm cx = ab \quad$ with h and c arbitrary. Take

$$h = \varepsilon b \cos\phi \quad , \quad c = \varepsilon b \sin\phi \quad , \quad y = r\sin\vartheta \quad , \quad x = r\cos\vartheta \quad .$$

Then $\quad r = a/1 \pm \varepsilon \sin(\vartheta \pm \phi)$.

By introducing the arbitrary constants h and c we have introduced ε and ϕ, which specify the eccentricity and the direction of the line of apsides.

Problem 3.23. Show that

$$A^2 = \frac{2E}{m}L^2 + \gamma^2 \quad . \tag{3.48}$$

It has been mentioned earlier that the Kepler problem is a fourth-order problem, requiring four constants of integration. Now we can count the constants that have been found. Since the orbit lies in a plane, L is perpendicular to it: one integral. E is the second one. A lies in the plane, but since its length is given by L and E, only one component is independent: the third integral. And as already mentioned, the last one is trivial: t_0, the time the clock started. The reason that $r(\vartheta)$ could be obtained in (3.46) by a purely algebraic process is that all the integrals had been found. This was relatively easy because the orbit is a figure stationary in space; in general it is very difficult or impossible.

In quantum mechanics the situation is similar. If one characterizes states of the hydrogen atom by angular momentum and energy as is usually done, a considerable amount of calculation remains. But if one introduces the quantum operator that corresponds to A [see (3.53)] the solution is reduced to quantum algebra. This is, in fact, how *Pauli* first solved the problem [35] in the days before Schrödinger's equation had simplified the calculation.

Hyperbolic Orbits. When the energy is positive and by (3.28) $e > 1$, the denominator of (3.27) vanishes or becomes negative for certain values of ϑ. The orbits are now hyperbolas, still with the sun at one focus, Fig. 3.8. The asymptotes are at the angles $\pm \vartheta$, where

$$\cos \vartheta = e^{-1} \quad . \tag{3.49}$$

By (3.28) this can also be written as

$$\tan \vartheta = \sqrt{\frac{2E}{m}} \frac{L}{|\gamma|} \quad .$$

At great distances from the origin, the energy is entirely kinetic and $(2E/m)^{1/2}$ is just the asymptotic velocity v_0, so that

$$\tan \vartheta = \frac{v_0 L}{|\gamma|} \quad . \tag{3.50}$$

If $\gamma > 0$ (attractive force) we must have $\cos \vartheta > e^{-1}$ and ϑ is bounded below. This is orbit 1 in Fig. 3.9. When the force is repulsive the inequality is the other way, and orbit 2 results.

Problem 3.24. Figure 3.10 shows (for $\gamma > 0$) the parameters used in calculations of scattering: s is the collision parameter and ϑ_s is the scattering angle. The incident particle's asymptotic linear momentum is $p_{as} = \sqrt{2mE}$. Show that the angular momentum is sp_{as}.

Problem 3.25. Show that these orbits are hyperbolas.

Problem 3.26. Show from the calculations just done that

$$s = \frac{\gamma}{2E}\cot\frac{1}{2}\vartheta_s \quad .$$

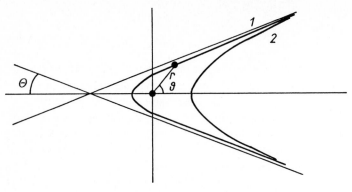

Fig. 3.9. Hyperbolic orbits corresponding to positive energies

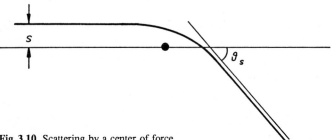

Fig. 3.10. Scattering by a center of force

The differential scattering cross section is given [1, p. 316] by

$$\frac{d\sigma}{d\Omega} = \frac{s}{\sin\vartheta_s}\frac{ds}{d\vartheta_s} \quad .$$

Derive the Rutherford scattering formula.

Problem 3.27. The vector integral A gives a simple derivation of ϑ_s. Derive (3.49) by forming the product $\boldsymbol{p}\cdot\boldsymbol{A}$ and considering the limit of large r.

3.6 Other Forces

Problem 3.28. A particle moves in a central attractive inverse-cube force given by the potential $V(r) = -\lambda r^{-2}$. Find and sketch $r(t)$, $\vartheta(t)$ and $r(\vartheta)$ for a particle projected in a tangential direction at $t = 0$.

Problem 3.29. Calculate the orbits of particles with $E > 0$ scattered by a repulsive inverse-cube force.

Problem 3.30. Show in quantum mechanics that a particle in an attractive inverse-cube potential has no stationary bound states.

Problem 3.31. The calculations in this chapter have all attempted to find orbits given the force. Show how to invert the argument to make inferences the other way.

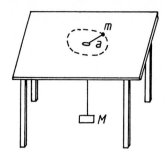

Fig. 3.11. To illustrate Problem 3.32

Suppose an astronomer finds a light object orbiting a motionless heavy one in a circle that has the heavy one located on its circumference instead of in the middle. Deduce the force law and the system's energy. Assume p_ϑ constant.

Problem 3.32. Figure 3.11 shows two masses connected by a cord that passes through a hole in a table top. The mass on the (frictionless) table moves in a circle with angular velocity $\omega = \sqrt{Mg/ma}$ so that the system is in dynamical equilibrium. The mass M is now given a sharp downward tug. Show that the orbit of m now oscillates around its original radius a and find the frequency of small oscillations.

Problem 3.33. A mass hangs from a fixed point by a cord of length l. It is set in motion so that it moves in a horizontal circle. If the motion is slightly disturbed, find the frequency of the resulting oscillations around the circular path and sketch the resulting motion.

Problem 3.34. How must the field of a central force vary with r in order that it be possible for a body to move in it describing a spiral $r = (c\vartheta)^{-1}$, with c a constant?

Problem 3.35. Several physical theories have proposed that the Newtonian G varies by a few parts in 10^{11} per year:

$$\dot{G} = -\lambda G \quad .$$

How will the radius, period and eccentricity of a circular planetary orbit vary if this is so? (Hint: for e, find $\dot{A}$.)

3.7 Bohr Orbits and Quantum Mechanics: Degeneracy

Planetary orbits and Bohr orbits are, in general, ellipses. The particle moves slowly when it is far from the center and tends to spend most of its time to one side of it. That is, in terms of expectation values, if A is in the direction of positive x,

$$\langle x \rangle > 0 \quad , \quad \langle y \rangle = \langle z \rangle = 0 \quad . \tag{3.51}$$

But (Problem 3.37) all these values are zero in any stationary state that is an eigenstate of the angular momentum. It is therefore not clear at first what quantum mechanics has to do with elliptical orbits. If the orbit is not re-entrant there is no difficulty, for the orbit will swing around and, on the average, spend equal times in all directions in its plane. Keplerian orbits, being stationary in space, are special, but how?

The solution to the puzzle is the degeneracy of the orbits. Degeneracy in quantum mechanics refers to the existence of different states of motion corresponding to the same energy, and the same definition will do for classical mechanics also. For example, the plane of an orbit can have different orientations: L, perpendicular to the plane, can point in any direction, and in quantum mechanics this is reflected in the $2l + 1$ values of L_z corresponding to any L^2. But a re-entrant orbit has still further degeneracy: with a given L there are still different orbits corresponding to different directions in which the vector A may point (always perpendicular to L). What is the additional degeneracy in quantum mechanics? In the hydrogen atom there is a degeneracy not only in L_z but also in L^2 itself: all orbits with the same value of the principal quantum number l are degenerate.

A general discussion is laborious; look only at the first excited level, which belongs to four different degenerate states:

$$\psi_0 = (8\pi)^{-1/2} a_0^{-3/2} \left(1 - \frac{r}{2a_0}\right) e^{-r/2a_0} \tag{3.52a}$$

$$\psi_{x,y,z} = (32\pi)^{-1/2} a_0^{-5/2} (x, y, z) e^{-r/2a_0} \quad . \tag{3.52b}$$

Problem 3.36. Show that all four of the functions (3.52) are eigenfunctions of L^2. Show that ψ_0 and ψ_z are also eigenfunctions of L_z, and that two further eigenfunctions can be formed from linear combinations of ψ_x and ψ_y.

Problem 3.37. Show that in eigenstates of E, L_z, and L^2, $\langle x \rangle = \langle y \rangle = \langle z \rangle = 0$.

Since the usual eigenfunctions of L^2 and L_z do not give expectation values satisfying (3.51), let us look at eigenfunctions of A. The classical formula for A,

$$A = \frac{\gamma}{r} r - \frac{1}{m} p \times L$$

is not quite ready to take over into quantum mechanics, since it involves the product of two noncommuting quantities, $\hat{p}$ and $\hat{L}$, and such a product is itself not hermitian. There is a general recipe to try in such cases: average the products taken in the two orders. Remembering that the vector product changes sign when the factors are interchanged, write

$$\hat{A} = \frac{\gamma}{r} r - \frac{1}{2m} (\hat{p} \times \hat{L} - \hat{L} \times \hat{p}) \quad . \tag{3.53}$$

Problem 3.38. Verify that this operator commutes with the hydrogen-atom Hamiltonian and therefore represents a conserved quantity.

It is now easy to see that the components of A do not commute, so that a state can diagonalize only one of them. Choose A_x. A somewhat tedious calculation using the wave functions (3.52) shows that

$$\hat{A}_x \psi_0 = -\tfrac{1}{2} \gamma \psi_x \quad , \qquad \hat{A}_x \psi_x = -\tfrac{1}{2} \gamma \psi_0$$

where γ is (electron charge)^2. Thus we can form two eigenfunctions of A_x by combining ψ_0 and ψ_x:

$$\hat{A}_x(\psi_0 + \psi_x) = -\tfrac{1}{2}\gamma(\psi_0 + \psi_x)$$
$$\hat{A}_x(\psi_0 - \psi_x) = \tfrac{1}{2}\gamma(\psi_0 - \psi_x) \quad . \tag{3.54}$$

By (3.47) we see that these quantum states correspond to oppositely oriented orbits with eccentricity $e = \tfrac{1}{2}$.

Thus the spatial degeneracy in the classical theory which allows us to have lopsided orbits corresponds exactly to a degeneracy in the quantum levels which allows us to combine states with different L to produce a similarly lopsided wave function. This happens to be the wave function for a hydrogen atom polarized by being immersed in an electric field, and there is a picture of it in [1, p. 277].

3.8 The Principle of Maupertuis and Its Practical Utility

In its primitive form, the principle stated by Maupertuis asserts that a body traces out its orbit between two given points in such a way that $\int v \cdot ds$ has a value smaller than it has for any other possible path *at the same energy*. For most purposes this proviso is an inconvenience, and there is no such restriction on the varied paths in Hamilton's principle, but it sometimes facilitates practical calculations.

In Cartesian coordinates

$$m \int v \cdot ds = m \int v \cdot \frac{ds}{dt} dt = \int mv^2 dt = \int 2T\, dt = \int (L + E)dt$$

and it is this integral, called the action, which is to be held stationary. E is not to vary, nor are the spatial endpoints of the path, though the time required for the motion may vary. In a schematic manner, showing only one coordinate q, Fig. 3.12 contrasts the principles of Hamilton and Maupertuis. We have now to prove that when the path is varied as in Fig. 3.12b the natural motion is described by

$$\delta \int (L + E)dt = \int_{t_0}^{t_2} L(q_2, \dot{q}_2)dt - \int_{t_0}^{t_1} L(q_1, \dot{q}_1)dt + \int_{t_0}^{t_2} E\, dt - \int_{t_0}^{t_1} E\, dt = 0 \quad .$$

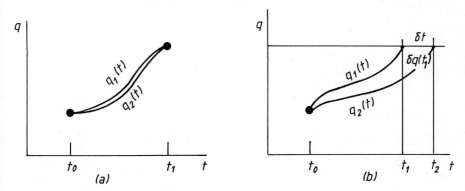

Fig. 3.12. Natural and varied paths for the variational principles of (a) Hamilton; (b) Maupertuis

The second pair of integrals is

$$E(t_2 - t_0 - t_1 + t_0) = E\delta t$$

and we write the first pair as

$$\int_{t_0}^{t_1} [L(q_2, \dot{q}_2) - L(q_1, \dot{q}_1)]dt + \int_{t_1}^{t_2} L(q, \dot{q})dt \quad .$$

The second integral is infinitesimal; write it as $L(t_1)\delta t$. Then treat the first integral as in Sect. 1.5 to get

$$\int_{t_0}^{t_1} \frac{\delta L}{\delta q_n} \delta q_n dt + \left[\frac{\partial L}{\partial \dot{q}_n} \delta q_n \right]_{t_0}^{t_1}$$

in which $\delta q_n = q_{2n} - q_{1n}$ and $\delta q_n = 0$ at t_0. At t_1, it is clear from the figure that

$$\delta q_n(t_1) = -\dot{q}_n(t_1)\delta t \quad .$$

Putting the integrals together gives

$$\delta \int (L + E)dt = \int \frac{\delta L}{\delta q_n} \delta q_n dt + \left[E + L - \frac{\partial L}{\partial \dot{q}_n} \dot{q}_n \right]_{t_1} \delta t \quad .$$

The first term vanishes if the equations of motion (3.5) are obeyed; the second vanishes from the definition of energy in (3.7). Thus we conclude that if the variation is performed as specified above,

$$\delta A = 0 \quad , \quad A = \int [L(q, \dot{q}) + E]dt \quad . \tag{3.55}$$

The fact that the endpoint of the integral is varied means that we can take the time of the motion as a variable, as we cannot in Hamilton's principle. This is especially useful in determining the period of an oscillating system.

Example 3.2. The Quartic Oscillator. Write the Lagrangian as

$$L = \tfrac{1}{2}(\dot{x}^2 - \mu x^4) \tag{3.56}$$

and approximate a half-period of the vibration with a parabola,

$$x(t) = at - bt^2 \quad .$$

The length of the half-period is a/b and the amplitude of the motion is $X = a^2/4b$. The action as calculated by integrating (3.55) over a half-period is

$$A = \frac{a^3}{6b} - \frac{\mu}{1260} \frac{a^9}{b^5} + E\frac{a}{b} \quad . \tag{3.57}$$

The parameters a and b are chosen so as to make A stationary:

$$\frac{\partial A}{\partial a} = \frac{\partial A}{\partial b} = 0$$

and they turn out to be

$$a = \sqrt{2E} \quad , \quad b = \left(\frac{64}{105} \mu E^3 \right)^{1/4} \quad .$$

The period is

$$T = \frac{420^{1/4}}{(\mu E)^{1/4}} = \frac{4.5270}{(\mu E)^{1/4}} \tag{3.58}$$

and the amplitude is

$$X = \left(\frac{105}{64} \frac{E}{\mu} \right)^{1/4} = 1.1318 \left(\frac{E}{\mu} \right)^{1/4} \quad .$$

Exact calculation of the coefficients in the last two expressions (Problem 2.5) gives 4.4098 and 1.1892, respectively. The accuracy is not striking but the form of $x(t)$ was chosen purely for ease of calculation. One can do better.

Problem 3.39. Try to improve the example by choosing a better form for $x(t)$.

Problem 3.40. Show that the Lagrangian for a simple pendulum can be written in the form

$$L = \frac{1}{2} \left[\dot{\vartheta}^2 - \omega_0^2 \left(\vartheta^2 - \frac{1}{12} \vartheta^4 + \ldots \right) \right] \quad , \quad \omega_0^2 = g/l$$

and use the variational method to show how the period depends on the amplitude of swing. I suggest you use more than one trial function. How can you tell which one is the best?

4. *N*-Particle Systems

Newton's first and second laws really refer only to particles, for they make no reference to properties like size, elasticity, and internal angular momentum. The objects described by these laws have only mass and position. The properties of extended objects must be included in the theory by a process of summation or integration. In physics generally there are two situations of special interest: where the bodies are few in number, say up to 4 or 5, and where there are about 10^{23} of them. The situations are very different, starting with the kind of question one wishes to ask — macroscopic quantities like pressure, temperature and density have no counterparts in a few-body system. If one is thinking about a rigid solid one must decide what rigidity means; if the solid is elastic there are assumptions to be made concerning the nature of its elasticity. In Sect. 4.6 an elementary calculation in hydrodynamics will require four assumptions in addition to the assumption that a small element of the fluid obeys Newton's second law. Except for Sects. 4.5 and 4.6 this chapter will be concerned with the simplest cases, in which the N particles can be considered one at a time, each one obeying Newton's laws. Rigid and elastic bodies will be discussed in Chaps. 8 and 9.

4.1 Center-of-Mass Theorems

A loaded beam balances at the center of mass. The condition for balance in Fig. 4.1 is

$$m_1 g(\overline{x} - x_1) + m_2 g(\overline{x} - x_2) = m_3 g(x_3 - \overline{x}) \quad \text{or}$$

$$\overline{x} = \frac{m_1 x_1 + m_2 x_2 + m_3 x_3}{m_1 + m_2 + m_3} . \tag{4.1}$$

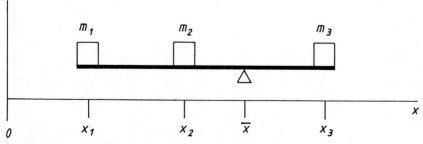

Fig. 4.1. A loaded weightless beam balanced at its center of mass

We take this as the definition of the center of mass and write the general formula as

$$M\overline{r} = \sum_{j=1}^{N} m_j r_j \quad M = \sum_{j=1}^{N} m_j \quad . \tag{4.2}$$

Linear Momentum. Suppose that the jth particle has two kinds of forces acting on it, an externally applied force F_j and a force f_j that is the sum of the forces exerted by all the other particles that compose the system,

$$f_j = \sum_{i \neq j} f_{ij} \tag{4.3}$$

where f_{ij} is the force that the ith particle exerts on the jth one. The total force on the jth particle is then $F_j + f_j$.

How does the center of mass of the whole system move, and how does the whole system turn? It will help to clarify some significant principles if these questions are answered first with a traditional and rather simple-minded argument based on Newton's Third Law. Newton says that "the mutual actions of two bodies upon each other are always equal and oppositely directed" but does not say what the direction is. We will have to amplify the statement with the following:

Axiom. When two bodies i and j interact, the forces f_{ij} and f_{ji} are equal and opposite and directed along the line joining the two bodies. Comment is reserved for a moment.

Now write the total momentum as

$$P = \sum_{j=1}^{N} m_j \dot{r}_j \quad . \tag{4.4}$$

By (4.2), this is

$$P = M\dot{\overline{r}} \tag{4.5}$$

and its rate of change is

$$\dot{P} = M\ddot{\overline{r}} = \sum_j f_j + \sum \sum_{i \neq j} f_{ij} \quad .$$

The last sum involves pairs of terms like $f_{12} + f_{21}$. According to the axiom, $f_{ij} = -f_{ji}$, and so the sum is zero. Further, the sum of the external forces exerted on all the particles of the system is the total force from outside, $\sum F_j := F$, and so we have

$$\dot{P} = M\ddot{\overline{r}} = F \quad . \tag{4.6}$$

Therefore if Newton's laws of motion hold for each particle (or element of volume) of a body or a group of bodies, they apply in this sense to the motion of the whole. Thus, for example, one can treat planets as if they were point particles.

Angular Momentum. If an extended object is pushed, its center of mass moves off, but also, unless the line of action of the force passes through the center of mass,

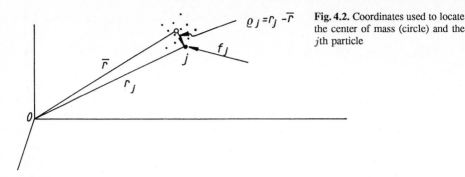

$\varrho_j = r_j - \bar{r}$

Fig. 4.2. Coordinates used to locate the center of mass (circle) and the jth particle

the object will begin to turn. Let l_j be the angular momentum of the jth particle around an origin of coordinate located arbitrarily, and let λ_j be that taken around the center of mass, Fig. 4.2:

$$l_j = m_j r_j \times \dot{r}_j \quad , \quad \lambda_j = m_j \varrho_j \times \dot{\varrho}_j \quad .$$

To see the relation between them, write

$$r_j = \varrho_j + \bar{r} \quad , \quad \dot{r}_j = \dot{\varrho}_j + \dot{\bar{r}} \quad . \tag{4.7}$$

Then the total angular momentum around 0 is

$$L = \sum l_j = \sum m_j (\varrho_j + \bar{r}) \times (\dot{\varrho}_j + \dot{\bar{r}})$$
$$= \sum m_j \varrho_j \times \dot{\varrho}_j + \bar{r} \times \sum m_j \dot{\varrho}_j - \dot{\bar{r}} \times \sum m_j \varrho_j + \bar{r} \times \dot{\bar{r}} \sum m_j \quad .$$

In the third sum,

$$\sum m_j \varrho_j = \sum m_j r_j - \bar{r} \sum m_j = 0 \tag{4.8}$$

by (4.2) and the second term involves the derivative of this sum which vanishes also, so that

$$L = \Lambda + \bar{L} \tag{4.9a}$$

$$\Lambda = \sum m_j \varrho_j \times \dot{\varrho}_j \quad , \quad \bar{L} = \bar{r} \times P \quad . \tag{4.9b}$$

The angular momentum is the sum of that around the center of mass and that of the entire system around the fixed point 0.

The rate of change of L is

$$\dot{L} = \frac{d}{dt} \sum r_j \times p_j = \sum r_j \times f_j + \sum \sum_{i \neq j} r_j \times f_{ij}$$

where the double sum represents angular momentum created by the system's internal stresses. Arrange the terms of the sum in pairs like

$$r_1 \times f_{12} + r_2 \times f_{21} = (r_1 - r_2) \times f_{12}$$

and according to the axiom this is zero. What remains is

$$\dot{L} = \sum \varrho_j \times f_j + \bar{r} \times F \quad . \tag{4.10}$$

84

If a force f acts on a particle at the point r it produces a torque, or moment, about the origin of coordinates defined as $r \times f$. The first term in (4.10) is the sum of the moments of all externally applied forces about the center of mass. Calling this M_c, we write (4.10) as

$$\dot{L} = M_c + \bar{r} \times F \quad . \tag{4.11}$$

It is often useful to choose the origin of coordinates at the center of mass. Then $\bar{r} = 0$ and only the first term of (4.11) survives.

A final remark will be useful later. Usually, in considering an N-particle system, one does not need to bother with motion *of* the center of mass and motion *about* the center of mass at the same time. In studying the behavior of a gas of interacting particles or a system of interacting stars and planets we do not consider translational motion of the whole system, for we naturally discuss it in coordinates with respect to which its center of mass is at rest. In such a system the angular momentum about a fixed point Q is independent of Q, and one can therefore speak of *the* angular momentum of such a system without specifying the point around which it is taken. The proof takes one line. If L_Q and $L_{Q'}$ are the two angular momenta, about Q and Q', and if Q' is at a constant distance r_0 from Q, then

$$L_{Q'} = \sum_j m_j[(r_j + r_0) \times \dot{r}_j] = L_Q + r_0 \times P = L_Q \tag{4.12}$$

because $P = 0$.

Problem 4.1. Prove that the force of attraction of a spherical sun for a spherical planet is directed along the line of centers and is exactly inverse-square. (This is a problem from elementary physics but a very important one.)

Newton's Third Law. Having used the amended Third Law to prove that the internal stresses of a system cannot be used to create linear or angular momentum, we can reasonably ask whether so restricted an axiom is necessary. Can internal stresses between two constituent bodies always reasonably be regarded as acting in pairs along the line joining them? If the bodies are of finite size, what is "the" line? If, say, they are two little magnets, the force will not usually be along any line joining them, and if two bodies that have some structure are close enough together so that (as with neighboring atoms) it is not clear where one ends and the other begins, it may be hard to imagine specifying their action on a third body in terms of two-body forces. We need a more general approach [36].

Consider the following argument. Without distorting the system, give the whole thing an arbitrary infinitesimal displacement δr. Work will have to be done against the external forces but not against the internal ones, since internally nothing has changed. Obviously, therefore,

$$\sum_j (m_j\ddot{r}_j - F_j - f_j) \cdot \delta r = \sum_j (m_j\ddot{r}_j - F_j) \cdot \delta r = (\dot{P} - F) \cdot \delta r = 0$$

from which, if δr is arbitrary, we find $\dot{P} = F$ as before. Or slightly rotate the body. Now $\delta r = \delta\vartheta \times r$, where $\delta\vartheta$ represents an infinitesimal rotation about an axis in the direction of $\delta\vartheta$ [this follows from (4.14) on multiplication by dt]. As before, since the rotation rearranges nothing,

$$\sum_j (m_j \ddot{\boldsymbol{r}}_j - \boldsymbol{F}_j) \cdot (\delta \vartheta \times \boldsymbol{r}_j) = \sum_j \boldsymbol{r}_j \times (m_j \ddot{\boldsymbol{r}}_j - \boldsymbol{F}_j) \cdot \delta \vartheta = 0 \quad .$$

With $\delta \vartheta$ arbitrary this gives

$$\sum_j m_j \boldsymbol{r}_j \times \ddot{\boldsymbol{r}}_j = \dot{\boldsymbol{L}} = \sum_j \boldsymbol{r}_j \times \boldsymbol{F}_j$$

and it comes out as before.

The axiom with which I have replaced Newton's Third Law is a good aid to intuition and it helps solve simple problems, but it is hardly general enough to deserve the name of law. In addition to the weaknesses mentioned above, the statement ignores the fact that fields, which are the agents that exert the real forces of nature, (a) propagate at finite speeds and also (b) carry energy and momentum. From (a) we see that the force exerted by i on j depends on where i was a little while ago and similarly for the action of j on i. Except in static situations, it would be very mysterious if at any given moment (even ignoring relativistic problems with simultaniety) f_{ij} were equal and opposite to f_{ji} and along the same straight line. In (b), Newton's law requires us to believe that all of that part of the energy and momentum delivered to the field by particle i which pertains to its interaction with j is delivered in turn to j, none of it remaining in the field. That this also is hard to believe is shown by the example illustrated in Fig. 4.3. Two charged particles move uniformly in a plane, and at a certain moment they are located as shown. Forget the effects of retardation; they only complicate matters. The electric forces, in this approximation, present no problem, but consider the magnetic forces. Particle

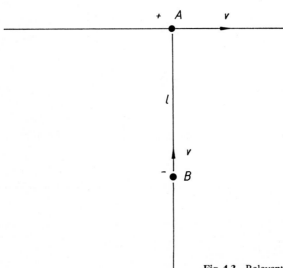

Fig. 4.3. Relevant to the law of action and reaction. Charged particles A and B move, each at velocity v, in the same plane. At the instant shown, A produces a magnetic field that acts on B to push it toward the right but no magnetic field from B acts on A

B produces no field at points directly ahead of it, so *A* is unaffected by *B*. But *A* produces a magnetic field at *B* that exerts a force acting toward the right. This force is not along the line from *A* to *B* and is not reciprocated by a force on *A*; therefore the axiom is violated twice. The system amounts to a rotating electric dipole that transfers energy and momentum to the field it generates. Of course, for most purposes in the macroscopic world Newton's law describes the phenomena very well, and that is what is was invented for.

Note on the First and Second Laws. Every student notices at some point that the First Law seems to be a special case of the Second, and wonders how Newton could have been so careless. It helps if one states the Second Law more carefully, perhaps "*In an inertial system,* the acceleration of a mass is proportional to the externally applied force", the proviso being necessary because in a non-inertial system the statement isn't even true. Next, note that if it is true in one inertial system then it is true in another system moving uniformly with respect to the first. Thus if there is one such system, there are many. But is there one? The First Law says yes, there is one, and that is why it is necessary.

Problem 4.2. Correct the argument concerning Fig. 4.3 for the effects of retardation and show that this does not save the axiom.

Problem 4.3. Can you think of reasons why one should dispute the correctness of Newton's first and second laws as we have done for the third one?

Kinetic Energy. The final theorem of this set concerns kinetic energy. If we write the kinetic energy of an N-particle system as

$$T = \frac{1}{2} \sum m_j \dot{r}_j^2 = \frac{1}{2} \sum m_j (\dot{\varrho}_j + \dot{\bar{r}})^2$$

it breaks up according to

$$T = \frac{1}{2} \sum m_j \dot{\varrho}_j^2 + \dot{\bar{r}} \cdot \sum m_j \dot{\varrho}_j + \frac{1}{2} \dot{\bar{r}}^2 \sum m_j$$

or, using (4.8),

$$T = \frac{1}{2} \sum m_j \dot{\varrho}_j^2 + \frac{1}{2} M \dot{\bar{r}}^2 \quad . \tag{4.13}$$

Once more, there is a term corresponding to the motion of the object as a whole and another corresponding to its internal motions.

The results just proved lead to simplifications in solving dynamical problems, for they tell us what we suspected already: that if we are interested in the internal motions of a system we need not bother about its motion as a whole and vice versa, and that a coordinate system attached to a system's center of mass has special advantages.

The formulas developed above are very general, applying to a large molecule, a container of gas, or a spinning baseball. They begin to look more familiar if we apply them to special cases. Consider the baseball for example, spinning around the z axis through its center with angular velocity ω. If one of its particles with coordinates ϱ moves with angular velocity ω, its linear velocity is

$$\dot{\varrho} = \omega \times \varrho \quad . \tag{4.14}$$

If ω is along the z axis, then

$$\dot{\varrho}_x = -\omega \varrho_y \quad , \qquad \dot{\varrho}_y = \omega \varrho_x$$

and the baseball's total kinetic energy is

$$T = \frac{1}{2} \sum m_j (\dot{\varrho}_{jx}^2 + \dot{\varrho}_{jy}^2) + \frac{1}{2} M \dot{\bar{r}}^2 \quad \text{or}$$

$$T = \frac{1}{2} I \omega^2 + \frac{1}{2} M \dot{\bar{r}}^2 \quad \text{where} \tag{4.15a}$$

$$I = \sum m_j (\varrho_{jx}^2 + \varrho_{jy}^2) \tag{4.15b}$$

is the baseball's moment of inertia about an axis through its center of mass. The internal part of the angular momentum is

$$\Lambda = \sum m_j \varrho_j \times \dot{\varrho}_j = \sum m_j \varrho_j \times [\omega \times \varrho_j]$$
$$= \sum m_j [\varrho_j^2 \omega - (\omega \cdot \varrho_j) \varrho_j]$$

by (3.43). Working out the components of Λ with ω parallel to the z axis gives

$$\Lambda_x = -\omega \sum m_j \varrho_{jz} \varrho_{jx} \quad , \qquad \Lambda_y = -\omega \sum m_j \varrho_{jz} \varrho_{jy}$$

$$\Lambda_z = \omega \sum m_j (\varrho_{jx}^2 + \varrho_{jy}^2 + \varrho_{jz}^2 - \varrho_{jz}^2) \quad \text{or}$$

$$\Lambda_z = I \omega \quad . \tag{4.16}$$

For a spherical distribution of mass Λ_x and Λ_y vanish by symmetry so that Λ is in this case parallel to the angular velocity ω. Note that this is not always the case, however (see Chap. 9). The moment of inertia of a sphere of mass M, density ϱ, and radius R is

$$I = \int \varrho (x^2 + y^2) dv \quad M = \frac{4}{3} \pi R^3 \varrho \quad .$$

If the density is uniform, this is

$$I = \varrho \int_0^R dr \int_0^\pi d\vartheta \int_0^{2\pi} d\phi (r^2 \sin^2 \vartheta) r^2 \sin \vartheta = \frac{2}{5} M R^2 \quad . \tag{4.17}$$

Example 4.1. A solid sphere rolls down an incline of height h, starting from rest. How fast is it moving when it reaches the bottom?

Solution. Energy is conserved. The potential energy at the top of the slope is Mgh. The kinetic energy at the bottom, by (4.14) is $I\omega^2/2 + M\bar{v}^2/2$, where $\bar{v} = \omega R$. With (4.17) this gives $\bar{v}^2 = 10gh/7$.

Problem 4.4. Show that Λ_x and Λ_y are zero for the ball discussed above.

Problem 4.5. A hoop and a solid cylinder of the same diameter are rolled down the same slope. Where is the hoop when the cylinder reaches the bottom?

Problem 4.6. A uniform bar of length l is used as a pendulum. Find the period if it is suspended (a) at the end and (b) at a point $\frac{2}{3}$ the way between the ends.

Problem 4.7. A rope of negligible mass supporting a mass m at the end is wrapped around a horizontal log of mass M with a frictionless axle down the middle. The mass is allowed to fall and the rope unwinds. Find the acceleration of the mass.

Problem 4.8. Find the position of the mass m in the preceding problem as a function of time if the mass of the rope is not negligible.

Problem 4.9. A small cylinder rolls inside a large one, Fig. 4.4. What is the period of small oscillations about equilibrium? ϑ will serve as a useful generalized coordinate.

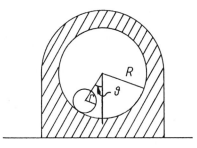

Fig. 4.4. To illustrate Problem 4.9

Problem 4.10. A uniform ladder of length $2l$ leans against a wall making an angle φ_0 with the vertical. Wall and floor are perfectly frictionless. At time $t = 0$ the ladder is allowed to start sliding. Analyze its motion, using the method of Lagrangian constraints (Sect. 3.1) or otherwise. You may want to stop with the small-angle approximation, but an exact treatment leads to interesting transformations of elliptic functions [37, Vol. 2].

4.2 Two-Particle Systems

We treated planetary motion in Chap. 3 on the hypothesis that the sun is clamped. Now relax the assumption. Let the center of mass of the two bodies be at rest at the origin, and let r be a vector running, as usual, from the sun to the planet. Then by (4.2), if M and m are the masses of sun and planet,

$$M r_s + m r_p = 0 \quad , \quad r_p - r_s = r \quad . \tag{4.18}$$

Solving these for r_s and r_p, calculate the total kinetic energy,

$$T = \frac{1}{2} M \dot{r}_s^2 + \frac{1}{2} m \dot{r}_p^2 = \frac{1}{2} \frac{Mm}{M+m} \dot{r}^2 \quad .$$

This is written as

$$T = \frac{1}{2} \mu \dot{r}^2 \quad \frac{1}{\mu} = \frac{1}{M} + \frac{1}{m} \tag{4.19}$$

where μ is called the reduced mass. We can write it also as

$$\mu = m\left(1 - \frac{m}{M+m}\right)$$

and when, as with atoms and planetary systems, $m \ll M$, we see that μ is slightly less than the smaller mass. The Lagrangian is

$$L = \tfrac{1}{2}\mu \dot{r}^2 - V(r) \tag{4.20}$$

and the only change from (3.14) is the substitution of μ for m.

Problem 4.11. The two-body Hamiltonian in quantum mechanics is

$$-\frac{\hbar^2}{2}\left(\frac{1}{m_2}\nabla_1^2 + \frac{1}{m_2}\nabla_2^2\right) + V(r) \quad .$$

Change the variables to r and $\bar{r}$ and show that the wave function factors into a term describing the motion around the center of mass and a term describing the motion of the system as a whole. Show that the effective mass enters here just as it does in classical theory.

4.3 Vibrating Systems

In several branches of study, from structural engineering to molecular chemistry, it is important to know how to analyze the motions of a system whose parts are coupled together so that they can vibrate. There was an example of this in the double pendulum of Sect. 3.2, where we found the modes of motion in which it moves periodically. In this section we shall study the periodic motions in more general situations, and then see how to deal with motions that are not periodic. The first part of the discussion will be only a generalization of what has already been done.

Suppose there are f degrees of freedom with coordinates q_n, and that the system has some position of stability around which small oscillations take place. The potential energy of the entire system is $V(q)$, and at the position of stability, where $q = q_0$, $V(q)$ has an absolute minimum or at least a local one, at which it is smaller than at any neighboring points. Since V is a minimum at q_0,

$$\left.\frac{\partial V}{\partial q_n}\right|_0 = 0 \quad n = 1, \ldots f \tag{4.21}$$

and if we expand V in the neighborhood of q_0 (as was done with the double pendulum) the linear terms in the Taylor series vanish and

$$V(q_0 + y) = V(q_0) + \frac{1}{2}\sum_{m=1}^{f}\sum_{n=1}^{f}\left.\frac{\partial^2 V}{\partial q_m \partial q_n}\right|_0 y_m y_n + \ldots$$

where the y's measure the system's departure from equilibrium. The first term is a constant and we can ignore it. The second term involves a set of constant coefficients

90

which we shall call V_{mn}, so that

$$V(q_0 + y) = \tfrac{1}{2} V_{mn} y_m y_n + \dots \tag{4.22}$$

in which the summation convention is used.

The kinetic energy is

$$T = \frac{1}{2} \sum m_i \dot{x}_i^2$$

in which the Cartesian components x_i are some functions of the q's and therefore of the y's. Since

$$x_i = x_i(y_1 \dots y_f)$$

we have

$$\dot{x}_i = \frac{\partial x_i}{\partial y_n} \dot{y}_n$$

and the kinetic energy will be some quadratic function of the $\dot{y}$'s,

$$T = \tfrac{1}{2} T_{mn} \dot{y}_m \dot{y}_n \quad . \tag{4.23}$$

In (4.23), the array T_{mn} is not necessarily symmetric in m and n to begin with, but

$$T_{mn} \dot{y}_m \dot{y}_n = T_{nm} \dot{y}_n \dot{y}_m = T_{nm} \dot{y}_m \dot{y}_n$$

so that

$$T_{mn} \dot{y}_m \dot{y}_n = \tfrac{1}{2}(T_{mn} + T_{nm}) \dot{y}_n \dot{y}_m \quad .$$

The coefficient $(T_{mn} + T_{nm})/2$, the symmetric part of T_{mn}, is all that enters the calculation, and so we may assume that both V_{mn} and T_{mn} are symmetrical to start with.

For small oscillations, the Lagrangian of the system is

$$L = \tfrac{1}{2} T_{mn} \dot{y}_m \dot{y}_n - \tfrac{1}{2} V_{mn} y_m y_n \quad .$$

To take the derivatives, vary the variables one at a time,

$$\begin{aligned} dL &= \tfrac{1}{2}(T_{mn} \dot{y}_m d\dot{y}_n + T_{mn} \dot{y}_n d\dot{y}_m - V_{mn} y_m dy_n - V_{mn} y_n dy_m) \\ &= T_{mn} \dot{y}_n d\dot{y}_m - V_{mn} y_n dy_m \end{aligned}$$

so that the equations of motion $\delta L / \delta y_m = 0$ are

$$T_{mn} \ddot{y}_n + V_{mn} y_n = 0 \quad m = 1, \dots f \quad . \tag{4.24}$$

This is a set of linear homogeneous equations with constant coefficients. Their solutions depend on time through expressions of the form $A \cos(\omega t + \phi)$. Since boundary conditions will involve us with different values of ϕ, it is strategic to consider y_n as the real part of an $a_n e^{i(\omega t + \phi_n)}$ with complex a_n. Now $e^{i\phi_n}$ can be absorbed into the constant a_n and all motions of the system have the same time dependence.

Substituting this form of y_n into (4.24) gives

$$(V_{mn} - \omega^2 T_{mn}) a_n = 0 \tag{4.25}$$

which is a set of f linear homogeneous equations in f unknowns. It has solutions only if the determinant of the coefficients vanishes

$$|V_{mn} - \omega^2 T_{mn}| = 0 \quad . \tag{4.26}$$

This is called the characteristic equation of the vibrating system. Its solutions are the values of ω^2 that correspond to the system's periodic motions. They are called characteristic values or, more commonly, eigenvalues. There are in general f different eigenvalues ω^2 (though some may happen to be equal) and since negative values of ω give nothing new, there are f solutions of the equations of motion. These solutions are called the system's normal modes of motion, and the corresponding frequencies are the normal mode frequencies, or characteristic frequencies of the system.

Equation (4.25) can be simplified. If we assume that none of the masses m_i are zero, it can be shown that the matrix T_{mn} is nonsingular, that is, it has an inverse. Multiplying (4.25) by this inverse gives

$$[(T^{-1}V)_{mn} - \omega^2 \delta_{mn}]a_n = 0 \tag{4.27}$$

and the characteristic equation is

$$|(T^{-1}V)_{mn} - \omega^2 \delta_{mn}| = 0 \quad . \tag{4.28}$$

The simplification is that ω^2 now occurs only along the main diagonal. Further, this standard form of eigenvalue equation is solved by readily available computer programs. The eigenvalues ω^2 are all real and positive. This follows from the conservation of energy: if ω in $e^{i\omega t}$ had an imaginary part the energy would mysteriously augment or disappear. It can also be proved from the mathematical properties of the matrices T_{mn} and V_{mn} [38].

Example 4.2. Suppose the double pendulum of Fig. 3.1 has $m_1 = 3$, $m_2 = 2$, $l_1 = 5$, $l_2 = 8$. To find the frequencies of the normal modes, one easily calculates

$$T = \frac{125}{2}\dot{\vartheta}_1^2 + 40\dot{\vartheta}_1\dot{\vartheta}_2 + 64\dot{\vartheta}_2^2$$

$$V = g\left(\frac{25}{2}\vartheta_1^2 + 8\vartheta_2^2\right)$$

so that the matrices for T and V are

$$T_{mn} = \begin{pmatrix} \frac{125}{2} & 20 \\ 20 & 64 \end{pmatrix} \quad , \quad V_{mn} = g\begin{pmatrix} \frac{25}{2} & 0 \\ 0 & 8 \end{pmatrix}$$

and

$$T_{mn}^{-1} = \frac{1}{3600}\begin{pmatrix} 64 & -20 \\ -20 & \frac{125}{2} \end{pmatrix} \quad , \quad (T^{-1}V)_{mn} = g\begin{pmatrix} \frac{2}{9} & -\frac{2}{45} \\ -\frac{5}{72} & \frac{5}{36} \end{pmatrix} \quad .$$

The characteristic equation (4.28) becomes

$$\begin{vmatrix} \frac{2}{9} - \frac{\omega^2}{g} & -\frac{2}{45} \\ -\frac{5}{72} & \frac{5}{36} - \frac{\omega^2}{g} \end{vmatrix} = 0$$

of which the solutions are

$$\frac{\omega^2}{g} = \frac{1}{4}, \frac{1}{9}$$

as can be verified from the earlier calculation.

Once the eigenvalues have been found, we can put them back into (4.25) and determine the relative sizes of the amplitudes a_n as in Sect. 3.2. (There is an arbitrary scale factor that is determined by the initial conditions.) Since all the coefficients in (4.25) are real, all the amplitudes are real or, if not, have the same complex phase. Thus all parts of a system in a normal mode vibrate exactly in or exactly out of phase with each other.

Problem 4.12. Figure 4.5 shows a model of linear triatomic molecule with masses and spring constants as shown. Find the frequencies of longitudinal oscillations.

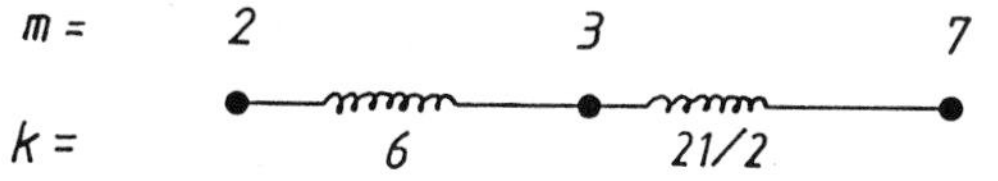

Fig. 4.5. To illustrate Problem 4.12

4.4 Coupled Oscillators

If a system is oscillating in one of its normal modes, its components move in or out of phase with each other with definite relations between the amplitudes. What happens if we start the system so that these conditions are not satisfied? A pessimist might conclude that we have to begin all over again, but a glance at an example already studied shows things aren't that bad.

The double pendulum of Sect. 3.1 had two characteristic frequencies

$$\omega_1 \simeq \sqrt{\frac{g}{l_1}} \left(1 + \frac{m_2}{4m_{12}} \frac{l_2}{l_2 - l_1} \right)$$
$$\omega_2 \simeq \sqrt{\frac{g}{l_2}} \left(1 - \frac{m_2}{4m_{12}} \frac{l_1}{l_2 - l_1} \right) \tag{4.29}$$

and, correspondingly, two alternative conditions to be satisfied by the amplitudes of motion,

$$a_2^{(1)} \simeq - \frac{l_1}{2(l_2 - l_1)} a_1^{(1)} := -\lambda_1 a_1^{(1)}$$
$$a_2^{(2)} \simeq \frac{2m_{12}}{m_2} \frac{l_2 - l_1}{l_2} a_1^{(2)} := \lambda_2 a_1^{(2)} . \tag{4.30}$$

The time behavior assumed there was of the form $\cos(\omega t + \phi)$; it could just as well have been $e^{\pm i\omega t}$. The point is that the equations of motion are linear; any linear combination of solutions is again a solution, and $e^{\pm i\omega t} = \cos \omega t \pm i \cos(\omega t - \pi/2)$. But we can go further and superpose solutions corresponding to different values of ω, the most general such superposition being of the form

$$\vartheta_1(t) = Ae^{i\omega_1 t} + Be^{-i\omega_1 t} + Ce^{i\omega_2 t} + De^{-i\omega_2 t}$$
$$\vartheta_2(t) = -\lambda_1(Ae^{i\omega_1 t} + Be^{-i\omega_1 t}) + \lambda_2(Ce^{i\omega_2 t} + De^{-i\omega_2 t}) \tag{4.31}$$

93

where A, B, C, D are arbitrary constants. The original equations (3.12a) are a set of the fourth order, which we know needs four arbitrary constants for its complete solution. Thus (4.31) is the complete solution; nothing more than the normal modes need be found.

Suppose for example that we hold pendulum 1 vertical with one hand, displace pendulum 2 through an angle ϑ_0 with the other, and release them both at $t = 0$. This gives four initial conditions,

$$A + B + C + D = 0 \quad , \quad -\lambda_1(A + B) + \lambda_2(C + D) = \vartheta_0$$
$$\omega_1(A - B) + \omega_2(C - D) = 0 \quad , \quad -\lambda_1\omega_1(A - B) + \lambda_2\omega_2(C - D) = 0$$

with solution

$$A = B = -C = -D = -\frac{\vartheta_0}{2(\lambda_1 + \lambda_2)}$$

and so the motion satisfying the given boundary conditions is

$$\vartheta_1(t) = -\frac{\vartheta_0}{\lambda_1 + \lambda_2}(\cos \omega_1 t - \cos \omega_2 t)$$

$$\vartheta_2(t) = \frac{\vartheta_0}{\lambda_1 + \lambda_2}(\lambda_1 \cos \omega_1 t + \lambda_2 \cos \omega_2 t) \quad . \tag{4.32}$$

The pendulums swing with varying amplitudes, and the upper one returns periodically to rest with an angular frequency $(\omega_1 - \omega_2)/2$. The two modes of motion take place simultaneously, and we are seeing the phenomena of beats.

Problem 4.13. Figure 4.6 shows an arrangement of coupled torsion oscillators that can very easily be made. Assigning suitable parameters, calculate the two basic frequencies when one bar is held still and the other is allowed to oscillate. In terms of these and other needed parameters, discuss the motion that begins when both bars are initially at rest and one is given a sudden displacement. Then build the thing and see what it does.

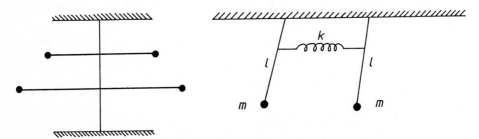

Fig. 4.6. To illustrate Problem 4.13. A torsion oscillator consisting of two weighted staffs supported by a wire

Fig. 4.7. To illustrate Problem 4.14

Problem 4.14. Figure 4.7 shows two identical pendulums coupled with a weak spring. Analyze the motion when they are started (a) in phase with the same amplitude, (b) out of phase with the same amplitude, (c) with both vertical but one in motion.

Fig. 4.8. To illustrate Problem 4.15

Problem 4.15. Three equal masses are connected by identical elastic cords, Fig. 4.8. Find the frequencies of normal oscillations and the corresponding normal modes.

Problem 4.16. A mass m is suspended from a spring of stiffness k and swings like a pendulum through small angles. Find the motion, (a) when the pendulum moves in a plane, (b) when it moves in three dimensions. Comment on special cases that may arise when the spring frequency is equal or nearly equal to the pendulum frequency. (c) Verify your main conclusions by experiment.

Normal Coordinates. The theory in Sect. 4.3 was set up in terms of generalized coordinates, and there was no necessity of starting with any particular set of q's or y's. But some choices are better than others. Suppose for example that instead of ϑ_1 and ϑ_2 in the double pendulum we had chosen linear combinations of the two solutions in (4.32) designed to produce quantities each of which oscillates at a single frequency:

$$
\begin{aligned}
u_1 &:= \vartheta_2 - \lambda_2 \vartheta_1 = \vartheta_0 \cos \omega_1 t \\
u_2 &:= \vartheta_2 + \lambda_1 \vartheta_1 = \vartheta_0 \cos \omega_2 t
\end{aligned}
\tag{4.33a}
$$

with λ's as in (4.30). With these variables,

$$
\vartheta_1 = \frac{u_2 - u_1}{\lambda_1 + \lambda_2} \quad , \quad \vartheta_2 = \frac{\lambda_1 u_1 + \lambda_2 u_2}{\lambda_1 + \lambda_2} \quad .
\tag{4.33b}
$$

We would have found u_1 to have the single frequency ω_1 and similarly for u_2, as if there were no coupling between them. How this occurs can be seen if we use (4.33b) to express the Lagrangian (3.11) in terms of u_1 and u_2. Omitting small terms, we find

$$
\begin{aligned}
L = \frac{m_{12}}{2(\lambda_1 + \lambda_2)^2} & \left\{ l_1^2 \left[\left(1 + \frac{m_2 l_2 (2l_1 - l_2)}{4 m_{12}(l_2 - l_1)^2} \right) \dot{u}_1^2 - \frac{g}{l_1} \left(1 + \frac{m_2 l_1 l_2}{4 m_{12}(l_2 - l_1)^2} \right) u_1^2 \right] \right. \\
& + 4 \frac{m_{12}}{m_2} (l_2 - l_1)^2 \left[\left(1 + \frac{m_2 l_1 (2l_2 - l_1)}{4 m_{12}(l_2 - l_1)^2} \right) \dot{u}_2^2 \right. \\
& \left. \left. - \frac{g}{l_2} \left(1 + \frac{m_2 l_1 l_2}{4 m_{12}(l_2 - l_1)^2} \right) u_2^2 \right] \right\}
\end{aligned}
$$

like the Lagrangian of two uncoupled oscillators. We now easily find that u_1 and u_2 oscillate with frequencies ω_1 and ω_2 calculated earlier.

The new variables u_1 and u_2 are called normal coordinates, and there is a mathematical theorem stating that it is always possible to define such coordinates for any system of particles oscillating around a potential minimum where the potential function is quadratic in the displacements [39, Chap. 3]. Calculations can be simplified if one uses the system's symmetries [40]. The procedure followed in the discussion above went backwards — first the problem was solved and then the normal coordinates were found, but for more complicated systems it is easier to do it the other way by algebraic methods. We shall encounter normal coordinates again in Sect. 10.2 in discussing the oscillations of a continuous system.

Problem 4.17. Find the normal coordinates of Problem 4.13, use them to reduce the Lagrangian to a sum of squares, and verify the frequencies.

Problem 4.18. Do the same for Problem 4.14.

Problem 4.19. Find the general solution for the system of Fig. 4.9 and discuss the motion that occurs if the system is initially at rest and the right-hand cart is struck toward the right with a hammer. What are the corresponding solutions when the left-hand spring is missing, i.e., what are the limiting forms of your solutions when $K = 0$?

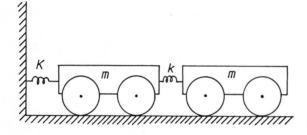

Fig. 4.9. To illustrate Problem 4.19

Problem 4.20. Figure 4.10 shows a scheme for the linear vibrations of the CO_2 molecule. The following steps will lead to a determination of the normal modes and thus to the entire solution. You should work them out in detail.

1. The Lagrangian is

$$L = \tfrac{1}{2}(M + 2m)\dot{x}^2 + m(\dot{\xi}_2 - \dot{\xi}_1)\dot{x} + \tfrac{1}{2}m(\dot{\xi}_1^2 + \dot{\xi}_2^2) - \tfrac{1}{2}k(\xi_1^2 + \xi_2^2) \quad . \qquad (4.34)$$

2. Since the momentum of the system is conserved, one can guess that the coordinate $\bar{x}$ will have special significance. The transformation

$$x = \bar{x} + \frac{m}{M + 2m}(\xi_1 - \xi_2)$$

eliminates x and gives

$$L = \frac{1}{2}(M + 2m)\dot{\bar{x}}^2 + \frac{1}{2}\frac{m^2}{M + 2m}(\dot{\xi}_1 - \dot{\xi}_2)^2 + \frac{1}{2}m(\dot{\xi}_1^2 + \dot{\xi}_2^2) - \frac{1}{2}k(\xi_1^2 + \xi_2^2) \quad .$$

3. The terms in $\dot{\xi}_2\dot{x}$ and $\dot{\xi}_1\dot{x}$ are gone but $\dot{\xi}_1\dot{\xi}_2$ remains. To remove it, rotate the ξ axes by $\pi/4$,

$$u_1 = 2^{-1/2}(\xi_1 + \xi_2) \quad , \quad u_2 = 2^{-1/2}(\xi_1 - \xi_2)$$

to get

$$L = \frac{1}{2}(M + 2m)\dot{\bar{x}}^2 + \frac{1}{2}m\dot{u}_1^2 + \frac{1}{2}\frac{Mm}{M + 2m}\dot{u}_2^2 - \frac{1}{2}k(u_1^2 + u_2^2) \quad .$$

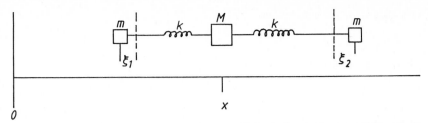

Fig. 4.10. To illustrate Problem 4.20. Mechanical model for the linear vibrations of CO_2. The ξ's measure the stretch of the two springs, to left and right respectively, from their equilibrium positions

4. The Lagrangian is now decoupled. Find the frequencies, describe the corresponding normal modes, and find the transformations from (x, ξ_1, ξ_2) to $(\overline{x}, u_1, u_2)$.

Problem 4.21. Solve the foregoing problem directly from the Lagrangian (4.34) and compare the frequencies with those just found.

Problem 4.22. Write down Schrödinger's equation for the CO_2 vibrations just discussed and transform it to the variables $\overline{x}$, u_1, u_2. Find the energy levels of the stationary molecule according to this model.

4.5 The Virial Theorem

The methods described in this chapter become useless when N is very large unless the particles have some specially simple spatial organization, as they have in a solid body. They are useless because the calculations are too complicated and because nobody wants the results anyhow. If the system is a gas, it would not be interesting to know the positions and velocities of all the particles, even if one could, because what we want to learn are the properties of gases in general, not those of some sample in which the molecules start in a particular way. We want a statistical theory, not a theory of specific systems. We cannot in this book start a serious discussion of statistical mechanics, but there is a simple statistical theorem, the virial theorem, that illustrates the changed point of view one adopts when N is large and permits some useful averages to be obtained.

Consider a finite isolated, bounded N-particle system at rest as a whole with respect to the origin, and define the quantity

$$G := \sum_{i=1}^{N} p_i \cdot r_i \quad . \tag{4.35}$$

Since no particle ever reaches an infinite value of momentum or wanders infinitely far away, G remains finite. Further, if we shift the origin by r_0 the new G is

$$G' = \sum p_i \cdot (r - r_0) = G - r_0 \cdot \sum p_i = G$$

and so G does not depend on the choice of origin. The time derivative of G is

$$\dot{G} = \sum(\boldsymbol{p}_i \cdot \dot{\boldsymbol{r}}_i + \dot{\boldsymbol{p}}_i \cdot \boldsymbol{r}_i)$$

$$= 2T + \sum \boldsymbol{F}_i \cdot \boldsymbol{r}_i$$

where T is the kinetic energy and the quantity

$$W := \sum \boldsymbol{F}_i \cdot \boldsymbol{r}_i$$

was called by Clausius the virial of the system. If the system as a whole remains at rest, W, like G, is independent of the choice of origin of coordinates.

Let us find the time average of $2T + W$. It is

$$\lim_{\tau \to \infty} \frac{1}{\tau} \int_0^\tau \dot{G} dt = \lim \frac{1}{\tau} [G(\tau) - G(0)] \quad .$$

But we have noted above that $G(\tau) - G(0)$ always remains finite; thus the limit is zero, and the relation between time averages is

$$\overline{T} = -\tfrac{1}{2}\overline{W} \quad . \tag{4.36}$$

This is the virial theorem [41].

The theorem has a useful application to systems bound by inverse-square forces – for example, an atom, certain plasmas, or slowly-evolving clusters of stars. Introducing the interparticle forces as in (4.3), we have

$$W = \sum_{i \neq j} \sum f_{ij} \cdot \boldsymbol{r}_j \quad .$$

A typical pair of terms is

$$\boldsymbol{f}_{12} \cdot \boldsymbol{r}_2 + \boldsymbol{f}_{21} \cdot \boldsymbol{r}_1 = \boldsymbol{f}_{12} \cdot (\boldsymbol{r}_2 - \boldsymbol{r}_1) \quad .$$

If the interparticle potential is

$$V_{12} = -\frac{\gamma}{|\boldsymbol{r}_2 - \boldsymbol{r}_1|}$$

the force is

$$\boldsymbol{f}_{12} = -\frac{\gamma(\boldsymbol{r}_2 - \boldsymbol{r}_1)}{|\boldsymbol{r}_2 - \boldsymbol{r}_1|^3} \quad \text{and}$$

$$\boldsymbol{f}_{12} \cdot (\boldsymbol{r}_2 - \boldsymbol{r}_1) = -\frac{\gamma}{|\boldsymbol{r}_2 - \boldsymbol{r}_1|} = V_{12} \quad .$$

Thus $W = V$, the total potential energy of the system, and

$$\overline{T} = -\tfrac{1}{2}\overline{V} \quad . \tag{4.37}$$

Since by hypothesis the system's total energy $\overline{T} + \overline{V} = E$ is conserved, this gives also

$$E = -\overline{T} = \tfrac{1}{2}\overline{V} \quad . \tag{4.38}$$

Clusters of Galaxies. The virial theorem has an important application in finding the mass of a cluster of galaxies. If the cluster has a mass M and a size of order R, the gravitational potential energy is of order

$$\overline{V} \simeq -\frac{GM^2}{R} \quad .$$

(4.39)

If $\overline{v^2}$ is the mean square velocity of a member of the cluster, then

$$\overline{T} = \tfrac{1}{2}M\overline{v^2}$$

and the virial theorem gives

$$M \simeq \frac{R\overline{v^2}}{G} \quad .$$

(4.40)

The numbers on the right can be measured; the masses come out much larger than would be expected from the luminosity of the clusters. Apparently a large amount of nonluminous matter is present, but one does not yet know what form it takes.

Ideal Gas. The virial theorem also gives insight into the behavior of gases. In an ideal gas the only force on a molecule is that exerted by the walls of the container. In magnitude and direction this is given by $d\mathbf{F} = -Pd\mathbf{A}$ where P is the pressure and $d\mathbf{A}$ is a directed element of area. The virial is

$$W = -P \int \mathbf{r} \cdot d\mathbf{A} \quad .$$

Evaluate this by Gauss's theorem:

$$W = -P \int \nabla \cdot \mathbf{r} dV = -3PV$$

where V is the volume of the container. The virial theorem now gives

$$\overline{T} = \tfrac{3}{2}PV$$

(4.41)

a relation familiar from elementary physics but here proved rather neatly.

Problem 4.23. How is the result (4.38) generalized if the force follows an inverse-nth power law?

Problem 4.24. Let the system be a hydrogen atom ($N = 2$) and the electron and proton in circular orbits around their common center of mass. Verify that the virial theorem is satisfied.

Problem 4.25. If the system considered above is immersed in a uniform magnetic field B, the force on each particle is augmented by $e\mathbf{v} \times \mathbf{B}$. Show that the virial theorem for charged particles now reads

$$2\overline{T} = -\frac{e}{m}\overline{\mathbf{B} \cdot \mathbf{L}} - \overline{V}$$

where $\mathbf{L}$ is the system's total angular momentum.

Problem 4.26. It is generally true that expectation values in quantum mechanics correspond to time averages in classical mechanics. State and prove a virial theorem for quantum mechanics. What form does (4.41) take in the quantum-mechanical theory of an ideal gas?

Problem 4.27. Three equal gravitating masses m are arranged to form an equilateral triangle of side a. Use the virial theorem to find out how fast the triangle must rotate (in its own plane) to keep the masses in equilibrium. You may want to verify your result by doing the calculation in another way.

Problem 4.28. The virial theorem was derived above from Newtonian mechanics. What is the relativistic form? What is the relation between pV and $\overline{T}$? Verify that the low-energy limit coincides with (4.41). What is the high-energy limit? To what commonly available substance does this apply?

4.6 Hydrodynamics

Rather than consider a fluid as an assemblage of particles, one can start out by picturing it as a continuous system. Various steps leading to the formulation and solution of problems in fluid mechanics can be justified by appealing to the theory of many-particle systems, but the equations themselves do not originate in arguments of that kind. In general, the flow of a fluid is a very complicated matter. It is reasonably simple to analyze if the fluid moves in a uniform way, but as soon as it breaks up into small regions of more or less independent movement (a state known as turbulence) statistical considerations take over and the analysis becomes much more difficult. The following discussion is therefore restricted to nonturbulent motion.

The basic equation for fluid flow can be derived as follows. Follow a small volume of the fluid, τ, as it moves along with velocity v. If p is the pressure of the fluid, the force this pressure exerts on τ equals $-\oint pd\boldsymbol{A}$, where $d\boldsymbol{A}$ is an outward-directed element of the surface of τ. This is transformed by a simple vector identity[1] into $-\int \nabla pdV$. That is to say, the force on the element τ is $-\nabla p$ per unit volume. We now make a series of assumptions.

1) Assume that τ moves according to Newton's second law. If ϱ is the density of the fluid, and if a gravitational force is acting, this gives

$$\varrho\frac{d\boldsymbol{v}}{dt} = -\nabla p + \varrho\boldsymbol{g} \quad .$$

But it is more convenient to study the motion of the fluid at a given point in space than to follow the element τ as it moves around. If the velocity of τ is considered as a function of position and time, it changes because v at the given point depends on time and also because τ moves from one place to another. Thus

---

[1] Let $\boldsymbol{a}$ be any constant vector. Then $\nabla \cdot (p\boldsymbol{a}) = \boldsymbol{a} \cdot \nabla p$ and so Gauss's theorem gives

$$\int \nabla \cdot (p\boldsymbol{a})dV = \int p\boldsymbol{a} \cdot d\boldsymbol{A} = \boldsymbol{a} \cdot \int pd\boldsymbol{A} = \boldsymbol{a} \cdot \int \nabla pdV$$

from which, since $\boldsymbol{a}$ is arbitrary, the result follows.

$$\frac{d\boldsymbol{v}}{dt} = \frac{\partial \boldsymbol{v}}{\partial t} + (\boldsymbol{v} \cdot \nabla)\boldsymbol{v}$$

so that the equation of motion becomes

$$\frac{\partial \boldsymbol{v}}{\partial t} = -(\boldsymbol{v} \cdot \nabla)\boldsymbol{v} - \frac{1}{\varrho}\nabla p + \boldsymbol{g} \quad . \tag{4.42}$$

A relation equivalent to this was given by Euler in 1755. Note that even this basic equation, which is no more than Newton's law, is nonlinear. Except in special situations and approximations it cannot be solved with pencil and paper.

As a very simple example to show what this equation does and does not say, look at the propagation of sound through a fluid. We need first an assumption as to the nature of sound:

2) Sound is a pressure wave that does not involve the fluid in any rotational motion. The mathematical statement of this is that[2] $\nabla \times \boldsymbol{v} = 0$, and it allows us to represent $\boldsymbol{v}$ in terms of a *velocity potential,* ϕ, by

$$\boldsymbol{v} =: \nabla \phi \quad .$$

Let $p = p_0 + p'$ and $\varrho = \varrho_0 + \varrho'$, where p_0 and ϱ_0 are the ambient values. Let the amplitude be small enough so that second-order terms in p, ϱ, and $\boldsymbol{v}$ can be neglected, and also neglect $\boldsymbol{g}$. Then Euler's equation gives

$$\nabla \left(\frac{\partial \phi}{\partial t} + \frac{p'}{\varrho_0} \right) = 0 \quad .$$

At some remote point where there is no sound the bracketed expression is zero; therefore it is zero everywhere and

$$\varrho_0 \frac{\partial \phi}{\partial t} + p' = 0 \quad . \tag{4.43}$$

Take the time derivative of this equation,

$$\varrho_0 \frac{\partial^2 \phi}{\partial t^2} + \frac{\partial p'}{\partial t} = 0 \quad .$$

Euler's equation tells about the dynamics of an element of fluid but says nothing about whether the fluid is conserved.

3) Conservation is expressed by

$$\nabla \cdot (\varrho \boldsymbol{v}) + \frac{\partial \varrho}{\partial t} = 0 \tag{4.44}$$

and to first order this gives

$$\varrho_0 \nabla \cdot \boldsymbol{v} = \varrho_0 \nabla^2 \phi = -\frac{\partial \varrho'}{\partial t} \quad .$$

Now we need a fourth hypothesis.

---

[2] To see this, suppose that $\nabla \times \boldsymbol{v} \neq 0$ in some region. Pass a surface S through this region and integrate $\nabla \times \boldsymbol{v}$ over any area of this surface. Except for special placement of the surface the result, by Stokes's theorem, is $\oint \boldsymbol{v} \cdot d\boldsymbol{l} \neq 0$: the fluid velocity has a component that flows around the boundary of the area.

4) The fluid obeys an equation of state connecting p with ϱ. With this, we can write

$$\frac{\partial p'}{\partial t} = \frac{\partial p}{\partial \varrho} \frac{\partial \varrho'}{\partial t} \quad .$$

How is the derivative $\partial p / \partial \varrho$ to be taken? Compression heats a fluid, and for sound waves the frequency is too high for the temperature to equalize itself before the next fluctuation; therefore

5) The derivative is to be evaluated isentropically; i.e. adiabatically.

The result of these five assumptions is that sound propagates at speed c according to d'Alembert's equation

$$\nabla^2 \phi - \frac{1}{c^2} \frac{\partial \phi^2}{\partial t^2} = 0 \quad , \quad c^2 := \left(\frac{\partial p}{\partial \varrho} \right)_{ad} \quad . \tag{4.45}$$

It would have been very hard to obtain this result starting from a molecular theory concentrating on the motions of individual particles. This brief discussion was intended to show that even for a simple process described finally by a simple equation, if we do not start with the motions of the particles of the fluid we must introduce along the way a number of separate physical hypotheses. This is the reason for the conceptual difficulties that beset continuum mechanics. The mathematical difficulties are legendary. In Chap. 10 the theory of sound vibrations (this time in a crystal) will be quantized so as to show how taking account of the discrete nature of these collective excitations leads to an explanation of the Mössbauer effect.

Problem 4.28. It is shown by thermodynamic arguments that in an adiabatic process the pressure and volume of a sample of gas are related by $pV^\gamma = \text{const}$, where γ equals the ratio of the specific heat of the gas at constant pressure to that at constant volume. For a monatomic gas $\gamma = \frac{5}{3}$ (a crude derivation is in Problem 7.21). Assuming also that the gas satisfies the ideal gas law, calculate c^2 in terms of readily available quantities. Equation (4.41) allows one to find $\overline{v^2}$, the mean squared velocity of an individual molecule. Calculate this and compare it with the c^2 just found.

5. Hamiltonian Dynamics

The Lagrangian procedure is helpful in setting up equations of motion and simplifying them through the use of cyclic coordinates and their constant conjugate momenta. But once this has been done, it is up to us to solve the equations. This chapter will develop methods which will later lead to solutions with a minimum of equation-solving; in fact a large class of calculations will be reduced in Chap. 6 to the evaluation of a few integrals. (Problems not so reducible are of an altogether different order of difficulty, and except for the chapter on perturbation theory we shall not get involved in solving them.)

It is important to choose the right variables for solving problems in dynamics – the solution of the Kepler problem, for example, uses polar coordinates to give the cyclic angular coordinate. A solution in Cartesian coordinates would be much harder. We need a technique for changing variables so as to find as many cyclic coordinates as possible. This will be introduced in Sect. 5.3 and developed in the following sections.

5.1 The Canonical Equations

There are two general ways to depict the time behavior of a dynamical system: a q, t diagram and a q, p diagram, where p_n, the momentum conjugate to a given q_n, is defined as

$$p_n := \frac{\partial L(q, \dot{q})}{\partial \dot{q}_n} \quad . \tag{5.1}$$

Lagrange's equations (3.3) then become

$$\dot{p}_n = \frac{\partial L(q, \dot{q})}{\partial q_n} \quad . \tag{5.2}$$

Figure 5.1 shows how a simple harmonic oscillator is described in each. Diagram (a) is a graph of $x = a \cos(\omega t + \phi)$. Diagram (b) is a graph of the energy equation (2.10), the ellipse

$$\frac{p^2}{2mE} + \frac{x^2}{2Ek^{-1}} = 1 \quad , \tag{5.3}$$

It represents time behavior because p itself represents time behavior. The point P representing the oscillator's state moves to the right on (a); on (b) it moves clockwise (why?) around the ellipse. Lagrangian dynamics, and the Hamilton-Jacobi theory to

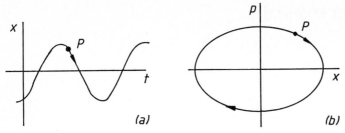

Fig. 5.1. Trajectory of an oscillator, (a) in q, t space; (b) in q, p space

be developed in Chap. 6, use the q, t representation. Hamiltonian dynamics uses q and p. Figure (b), corresponding to a system with one degree of freedom, represents the oscillator's motion in what is called phase space. It is easy to look at if it has only two dimensions, but for higher dimensionality it must be imagined. The coordinates of the phase point P specify the value of every coordinate and momentum and so specify the state of the system. The phase point does not usually wander into every part of the space. For the oscillator it is kept on an ellipse by the conservation of energy, and in general if energy is conserved, it moves over a surface of constant energy in the full space. If there are other conserved quantities such as angular momentum it moves on the surfaces they define, and if there are several of them it moves in the subspace in which all these surfaces intersect. This will be further discussed in Sect. 6.5.

The Hamiltonian function corresponding to a given Lagrangian is defined [compare (3.7)] as

$$H(p, q) := p_n \dot{q}_n - L(q, \dot{q}) \tag{5.4}$$

(summed). To see why it is called a function of p and q, imagine that the state of the system is varied, i.e., changed so that p, q, and $\dot{q}$ all change slightly:

$$\delta H = p_n \delta \dot{q}_n + \dot{q}_n \delta p_n - \frac{\partial L}{\partial q_n} \delta q_n - \frac{\partial L}{\partial \dot{q}_n} \delta \dot{q}_n \quad .$$

By (5.1) and (5.2), this is

$$\delta H = \dot{q}_n \delta p_n - \dot{p}_n \delta q_n \tag{5.5}$$

so that we need specify only the δp's and δq's in order to evaluate the change in H. Changing one p at a time and one q at a time gives Hamilton's equations [5, p. 162] (which Lagrange had published many years earlier without recognizing their value),

$$\dot{q}_n = \frac{\partial H}{\partial p_n} \quad , \quad \dot{p}_n = -\frac{\partial H}{\partial q_n} \quad n = 1, \dots f \quad . \tag{5.6}$$

The first of these, anticipated in (1.24), tells how $\dot{q}$ is related to q and p; we need this in order to write $H(p, q)$. The second is essentially (5.2) again, the generalized form of Newton's second law. Computationally, these equations are no great advance over Lagrange's equations. Formally however, by virtue of their transparent structure, being $2f$ first-order equations, they open the way for a profound study of the laws of dynamics. To express the fact that all further studies start from here, they are called the canonical equations.

Example 5.1. Set up the problem of plane orbits by Hamilton's equations.

Solution. Use the coordinates r, ϑ. Then

$$L(q, \dot{q}) = \tfrac{1}{2}m(\dot{r}^2 + r^2\dot{\vartheta}^2) - V(r) \quad .$$

The momenta are $\partial L/\partial \dot{q}_n$, or

$$p_r = m\dot{r} \quad , \quad p_\vartheta = mr^2\dot{\vartheta} \quad .$$

The Hamiltonian function (3.19) expressed in terms of p and q is

$$H = \frac{1}{2m}(p_r^2 + p_\vartheta^2/r^2) + V(r) \tag{5.7}$$

and Hamilton's equations are

$$\dot{r} = \frac{p_r}{m} \quad , \quad \dot{\vartheta} = \frac{p_\vartheta}{mr^2} \quad , \quad \dot{p}_\vartheta = 0 \quad , \quad \dot{p}_r = \frac{p_\vartheta^2}{mr^3} - \frac{\partial V}{\partial r} \quad .$$

Since ϑ is cyclic, p_ϑ is constant. Further solution leads along essentially the same path that was travelled before.

To investigate the rate at which energy in a system changes, consider the general situation in which the external interactions are time-dependent and H is a function of p, q, and, explicitly, t. We have

$$\frac{dH}{dt} = \frac{\partial H}{\partial p_n}\dot{p}_n + \frac{\partial H}{\partial q_n}\dot{q}_n + \frac{\partial H}{\partial t} \quad .$$

From (5.6), this is

$$\frac{dH}{dt} = \frac{\partial H}{\partial t} \tag{5.8a}$$

and if H is not explicitly time-dependent,

$$\frac{dH}{dt} = 0 \quad . \tag{5.8b}$$

The system then moves in such a way that its energy is conserved.

Example 5.2. In one dimension, start with the expression for kinetic energy in special relativity and find the appropriate Lagrangian for a free particle.

Solution. The kinetic energy is

$$E = mc^2[(1 - v^2/c^2)^{-1/2} - 1] \quad , \quad m = \text{rest mass} \quad . \tag{5.9}$$

Its relation with L is

$$E = \frac{dL}{dv}v - L = v^2\frac{d}{dv}\left(\frac{L}{v}\right)$$

so that

$$L = v\int\frac{Edv}{v^2} = mc^2[1 - \sqrt{1 - v^2/c^2}] \tag{5.10}$$

plus a constant multiple of v which we can omit (Problem 5.1). There is an important point that can be made if we drop the irrelevant additive constant in L and write the action as

$$S = -mc^2 \int \sqrt{1 - v^2/c^2}\, dt = -mc^2 \int \sqrt{dt^2 - c^{-2}dx^2}$$

This is

$$S = -mc^2 \int ds \tag{5.11a}$$

and the law of motion becomes

$$\delta S = \delta \int ds = 0 \tag{5.11b}$$

where ds is the invariant element of proper time.

Let us suppose that the integrand to be varied in Hamilton's principle (3.6) were not a relativistic invariant. Then we could make the integral larger or smaller by evaluating it in one coordinate system or another, and setting $\delta S = 0$ would pick out some preferred coordinate system. The only way to have results that are covariant (independent of the coordinate system) is to have the integrand an invariant, as has happened here automatically. If there is an interaction term, it too must be an invariant. (This fact will be noted again following Eq. (5.15).)

Problem 5.1. Show that no generality is lost by omitting the constant multiple of v that appears in L. Set up the equations for an actual problem and sketch their solution to the point where it can be seen that the extra term does not matter.

Problem 5.2. Show that H and L are equal in the low-velocity limit of the relativistic formulas just given.

Problem 5.3. (This and the following problems are designed to allow the reader to explore one aspect of the relation between classical and quantum mechanics.) Suppose that the equation of motion for matter waves is taken to be of the general form

$$i\hbar \frac{\partial \psi}{\partial t} = H(p_n, q_n)\psi \quad , \quad n = 1 \ldots f$$

where the p's and q's are q-numbers and H is a function of them. Why does the ordinary probabilistic interpretation of ψ require H to be hermitian?

Problem 5.4. Assume that

$$[p_m, q_n] = -i\hbar \delta_{mn} \quad , \quad [p_m, p_n] = 0$$

and prove that

$$[q_n^k, p_n] = k i\hbar q_n^{k-1} = i\hbar \frac{\partial}{\partial q_n} q_n^k \quad .$$

Problem 5.5. Show via Taylor's series that

$$[f(q_n), p_n] = i\hbar \frac{\partial f(q_n)}{\partial q_n} \quad .$$

What if the coefficients in the Taylor series are functions of p? Give a recipe for finding $[f(p, q), p_n]$.

Problem 5.6. Prove that

$$\frac{dp_{Hn}}{dt} = \frac{i}{\hbar}[H, p_{Hn}] = -\frac{\partial H_H}{\partial q_{Hn}}$$

provided that the differentiations are carried out properly.

Problem 5.7. Show that

$$\frac{dq_{Hn}}{dt} = \frac{\partial H_H}{\partial p_{Hn}} \quad .$$

(A general recipe for carrying out such differentiations will be given in Sect. 5.7.)

Problem 5.8. A crazy oscillator (mass=1) is defined by the Hamiltonian

$$H = \tfrac{1}{2}(p^2/x^2 + \mu^2 x^2) \quad .$$

Discuss its motion and calculate its period. (The *Note on Arc-Sines* in Sect. 2.2 may be useful.)

5.2 Magnetic Forces

We have not yet looked at the effect of an applied magnetic field because it does not fit obviously into considerations based on energy. This is because the force on a charged particle

$$F = ev \times B \tag{5.12}$$

(e/c in Gaussian units) is perpendicular to the particle's motion and hence cannot affect its energy. Further, since we do not introduce electric forces into the Lagrangian formalism by means of the field, we should not expect to do so for the magnetic forces. For both kinds of force, it is the potential that we must use.

Electrostatic fields are given in terms of the scalar potential Φ by $E = -\nabla\Phi$. Magnetic fields are formed from a vector potential A by

$$B = \nabla \times A \quad . \tag{5.13a}$$

If the potentials are time-dependent, (5.13a) still holds and E becomes (in MKS units)

$$E = -\nabla\Phi - \frac{\partial A}{\partial t} \quad . \tag{5.13b}$$

Since the Lagrangian is a scalar, A must enter as a dot product, and since the forces are velocity-dependent we try the product with the velocity $\dot{x}$. Let λ be a constant to be determined later. Suppose that

$$L = \tfrac{1}{2}m\dot{x}^2 - e(\Phi - \lambda A \cdot \dot{x}) \quad .$$

Then

$$p_i = \frac{\partial L}{\partial \dot{x}_i} = m\dot{x}_i + e\lambda A_i \quad , \quad \frac{\partial L}{\partial x_i} = -e\left(\frac{\partial \Phi}{\partial x_i} - \lambda\frac{\partial A_j}{\partial x_i}\dot{x}_j\right) \quad .$$

In taking the time derivative of $\partial L/\partial \dot{x}_i$, remember that a moving particle encounters a changing A even if A is time-independent, so consider A_i as a function of $x_j(t)$ and t:

$$\frac{d}{dt}\frac{\partial L}{\partial \dot{x}_i} = m\ddot{x}_i + e\lambda\left(\frac{\partial A_i}{\partial x_j}\dot{x}_j + \frac{\partial A_i}{\partial t}\right)$$

and the Lagrange equations are

$$m\ddot{x}_i + e\lambda\left(\frac{\partial A_i}{\partial x_j}\dot{x}_j + \frac{\partial A_i}{\partial t} - \frac{\partial A_j}{\partial x_i}\dot{x}_j\right) + e\frac{\partial \Phi}{\partial x_i} = 0 \quad.$$

Let $i = 1$. Then in this expression

$$\left(\frac{\partial A_1}{\partial x_j} - \frac{\partial A_j}{\partial x_1}\right)\dot{x}_j = \left(\frac{\partial A_1}{\partial x_2} - \frac{\partial A_2}{\partial x_1}\right)\dot{x}_2 + \left(\frac{\partial A_1}{\partial x_3} - \frac{\partial A_3}{\partial x_1}\right)\dot{x}_3$$

$$= -B_3\dot{x}_2 + B_2\dot{x}_3$$

$$= (B \times \dot{x})_1 = -(\dot{x} \times B)_1$$

and the equation of motion is

$$m\ddot{x} = e\lambda\dot{x} \times B - e\left(\nabla\Phi + \lambda\frac{\partial A}{\partial t}\right) \quad.$$

Take $\lambda = 1$ and remember (5.13b); this becomes the equation for a charged particle in MKS units,

$$m\ddot{x} = e\dot{x} \times B + eE \tag{5.14}$$

whereas Gaussian units come from taking $\lambda = c^{-1}$.

The nonrelativistic Lagrangian of a particle in an electric and magnetic field is therefore

$$L = \tfrac{1}{2}m\dot{x}^2 - e(\Phi - A \cdot \dot{x}) \quad. \tag{5.15}$$

In special relativity the first term becomes $-mc^2\sqrt{1 - \dot{x}^2/c^2}$, which we know gives a relativistically invariant contribution to S. The contribution of the interaction term is

$$S_{\text{int}} = -e\int(\Phi - A \cdot \dot{x})dt = -e\int(\Phi\,dt - A \cdot d x) \quad. \tag{5.16}$$

Since (Φ, A) forms one four-vector [42, Chap. 11] and $(dt, d x)$ is another, S_{int} is already invariant in form and need not be changed in writing down the relativistic version of (5.15).

To find the Hamiltonian corresponding to (5.15) calculate

$$\frac{\partial L}{\partial \dot{x}} = p = m\dot{x} + eA \tag{5.17}$$

so that the energy is

$$E = \dot{x} \cdot (m\dot{x} + eA) - \tfrac{1}{2}m\dot{x}^2 + e(\Phi - A \cdot \dot{x}) \quad \text{or}$$

$$E = \tfrac{1}{2}m\dot{x}^2 + e\Phi \quad. \tag{5.18}$$

The magnetic field seems to have disappeared from the theory, but this expression for E merely repeats that a magnetic field does not change the energy. To find the

Hamiltonian use (5.17) to express E in terms of p:

$$H = \frac{1}{2m}(p - eA)^2 + e\Phi \quad . \tag{5.19}$$

Here p contains A though the energy does not. One of the difficulties in the historical development of the concept of mechanical energy was that it turned out to have a visible (kinetic) and an invisible (potential) part. We can see now that the same is true of momentum.

Problem 5.9. Show that the angular momentum of a particle in a magnetic field is $r \times p$ (and not, as one might perhaps think, $r \times (p - eA)$).

The Relativistic Lagrangian. It would appear from (5.11) and (5.16) that the action integral for relativistic electrodynamics looks like

$$S = \int [-mc^2 ds - e(\Phi\, dt - A \cdot dx)] \tag{5.20}$$

(remember that here and below, m is the particle's *rest* mass), but this is unsatisfactory in several ways: the first term represents the kinetic energy but does not contain any dynamical variables; it is not clear what is or should be the variable of integration, and most of the motions it allows are impossible. To say in the simplest language why that is so it will be useful to introduce some relativistic notation. Define the four coordinates as $dx^0 := dt$, $dx^1 := dx$, $dx^2 := dy$, $dx^3 := dz$ (it is standard notation to use the superscripts) and the corresponding four-velocities as

$$u^i := dx^i/ds \quad (i = 0, \ldots, 3) \quad .$$

(Gothic letters will be used to denote quantities in which proper time is the independent variable if there is any risk of confusion.) Further, to form scalar products such as that in (5.16), introduce quantities η_{ij} defined as

$$\eta_{00} := 1 \quad , \quad \eta_{11} := \eta_{22} := \eta_{33} := -c^{-2} \quad , \quad \text{all others} = 0$$

in terms of which the relation

$$c^2 dt^2 - dx^2 - dy^2 - dx^2 = c^2 ds^2$$

becomes

$$\eta_{ij} u^i u^j = 1 \tag{5.21}$$

(summed). Finally, define subscripted coordinates by

$$u_i := \eta_{ij} u^j$$

and similarly for other four-vectors, so that (5.21) can be written

$$u_i u^i = 1 \quad . \tag{5.22}$$

(Vectors such as u_i and u^i are called covariant and contravariant, respectively.) With these notations established, the trouble with (5.20) is that however a particle moves its coordinates must satisfy (5.22), but that nothing in (5.20) requires this to be so.

The requirement must be introduced with a Lagrangian multiplier, and when this is done the other difficulties of (5.20) will also disappear.

Choose s as the variable of integration, let $\mu(s)$ be the Lagrangian multiplier, and write the integrand in (5.20) as

$$\mathcal{L} := -mc^2 - eA_i u^i + \mu(s)(\eta_{ij} u^i u^j - 1) \quad . \tag{5.23}$$

Lagrange's equations are

$$\frac{\partial \mathcal{L}}{\partial x^k} - \frac{d}{ds}\frac{\partial \mathcal{L}}{\partial u^k} = 0 \quad (k = 0, \dots, 3) \quad . \tag{5.24}$$

Look ahead for a moment at the solutions that will result from these equations. They will specify the four x^i as functions of s; that is, they will tell the values of the three spatial coordinates and the reading t of the clock on the laboratory wall as functions of the proper time measured by the particle's own watch. If one wants to know $x(t)$, say, instead of $x(s)$, one must eliminate s between $x(s)$ and $t(s)$, or alternatively (see Problem 5.12) set up the Lagrangian formalism with t instead of s as the independent variable. This destroys the formal relativistic symmetry but corresponds with the facts of human experience: ordinarily it is t that we know, not s.

The equations that come from (5.24) are

$$eA_{i,k} u^i - \frac{d}{ds}(eA_k - 2\mu\eta_{ik} u^i) = 0 \quad \text{where}$$

$$A_{i,k} := \partial A_i/\partial x^k$$

and please note carefully how the factors of 2 arise. Perform the differentiation:

$$eA_{i,k} u^i - eA_{k,i} u^i + 2\mu\eta_{ik} u^{i\prime} + 2\mu'\eta_{ik} u^i = 0$$

where the prime differentiates with respect to s. Now multiply by u^k and sum:

$$e(A_{i,k} - A_{k,i})u^i u^k + \mu(\eta_{ik} u^i u^k)' + 2\mu'\eta_{ik} u^i u^k = 0 \quad .$$

Relabelling indices and using (5.21) reduces this to $\mu' = 0$, and so μ is a constant. Thus, when irrelevant constants are omitted, $\mathcal{L}$ becomes

$$\mathcal{L} = (\mu u_i - eA_i)u^i$$
$$= (\mu/c^2)(c^2 u^{02} - u^2) - e(\Phi u^0 - A \cdot u) \quad .$$

At nonrelativistic velocities u^0 is close to 1. If $\mu = -mc^2/2$ this Lagrangian reduces to that in (5.15); thus

$$\mathcal{L} = (-\tfrac{1}{2}mc^2 u_i + eA_i)u^i \quad . \tag{5.25}$$

The canonical momenta are $\partial \mathcal{L}/\partial u^i$:

$$\mathfrak{p}_i = -mc^2 u_i + eA_i \quad . \tag{5.26}$$

Since (Problem 5.11) u^0 is the usual γ, $\mathfrak{p}_0$ is the negative of the total energy including mc^2, while $\boldsymbol{p}$ is as in (5.17).

Problem 5.10. Calculate the quantity

$$\mathfrak{H}(\mathfrak{p}, x) := \mathfrak{p}_i u^i - \mathcal{L}$$

that corresponds to the Hamiltonian function in this formalism. According to dynamical principles $\mathfrak{H}$ should be independent of s. Calculate it; is it?

Problem 5.11. Show that u^0 is the usual γ.

Problem 5.12. In relativistic mechanics it is neat to use s as the independent variable but not necessary. Show how to do it using t as in the nonrelativistic theory. Find the Hamiltonian and compare the resulting equations of motion with those just derived.

Answer:

$$H = [(\mathbf{p} - e\mathbf{A})^2 + m^2 c^4]^{1/2} + e\Phi \quad . \tag{5.27}$$

Problem 5.13. A charged hollow vessel in an electric field E has a potential energy $e\Phi = -eEx$. It starts from rest at $t = 0$. Show that its coordinate distance x traveled in time t is given by

$$\left(x + \frac{mc^2}{eE}\right)^2 - c^2 t^2 = \left(\frac{mc^2}{eE}\right)^2 \quad .$$

Sketch the shape of this curve. What is the limiting value of x after a long time? Observers inside the vessel measure its acceleration with an accelerometer. What value do they find?

Problem 5.14. Suppose that to make its passengers comfortable a spaceship is driven with a constant acceleration equal to g ($= 9.8 \text{ m/s}^2$). What is g measured in the cosmic units of lt yr/yr^2? How fast, in the coordinates of observers left at home, will it be going at the end of a year?

An Electron Lens. The immediate reason why Hamilton's equations simplify some dynamical calculations is that they are ordinary differential equations of the first order in time. With luck, such an equation can be solved by a simple integration. The following simple and approximate calculation shows how this can happen. Figure 5.2 shows an electron lens made of a single circular loop of wire. Since the electron follows a spiral path, we use cylindrical coordinates, ϱ, ϑ, z. The vector potential inside the ring is given by an intractable expression [42, p. 177] which for points near the axis is approximated by

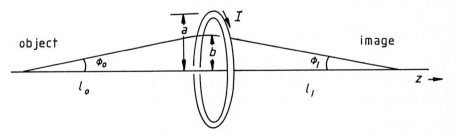

Fig. 5.2. Loop of current serving as an electron lens

$$A_\vartheta = I\pi a^2 \frac{\varrho}{[(a+\varrho)^2 + z^2]^{3/2}} \simeq I\pi a^2 \frac{\varrho}{(a^2+z^2)^{3/2}} \quad . \tag{5.28}$$

The Lagrangian is

$$L = \frac{1}{2}m(\dot\varrho^2 + \varrho^2\dot\vartheta^2 + \dot z^2) + \Lambda\frac{\varrho^2\dot\vartheta}{(a^2+z^2)^{3/2}} \quad , \qquad \Lambda := I\pi a^2 e$$

and the Hamiltonian is

$$H = \frac{1}{2m}(p_\varrho^2 + p_z^2) + \frac{1}{2m\varrho^2}\left[p_\vartheta - \frac{\Lambda\varrho^2}{(a^2+z^2)^{3/2}}\right]^2 \quad .$$

Here ϑ is a cyclic coordinate and p_ϑ = const. From $\dot\vartheta = dH/dp_\vartheta$ we have

$$\dot\vartheta = \frac{1}{m\varrho^2}\left[p_\vartheta - \frac{\Lambda\varrho^2}{(a^2+z^2)^{3/2}}\right] \quad .$$

Assume that far away at the object the beam originates on the axis. It then has $\dot\vartheta = 0$ and so $p_\vartheta = 0$. During the subsequent motion,

$$\dot\vartheta = -\frac{\Lambda}{m(a^2+z^2)^{3/2}} \quad .$$

Thus it spirals around the axis in order to conserve its angular momentum. Since the magnetic field does no work, the electron's speed remains constant, and since the rays stay close to the axis this means that $\dot z$ is very nearly constant.

To study the orbit

$$\frac{dp_\varrho}{dt} = -\frac{\partial H}{\partial\varrho} = -\frac{\Lambda^2\varrho}{m(a^2+z^2)^3}$$

and with $\dot z := u \simeq$ const this gives

$$\frac{dp_\varrho}{dz} = \frac{dp_\varrho}{dt}\frac{dt}{dz} = -\frac{\Lambda^2\varrho}{mu(a^2+z^2)^3} \quad .$$

Assume that while the electron passes through the ring, ϱ is about constant at b. Then the total change in p_ϱ is

$$\Delta p_\varrho = -\frac{\Lambda^2\varrho}{mu}\int_{-\infty}^{\infty}\frac{dz}{(a^2+z^2)^3} = -\frac{3\pi}{8}\frac{\Lambda^2 b}{mua^5} \quad .$$

From Fig. 5.2, $\tan\phi_0 = (p_\varrho/mu)_{\text{initial}}$ and $\tan\phi_i = (-p_\varrho/mu)_{\text{final}}$, so that

$$\tan\phi_0 + \tan\phi_i = -\frac{\Delta p_\varrho}{mu}$$

and since $\tan\phi_0 = b/l_0$, $\tan\phi_i = b/l_i$ this is

$$\frac{1}{l_0} + \frac{1}{l_i} = \frac{3\pi}{8}\frac{\Lambda^2}{m^2u^2a^5} = \frac{3\pi}{8a}\left(\frac{eI\pi}{mu}\right)^2 \quad .$$

This is the equation for a thin lens of focal length

$$f = \frac{8a}{3\pi}\left(\frac{mu}{eI\pi}\right)^2 \quad . \tag{5.29}$$

When the lens forms a real image it is not simply inverted, for the electrons all spiral at a rate

$$\frac{d\vartheta}{dz} = \frac{d\vartheta}{dt}\frac{dt}{dz} = -\frac{\Lambda}{mu(a^2 + z^2)^{3/2}}$$

so that the total rotation is

$$\vartheta = -\frac{\Lambda}{mu}\int_{-\infty}^{\infty}\frac{dz}{(a^2 + z^2)^{3/2}} = -\frac{2\Lambda}{mua^2} \quad .$$

In terms of the focal length, this is

$$\vartheta = -4\sqrt{\frac{2a}{3\pi f}} \quad .$$

If for example $f = 3a$, then $\vartheta = -61°$.

Note particularly that we have got these results not by solving differential equations but just by evaluating integrals. This is a first step in the direction we shall take in Chap. 6, which will give a systematic way of reducing a large class of calculations to more or less difficult integrations. But the principal merit of the canonical equations is that they form the basis for further formal developments, as the rest of this chapter will show.

Problem 5.15. A "thin" lens is made with two identical coils that carry equal currents in opposite directions and are separated by a distance l that is small compared with the diameter of the coils. Find the focal length and rotation as before.

Problem 5.16. According to (5.29) the focal length of the lens is proportional to the electrons' kinetic energy. How is this conclusion changed if one calculates relativistically?

Problem 5.17. Show that if a magnetic field B is uniform and in the z direction, a possible (but not unique) vector potential is

$$A_x = -\tfrac{1}{2}By \quad A_y = \tfrac{1}{2}Bx \quad A_z = 0 \quad . \tag{5.30}$$

Express the Hamiltonian in this case using the angular momentum.

Problem 5.18. To find the equations governing dynamical phenomena observed on a merry-go-round, let

$$x = x' \cos \omega t - y' \sin \omega t \quad , \quad y = x' \sin \omega t + y' \cos \omega t \quad , \quad z = z'$$

and find the transformed Lagrangian. Show that when written in terms of momenta the kinetic energy is unchanged, and show that the Hamiltonian becomes

$$H = \frac{1}{2m}p'^2 - \omega \cdot p'_\vartheta + V(r) \quad . \tag{5.31}$$

Problem 5.19. Comparing the results of the last two problems compare the behavior of a system of identical charged particles in a magnetic field with the behavior of the same system without a magnetic field but viewed from a suitably rotating system of reference. (This is the subject of Larmor's theorem.)

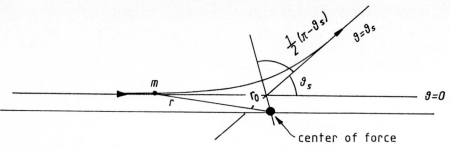

Fig. 5.3. To illustrate Problem 5.20. Path of a particle under a repulsive force

Problem 5.20. Calculate the relativistic version of Rutherford's formula for the differential cross-section for scattering by a fixed center of inverse-square force. A good way is to consider half the trajectory, from infinity to the point of closest approach. Find how much the trajectory has turned, then double it to find the total deflection.

Solution. Figure 5.3 gives the geometry. The object is to find out how ϑ changes as r changes. Note that $p_\vartheta = \text{const}$. Further,

$$\dot\vartheta = \frac{\partial H}{\partial p_\vartheta} \quad , \quad \dot p_r = -\frac{\partial H}{\partial r}$$

so that

$$\frac{\partial \dot\vartheta}{\partial r} = \frac{\partial^2 H}{\partial r \partial p_\vartheta} = -\frac{\partial \dot p_r}{\partial p_\vartheta}$$

and integrating this over time gives

$$\frac{\partial \vartheta}{\partial r} = -\frac{\partial p_r}{\partial p_\vartheta} + \text{const} \quad .$$

Now show from the figure that

$$\vartheta_s = \pi + 2 \int_{r_0}^{\infty} \frac{\partial p_r}{\partial p_\vartheta} dr$$

though it may be easier to integrate p_r before differentiating. Try it nonrelativistically for practice, then use the relativistic Hamiltonian.

5.3 Canonical Transformations

If a Hamiltonian dynamical problem contains cyclic coordinates their conjugate momenta are constants and can be used to lower the order of the remaining differential equations. Cyclic coordinates correspond to symmetries, but they only show up when appropriate coordinates are used. (In the Kepler problem, for example, circular symmetry leads to the conservation of angular momentum, but one has to use polar coordinates in order to get the cyclic coordinate.) It is therefore important to be able to study the symmetries of a Hamiltonian and to capitalize on them by shifting to appropriate coordinates when one is found.

With this motivation, let us see how to change variables in the canonical equations. The place to start is Lagrange's equations and the fact, mentioned above but not yet used, that they are derivable from Hamilton's variational principle

$$\delta \int_{t_0}^{t} L[q(t), \dot{q}(t), t] dt = 0 \tag{5.32}$$

provided that the variations in the δq's are made to vanish at the end points. The integral is known as Hamilton's principal function, S, and (for fixed lower limit) it is a function $S(q, t)$ of the point q at which all the paths arrive at time t. In terms of coordinates and momenta, the varied integral in (5.32) is

$$\delta \int_{t_0}^{t} (p_n \dot{q}_n - H[p(t), q(t), t]) dt = 0 \tag{5.33}$$

where we allow time-dependent interactions in H because for the moment it is no more difficult to do so. Let us vary this integral differently than was done in deriving Lagrange's equations. There, q was varied by δq and δq was required to vanish at the limits of integration. Further, $\delta \dot{q} = d(\delta q)/dt$. Since Hamilton's theory puts p and q on the same level, we will vary them independently, and drop for the moment the requirement that δq vanish at the limits. Figure 5.4 shows two paths starting at the same point and ending at points different in place and time. Comparing them we have

$$\delta S = \delta \int_{t_0}^{t} L \, dt = \int_{t_0}^{t} \delta[p_n \dot{q}_n - H(p, q)] dt + L(t) \delta t$$

$$= \int_{t_0}^{t} \left[\dot{q}_n \delta p_n + p_n \delta \dot{q}_n - \frac{\partial H}{\partial p_n} \delta p_n - \frac{\partial H}{\partial q_n} \delta q_n \right] dt + L \delta t \quad .$$

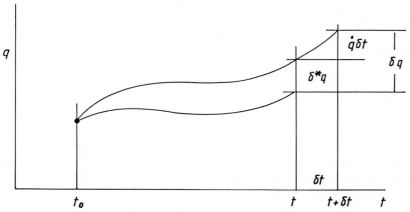

Fig. 5.4. Two paths in q, t space arriving at different places and times

115

The first and third terms cancel and

$$\delta S = \int_{t_0}^{t} (p_n \delta \dot{q}_n + \dot{p}_n \delta q_n) dt + L \delta t \quad .$$

With (5.4) this is

$$\delta S = \left[p_n(t) \delta^* q_n(t) + p_n(t) \dot{q}_n(t) \delta t - H(p, q) \delta t \right]_{t_0}^{t}$$

where $\delta^* q$ is the difference in q's at time t (Fig. 5.4). It is clear from the diagram that the difference in the q's at the ends of the two paths is

$$\delta q(t) = \delta^* q(t) + \dot{q}(t) \delta t$$

so that the final result can be written

$$\delta S = [p_n \delta q_n - H \delta t]_{t_0}^{t} \qquad (5.34)$$

independent of variations in p. Since Hamilton's equations (5.6) were used in deriving this result, it holds only for variations away from a *natural path* where the equations are valid. This simple relation, given by Hamilton in 1834, contains all of classical dynamics. If δq_n and δt are suitably restricted it yields the canonical equations (see the next problem). We will see that it offers profound insight into the relation of invariance to conservation via Noether's argument, and finally we will use it to derive the Hamilton-Jacobi theory.

Problem 5.21. Show that Hamilton's principle is a special case of (5.34). In Sect. 3.8 the Maupertuis principle was derived in several tedious steps. Show that if one thinks about the meaning of δq and δt it follows at once from (5.34).

Problem 5.22. Considering L as a function of the independently variable p's and q's as well as t, let the variations of all these quantities in (5.34) vanish at the limits of integration and show that the resulting Euler-Lagrange equations are the canonical equations.

Noether's Procedure. I have commented several times in increasingly general terms on the relation between invariance (i.e., symmetry) and conservation principles. Equation (5.34) permits a very general statement of this relation, using a kind of argument first given by Emmy Noether in 1918. Suppose we adopt a new set of coordinates to describe the system, differing infinitesimally from the old one in geometry and in dependence on time, and suppose this does not make any difference. That is, the system has the same behavior and obeys the same equations of motion in the new coordinates as in the old. The old equations of motion were expressed by specifying the final time in (5.34) as t and requiring that if δt and the δq's are initially and finally zero, $\delta S = 0$ (Problem 5.22). In the new coordinates δS will still be zero, and this allows one to see some of the conservation laws at once.

Example 5.3. Suppose for example we do not change the spatial coordinates but infinitesimally reset the clock by an amount δt which is small but arbitrary. It follows from (5.34) that $H(t) = H(t_0)$ so that if H is numerically equal to the energy,[1] energy is conserved. One can say either that energy is conserved if the Hamiltonian does not involve time explicitly, as already noted in (5.8a), or if the result of an experiment governed by these equations does not depend on when the experiment is performed.

Example 5.4. In the same spirit, displace Cartesian coordinates with respect to the physical system, or the system with respect to the coordinates, by a distance δx. It follows that p_x is constant. If we rotate it through an angle $\delta\vartheta$, then p_ϑ is constant.

Example 5.5. Equation (5.34) is one of the relations which occasionally emerge in Newtonian physics that are relativistic in form. Suppose a massive object, as seen by any inertial observer, moves through space without external interaction. Let us try a hyperbolic rotation of coordinates in the x-t plane:

$$x \rightarrow x + \varepsilon ct \quad , \quad t \rightarrow t + \varepsilon x/c$$

with which

$$x^2 - c^2 t^2 \rightarrow (1 - \varepsilon^2)(x^2 - c^2 t^2)$$

so that distance in spacetime is unchanged to the first order in ε: this is an infinitesimal Lorentz transformation. Put into (5.34) this gives

$$Hx - c^2 pt = \text{const} \quad , \quad x = \frac{c^2}{H} p(t - t_0)$$

so that the object moves through space at a constant rate given by well-known relativistic kinematics.

Transformations. Now let us adopt a new set of coordinates and momenta, supposing them to be functions of the old ones and the time, $P_n(p, q, t)$ and $Q_n(p, q, t)$. We assume that these relations are soluble to give $p_n(P, Q, t)$ and $q_n(P, Q, t)$. Further, we require that the equations of motion in P and Q shall again be canonical; that is, that when the changes of variable have been made we again have

$$\delta \int [P_n \dot{Q}_n - K(P, Q, t)] dt = 0 \tag{5.35}$$

where K is the new Hamiltonian and δQ_n, like δq_n, vanish at the limits of integration. Transformation of this very general kind which preserve the canonical form of the equations of motion are called *canonical transformations*. The subset consisting of transformations $Q(q)$ and $P(p, q)$ that arise from changing only the spatial coordinates are called *point transformations*.

The other requirement for a contact transformation is that every natural path in (p, q) should, when translated into (P, Q) again be a natural path. This will be so if, for corresponding choices of the initial conditions, the integrals in (5.33) and (5.35) are exactly the same for all values of t_0 and t_1, that is, if the integrands differ at most by the time derivative of a function whose variations vanish at the limits of

---

[1] There are several situations in which H is not an energy. One is in the treatment of dissipative systems, not discussed in this book; another is when there is a time-dependent external field, see Sect. 5.8.

integration. For example,

$$\delta \int_{t_0}^{t} \dot{F}_1(q, Q, t)\,dt = \delta[F_1(q, Q, t)]_{t_0}^{t} = 0 \quad . \tag{5.36}$$

Writing out the time derivative of $F_1(q, Q, t)$, we have

$$p_n \dot{q}_n - H(p, q, t) = P_n \dot{Q}_n - K(P, Q, t) - \frac{\partial F_1}{\partial q_n}\dot{q}_n - \frac{\partial F_1}{\partial Q_n}\dot{Q}_n - \frac{\partial F_1}{\partial t} \tag{5.37}$$

which is satisfied identically provided that

$$p_n = -\frac{\partial F_1(q, Q, t)}{\partial q_n} \quad , \quad P_n = \frac{\partial F_1(q, Q, t)}{\partial Q_n} \quad \text{and} \tag{5.38}$$

$$K(P, Q, t) = H(p, q, t) - \frac{\partial F_1(q, Q, t)}{\partial t} \quad . \tag{5.39}$$

This formula means: Solve (5.38) to find p and q as functions of P, Q, and possibly t; put them into (5.39) and this will give the Hamiltonian in the new variables. Thus given F_1 we find K. Using the methods of the next chapter it is also possible to start with a desired K and find the F_1 that produces it [43], but the argument will not be needed here. (We can also introduce scale factors by multiplying the two sides of (5.37) by different constants, but this only changes the units of measurement.)

There are other forms possible for F: provided it satisfies (5.36) it can be a function of any two of the variables p, q, P, Q, but not more than two because with a function $F(q, Q, P)$, for example, we could solve the resulting equation $P_n = P_n(q, Q, P)$ for P_n and reduce F to a function of q and Q.

To get a new transformation, replace $F_1(q, Q)$ by a new function

$$F_2(q, P) = F_1(q, Q) - P_n Q_n \quad \text{(summed)}$$

where P is considered as a function of q and Q and its variation vanishes at the endpoints. Then (5.37) is

$$p_n \dot{q}_n - H(p, q, t) = P_n \dot{Q}_n - K(P, Q, t)$$
$$- \frac{\partial F_2}{\partial P_n}\dot{P}_n - \frac{\partial F_2}{\partial q_n}\dot{q}_n - \frac{\partial F_2}{\partial t} - P_n \dot{Q}_n - Q_n \dot{P}_n \quad .$$

As before, this is satisfied provided that

$$p_n = -\frac{\partial F_2(q, P)}{\partial q_n} \quad , \quad Q_n = -\frac{\partial F_2(q, P)}{\partial P_n} \quad , \quad H = K + \frac{\partial F_2}{\partial t} \quad . \tag{5.40}$$

In a similar way, we can form two more functions and generate the transformations from

$$P_n = \frac{\partial F_3(p, Q, t)}{\partial Q_n} \quad , \quad q_n = \frac{\partial F_3(p, Q, t)}{\partial p_n} \quad , \quad H = K + \frac{\partial F_3(p, Q, t)}{\partial t} \tag{5.41}$$

and

$$q_n = \frac{\partial F_4(p, P, t)}{\partial p_n} \quad , \quad Q_n = -\frac{\partial F_4(p, P, t)}{\partial P_n} \quad , \quad H = K + \frac{\partial F_4(p, P, t)}{\partial t} \quad . \tag{5.42}$$

These are the four simple types but they produce a great variety of transformations since one can mix them, using different types on different pairs of variables in the same calculation.

Example 5.6. Let

$$F_3(p, Q, t) = p_n Q_n \quad . \tag{5.43}$$

Then (5.41) gives

$$P_n = p_n \quad , \quad Q_n = q_n \quad , \quad H = K$$

so that this is the identity transformation. (It is not as useless as it looks and will be needed later.)

Example 5.7. Let

$$F_1(q, Q) = q_n Q_n \quad .$$

We find

$$P_n = q_n \quad , \quad Q_n = -p_n \quad .$$

This is not very interesting computationally, but it shows that momenta and coordinates are merely labels we attach to certain variables for historical reasons. Nothing is changed if (remembering the sign) we switch the labels around.

Point transformations to new coordinates $Q(q)$ are often conveniently generated by

$$F_3(p, Q) = p_m f_m(Q) \tag{5.44}$$

since from (5.41),

$$q_n = f_n(Q) \quad , \quad P_n = p_m \frac{\partial f_m}{\partial Q_n}$$

Example 5.8. To transform from Cartesian to polar coordinates in a plane let

$$x = R \cos \vartheta \quad , \quad y = R \sin \vartheta \quad , \quad p_x = m\dot{x} \quad , \quad p_y = m\dot{y}$$

so that

$$F_3(p_x, p_y, R, \vartheta) = R(p_x \cos \vartheta + p_y \sin \vartheta)$$

and along with x and y we get

$$P_R = p_x \cos \vartheta + p_y \sin \vartheta = (x p_x + y p_y)/R$$

$$P_\vartheta = R(-p_x \sin \vartheta + p_y \cos \vartheta) = x p_y - y p_x$$

yielding the expressions for radial and angular momentum at once.

Problem 5.24. Find the transformation from rectangular to spherical polar coordinates.

Problem 5.25. Find the generator $F_3(p, X)$ that carries out a rotation of plane Cartesian axes through an angle α:

$$X = x \cos \alpha - y \sin \alpha \quad P_x = p_x \cos \alpha - p_y \sin \alpha$$
$$Y = x \sin \alpha + y \cos \alpha \quad P_y = p_x \sin \alpha + p_y \cos \alpha \quad . \tag{5.45}$$

The usual point transformations are expressed in terms of F_3's as are the infinitesimal transformations to be discussed in the next section, but we will encounter other forms in subsequent chapters. In particular, we will be interested in canonical transformations that leave the Hamiltonian invariant, i.e., the same function of P, Q as it was of p, q. That is what defines invariance.

Problem 5.26. The Hamiltonian of a simple harmonic oscillator is

$$H(p, x) = \frac{p^2}{2m} + \frac{1}{2} k x^2 \quad .$$

Carry out the contact transformation

$$F_1(x, X) = -\tfrac{1}{2} m \omega x^2 \cot X \quad , \quad \omega^2 = k/m \quad . \tag{5.46}$$

Find the new coordinate, momentum, and Hamiltonian, and solve the oscillator in the new variables.

Solution. $p = -\partial F_1/\partial x = m \omega x \cot X$, $P = \partial F_1/\partial X = \tfrac{1}{2} m \omega x^2 \csc^2 X$, $K = \tfrac{1}{2} m \omega^2 x^2 \csc^2 X = \omega P$. The equations of motion give $P = \text{const}$ and $X = \omega(t - t_0)$. The values of $x(t)$ and $p(t)$ can now easily be found.

Problem 5.27. Find the F representing a transformation to a system of coordinates rotating around the z axis, and verify that the Hamiltonian in the rotating system is (5.31).

Problem 5.28. A point transformation in the form (5.44) gives the old coordinates in terms of the new ones. Find the transformation that gives the new coordinates in terms of the old.

Problem 5.29. Show by comparing second derivatives of the generating functions that a transformation is canonical if it satisfies certain conditions among which are

$$\left(\frac{\partial p_m}{\partial Q_n} \right)_q = - \left(\frac{\partial P_n}{\partial q_m} \right)_Q \quad .$$

Find the rest of these conditions.

Problem 5.30. Show that the transformation

$$Q = \sqrt{q} \cos 2p \quad P = \sqrt{q} \sin 2p$$
$$q = P^2 + Q^2 \quad \tan 2p = P/Q$$

is canonical.

Problem 5.31. Find by integrating (5.38) the form of $F_1(q, Q)$ that generates the transformation of the preceding problem.

(**Answer:** $F_1(q, Q) = [\sqrt{Q(q - Q^2)} + q \sin^{-1} (Qq^{-1/2})]/2$.

Note that in all the examples given except that of Problem 5.27 the transformations are independent of t and therefore, by (5.39) and its later analogues,

$K(P, Q) = H(p, q)$. In the exception we are looking at a system from a moving frame of reference and would not expect the energy to be the same.

Dynamical Symmetry. Problem 5.26 offers a glimpse of what is to come. We have already encountered ignorable coordinates; by not appearing in the Hamiltonian they express a geometrical symmetry of the system and thus tell us that the corresponding momentum is constant. The new Hamiltonian in Problem 5.26 is independent of the coordinate X. Accordingly P is constant and the problem is easy to solve. But does this express a symmetry? In the new variables, yes, but not in space and time. There are other kinds. To see what is happening here, work backward from the solution. In terms of the energy,

$$x = \frac{1}{\omega}\sqrt{2E/m}\sin \omega t \quad , \quad p = \sqrt{2mE}\cos \omega t$$

and eliminating the amplitude (which belongs to the solution, not the equation of motion) gives $p/x = m\omega \cot \omega t$ as was found above, from which F_1 follows easily. But what is the symmetry? Follow the motion of the oscillator for a short time. The coordinate and momentum change according to

$$x \to x + \dot{x}\delta t = x + \frac{p}{m}\delta t \quad , \quad p \to p + \dot{p}\delta t = p - m\omega^2 x\,\delta t$$

and this represents what is essentially a rotation in x-p space. The Hamiltonian is invariant under such a rotation; that is how it stays constant. It is not a symmetry in the conventional sense but dynamical symmetries of this kind can be used to simplify a solution in the same way as the geometrical symmetries already encountered. Such symmetries lie buried and invisible in many dynamical systems, and it is the business of dynamics to uncover and use them. The Hamilton-Jacobi theory developed in Chap. 6 is a technique that helps in the process, but ingenuity is also required.

5.4 Infinitesimal Transformations

There are many transformations that can be carried out in a sequence of infinitesimal steps. Rotations and translations are familiar examples, and we will encounter others less familiar. Other transformations are discontinuous, for example that from Cartesian to polar coordinates. Infinitesimal transformations are of special interest because of their formal simplicity and closeness to analogous considerations in quantum mechanics, and because the criterion for invariance of the Hamiltonian under such transformations is very easy to state.

Infinitesimal coordinate transformations are familiar and simple cases of infinitesimal canonical transformations. Corresponding to a displacement of an object to the right (or a displacement of the origin of coordinates to the left) we have

$$x \to X = x + \varepsilon \tag{5.47a}$$

whereas for rotations, the infinitesimal form of (5.45) is

$$X = x - \varepsilon y \quad , \quad Y = y + \varepsilon x \quad . \tag{5.48a}$$

A function $f(x, y)$ is transformed by (5.47a) into

$$f(X, Y) = f(x, y) + \varepsilon \frac{\partial f}{\partial x} \tag{5.47b}$$

and by (5.48a) into

$$f(X, Y) = f(x, y) + \varepsilon \left(x \frac{\partial f}{\partial y} - y \frac{\partial f}{\partial x} \right) \quad . \tag{5.48b}$$

Quantum Mechanics. The application of these formulas in quantum mechanics is elementary. The last two equations can each be written in the form

$$f(X, Y) = (1 + \varepsilon \hat{D}) f(x, y) \tag{5.49a}$$

with the inverse

$$f(x, y) = (1 - \varepsilon \hat{D}) f(X, Y) \tag{5.49b}$$

where $\hat{D}$ is the appropriate differential operator. Suppose that $\hat{a}(x, y)$ is some dynamical variable in quantum mechanics and that it satisfies a relation of the form

$$\hat{a}(x, y) f(x, y) = g(x, y) \quad .$$

Operate with $(1 + \varepsilon \hat{D})$ on this equation:

$$(1 + \varepsilon \hat{D}) \hat{a}(x, y) f(x, y) = (1 + \varepsilon \hat{D}) g(x, y) = g(X, Y) \quad .$$

By (5.49b), this is

$$(1 + \varepsilon \hat{D}) \hat{a}(x, y)(1 - \varepsilon \hat{D}) f(X, Y) = g(X, Y) \quad .$$

Thus the operator playing the same role in (X, Y) that $\hat{a}(x, y)$ did in (x, y) is

$$\hat{A}(X, Y) := (1 + \varepsilon \hat{D}) \hat{a}(x, y)(1 - \varepsilon \hat{D}) \tag{5.50}$$

and to first order this is

$$\hat{A}(X, Y) = \hat{a}(x, y) + \varepsilon [\hat{D}, \hat{a}] \tag{5.51}$$

where $[\hat{D}, \hat{a}]$ is the commutator $\hat{D}\hat{a} - \hat{a}\hat{D}$. If $\hat{D}$ commutes with $\hat{a}$, then $\hat{A}(X, Y) = \hat{a}(x, y)$ and we say that $\hat{a}$ is invariant under the transformation. If $\hat{a}$ happens to be the Hamiltonian operator of some system, then $\hat{D}$ defines a dynamical variable that satisfies a conservation law. Adding factors of i to make $\hat{D}$ hermitian and $\hbar$ to introduce conventional units of measurement, we write (5.49a) as

$$f(X, Y) = \left(1 + \frac{i\varepsilon}{\hbar} \hat{G} \right) f(x, y) \tag{5.52}$$

where

$$\hat{G} = -i\hbar \hat{D} \tag{5.53}$$

and (5.51) becomes

$$\hat{A} - \hat{a} := \delta \hat{a} = \varepsilon \frac{i}{\hbar} [\hat{G}, \hat{a}] \quad . \tag{5.54}$$

(In contrasting classical and quantum mechanics I will generally use hats to designate operators.) The two examples above give

$$\hat{G} = -i\hbar\frac{\partial}{\partial x} = \hat{p}_x \quad \text{(displacement in } x\text{)} \tag{5.55}$$

and

$$\hat{G} = x\hat{p}_y - y\hat{p}_x \quad \text{(rotation in } xy \text{ plane)} \tag{5.56}$$

respectively, which are recognized as the operators of linear and angular momentum. If $\hat{H}$ is invariant, the quantity is conserved, and the operators $\hat{G}$, called the generators of the transformations, are the same as the operators that represent the conserved dynamical variables.

The situation is especially simple in quantum mechanics because a system is supposed to be specified as completely as possible by its wave function, which depends only on the coordinates and time, $f+1$ variables in general, while the more closely specified systems imagined in classical theory require coordinates, momenta, and time, $2f+1$ variables. Thus we can expect that the corresponding formalism in classical dynamics will be a little more complicated.

Problem 5.32. Formulas like (5.52) were derived from the consideration of wave functions that depend on spatial coordinates and so the arguments of operators like $\hat{a}$ and $\hat{A}$ were written as functions of coordinates. Show that if they are written as functions of coordinates and momenta the relations are still correct.

Unitary Transformations. In the two examples given above in (5.55) and (5.56), the generator $\hat{G}$ turned out to be a hermitian operator. Let us see why this was to be expected. If $\psi(x, y)$ describes a certain system, then $\psi' = (1 + i\varepsilon\hbar^{-1}\hat{G})\psi$ describes another identical system infinitesimally transformed away from the first. Clearly this can make no difference to the normalization of the wave function; we must have

$$1 = \int \psi'^*\psi' \, dx = \int [(1 + i\varepsilon\hbar^{-1}\hat{G})\psi]^*(1 + i\varepsilon\hbar^{-1}\hat{G})\psi \, dx$$

and only if, for all ψ in a suitable domain, $\hat{G}$ is hermitian so that

$$\int (\hat{G}\psi)^*\psi \, dx = \int \psi^*\hat{G}\psi \, dx$$

will we have

$$\int \psi'^*\psi' \, dx = \int \psi^*(1 - i\varepsilon\hbar^{-1}\hat{G})(1 + i\varepsilon\hbar^{-1}\hat{G})\psi \, dx$$

$$= \int \psi^*\psi \, dx = 1 \quad \text{to first order} \quad .$$

Transformations of the form (5.52) with $\hat{G}$ hermitian are called unitary, and their defining property is that they preserve normalization. A fundamental property of hermitian operators is that they have real eigenvalues, and so all dynamical variables with analogues in classical physics are represented by hermitian operators. Thus we have a reciprocal relation: to every dynamical variable represented by a hermitian operator corresponds an infinitesimal unitary transformation and vice versa. As we have seen, if the transformation leaves the Hamiltonian invariant, the corresponding dynamical variable is a constant of the motion. Of more interest nowadays is the converse statement. Physics, especially the physics of elementary particles, is rich in conservation laws: baryon number, electron number, muon number, charge, hypercharge, color, and so on. Each of these probably corresponds to some invariance of the underlying theory and so provides a clue as to its eventual structure.

Classical Mechanics. An infinitesimal transformation is defined as one which differs only infinitesimally from the identity transformation. In Example 5.6 the identity is $F_3 = p_n Q_n$ (it is also $F_2 = -q_n P_n$, but that leads to the same results). We write

$$F_3 = p_n Q_n - \varepsilon G(p, Q)$$

where ε is infinitesimal, noting that in the limit $\varepsilon \to 0$, G becomes a function of p and q. From (5.41), we have to order ε

$$P_n = p_n - \varepsilon \frac{\partial G}{\partial Q_n} \quad , \quad q_n = Q_n - \varepsilon \frac{\partial G}{\partial p_n} \quad , \quad H(p, q) = K(P, Q) \quad . \tag{5.57}$$

Again to order ε, this gives

$$\Delta p_n = P_n - p_n = -\varepsilon \frac{\partial G}{\partial q_n} \quad , \quad \Delta q_n = \varepsilon \frac{\partial G}{\partial p_n} \tag{5.58}$$

so that, for any function $a(p, q)$,

$$\Delta a = -\varepsilon (G, a) \tag{5.59}$$

where we define

$$(G, a) := \frac{\partial G}{\partial q_n} \frac{\partial a}{\partial p_n} - \frac{\partial G}{\partial p_n} \frac{\partial a}{\partial q_n} \quad . \tag{5.60}$$

This combination of derivatives was introduced into dynamics by S.D. Poisson in 1809 and is called the Poisson bracket of a and G. From (5.54)

$$[\hat{G}, \hat{a}] \quad \text{corresponds to} \quad i\hbar (G, a) \tag{5.61}$$

As a special example, invariance of $H(p, q)$ under the transformation generated by G means that

$$(G, H) = 0 \quad . \tag{5.62}$$

The result of Problem 5.34 shows that if this is so, the numerical value of G remains constant.

Problem 5.33. Verify the correspondence (5.61) using some pairs of dynamical variables familiar from classical and quantum mechanics: p_i (linear momentum) and x_j, L_i (angular momentum) and L_j, L_i and x_j, L_i and p_j.

Problem 5.34. In quantum mechanics

$$\frac{d\langle a \rangle}{dt} = \frac{i}{\hbar} \langle [\hat{H}, \hat{a}] \rangle \quad \text{(Schrödinger representation)}$$

$$\frac{d\hat{a}_\mathrm{H}}{dt} = \frac{i}{\hbar} [\hat{H} \, \hat{a}_\mathrm{H}] \quad \text{(Heisenberg representation)} \tag{5.63}$$

for any time-independent dynamical variable a. Show that for any classical dynamical variable $a(p, q)$,

$$\dot{a} = (a, H) \tag{5.64}$$

in agreement with the correspondence (5.61).

Problem 5.35. Show that (5.64) contains Hamilton's equations as special cases.

Problem 5.36. Show that the canonical coordinates and momenta satisfy

$$(p_m, p_n) = (q_m, q_n) = 0 \quad , \quad (q_m, p_n) = \delta_{mn} \quad . \tag{5.65}$$

Problem 5.37. If a dynamical variable $a(p, q)$ is re-expressed in terms of P and Q call it A:

$$A(P(p, q), Q(p, q)) = a(p, q) \quad . \tag{5.66}$$

Express $\partial A/\partial P$ and $\partial A/\partial Q$ in terms of the generator of the infinitesimal transformation.

Solution. Writing

$$\frac{\partial A}{\partial P_n} = \frac{\partial a}{\partial p_m} \frac{\partial p_m}{\partial P_n} + \frac{\partial a}{\partial q_m} \frac{\partial q_m}{\partial P_n}$$

and similarly for $\partial A/\partial Q_n$, we need $\partial q_m/\partial P_n$ and $\partial p_m/\partial P_n$. From (5.57), to order ε,

$$p_m = P_m + \varepsilon \frac{\partial G}{\partial q_m} \quad , \quad q_m = Q_m - \varepsilon \frac{\partial G}{\partial p_m}$$

so that, again to order ε,

$$\frac{\partial p_m}{\partial P_n} = \delta_{mn} + \varepsilon \frac{\partial^2 G}{\partial p_n \partial q_m} \quad , \quad \frac{\partial q_m}{\partial P_n} = -\varepsilon \frac{\partial^2 G}{\partial p_m \partial p_n} \quad \text{and}$$

$$\frac{\partial A}{\partial P_n} = \frac{\partial a}{\partial p_n} + \varepsilon \left(\frac{\partial a}{\partial p_m} \frac{\partial^2 G}{\partial p_n \partial q_m} - \frac{\partial a}{\partial q_m} \frac{\partial^2 G}{\partial p_n \partial p_m} \right)$$

$$\Delta \frac{\partial a}{\partial p_n} = \varepsilon \left(\frac{\partial G}{\partial p_n}, a \right) \quad . \tag{5.67a}$$

Finish the problem by showing that

$$\Delta \frac{\partial a}{\partial q_n} = \varepsilon \left(\frac{\partial G}{\partial q_n}, a \right) \quad . \tag{5.67b}$$

Problem 5.38. What transformation, for an N-particle system, is generated by the time-dependent generator

$$G = Pt - M\bar{r}$$

where P is the total momentum, M is the total mass, and $\bar{r}$ locates the center of mass?

Problem 5.39. Show that if the forces between particles in an n-particle system are inverse-cube, there are two more general integrals of the equations of motion

$$D = 2Ht - \sum_{i=1}^{n} \mathbf{r}_i \cdot \mathbf{p}_i \quad \text{and}$$

$$F = Ht^2 - \sum_{i=1}^{n} \left(t\mathbf{r}_i \cdot \mathbf{p}_i - \frac{1}{2} m_i r_i^2 \right) \quad .$$

Special Transformations. Up till now we have not specified any generator $G(p, q)$. Let us try some of the dynamical variables we have encountered so far.

1) $G = p_x$. We have

$$\Delta a = \varepsilon(a, p_x) = \varepsilon \frac{\partial a}{\partial x}$$

so that as in quantum mechanics, p generates the linear displacement, (5.47a).

2) $G = p_\vartheta = x p_y - y p_x$

$$\Delta a = \varepsilon \left(x \frac{\partial a}{\partial y} + p_x \frac{\partial a}{\partial p_y} - y \frac{\partial a}{\partial x} - p_y \frac{\partial a}{\partial p_x} \right)$$

corresponding to a rotation through an angle ε. This also is as in quantum mechanics, but note that the momentum variables are transformed separately.

3) $G = H(p, q)$

$$\Delta a = \varepsilon(a, H) = \varepsilon \dot{a} \tag{5.68}$$

by (5.64). Thus Δa is the change in a during a short time ε, and the natural motion of a system can be regarded as the continuous unfolding of a canonical transformation generated at each instant by H. This corresponds to Schrödinger's equation, which in a similar notation is

$$\Delta \psi(q, t) = -\frac{i}{\hbar} \varepsilon \hat{H} \psi(q, t) \tag{5.69}$$

so that $\hat{H}$ is the operator that takes the system from one moment to the next.

Problem 5.40. Because the vector A of (3.45) is constant, it must generate an infinitesimal transformation that leaves the Hamiltonian of the Kepler problem invariant. Choose a single component, say A_z, and find the transformations it generates in x, y, z, p_x, p_y, and p_z. You will find the new coordinates involve the old momenta, so that the transformation is of the kind discussed at the end of Sect. 5.3. If you now continue and prove the invariance of the Hamiltonian under these changes of variable, you will have some idea of the mathematical complexities that lie beneath the formulas of elementary mechanics.

Properties of Poisson Brackets. The Poisson brackets of dynamical variables satisfy the following relations:

$$(a, b) = -(b, a) \tag{5.70a}$$

$$(a, b + c) = (a, b) + (a, c) \tag{5.70b}$$

$$(a, bc) = (a, b)c + b(a, c) \tag{5.70c}$$

$$(a, (b, c)) + (b, (c, a)) + (c, (a, b)) = 0 \quad . \tag{5.70d}$$

All these formulas apply to commutators as well (the order of factors in (5.70c) is correct for commutators) and all are trivially easy to derive except that for Poisson brackets the last takes a little work. It is called Jacobi's identity.

A remarkable property of (a, b), established by Poisson in 1809, is that if the variables p, q are transformed to P, Q by a canonical transformation the numerical value of (a, b) remains unchanged.

To prove Poisson's theorem we must use the fact that not only is there a change of variables but that it leads to canonical equations of motion. Time-independent transformations are simplest. Start with the Poisson bracket (a, b) of two arbitrary dynamical variables expressed as functions of p and q. Make a canonical transformation from p and q to P and Q, but for the moment make it only in b:

$$(a, b) = \frac{\partial a}{\partial q_i} \left(\frac{\partial b}{\partial Q_j} \frac{\partial Q_j}{\partial p_i} + \frac{\partial b}{\partial P_j} \frac{\partial P_j}{\partial p_i} \right) - \frac{\partial a}{\partial p_i} \left(\frac{\partial b}{\partial Q_j} \frac{\partial Q_j}{\partial q_i} + \frac{\partial b}{\partial P_j} \frac{\partial P_j}{\partial q_i} \right)$$

$$= (a, P_j) \frac{\partial b}{\partial P_j} + (a, Q_j) \frac{\partial b}{\partial Q_j} \quad . \tag{5.71}$$

Now let b be the Hamiltonian H; in the new variables it is $K(P, Q)$. By Hamilton's equations,

$$\dot{a} = (a, H) = (a, P_j) \dot{Q}_j - (a, Q_j) \dot{P}_j \quad .$$

But a is arbitrary, so that if it is now considered as a function of P and Q we must have

$$(a, P_j) = \frac{\partial a}{\partial Q_j} \quad , \quad (a, Q_j) = -\frac{\partial a}{\partial P_j} \quad .$$

Putting these relations into (5.71) gives

$$(a, b) = \frac{\partial a}{\partial Q_j} \frac{\partial b}{\partial P_j} - \frac{\partial a}{\partial P_j} \frac{\partial b}{\partial Q_j} \quad . \qquad\qquad \Box \ (5.72)$$

Poisson's theorem results in a convenience of notation: in writing (a, b) we need not specify what independent variables are being used, and of course it leads to computational simplifications as well, as will be seen in Chap. 8, since some coordinates are much more convenient than others.

As an exercise for the reader, Problem 5.43 asks for a proof of Poisson's theorem valid for infinitesimal transformations, which can easily be constructed from (5.67).

Poisson Brackets and Canonical Transformations. This paragraph develops another definition of canonical transformations which is more useful for verifying that transformations are canonical than for finding new ones. According to Problem 5.29,

$$\frac{\partial^2 F_1}{\partial q_m \partial Q_n} = \frac{\partial P_n}{\partial q_m} = -\frac{\partial p_m}{\partial Q_n}$$

and similarly,

$$\frac{\partial p_m}{\partial P_n} = \frac{\partial Q_n}{\partial q_m} \quad , \quad \frac{\partial P_m}{\partial p_n} = \frac{\partial q_n}{\partial Q_m} \quad , \quad \frac{\partial q_m}{\partial P_n} = -\frac{\partial Q_n}{\partial p_m} \quad .$$

Suppose that P is a function $P(p, q)$ and similarly $Q(p, q)$ and $p(P, Q)$ and $q(P, Q)$. Thus we can consider Q as a function $Q[p(P, Q), q(P, Q)]$. Vary Q_n in this expression:

$$\delta Q_n = \frac{\partial Q_n}{\partial q_m} \frac{\partial q_m}{\partial Q_k} \delta Q_k + \frac{\partial Q_n}{\partial p_m} \frac{\partial p_m}{\partial Q_k} \delta Q_k \quad .$$

If the transformation is canonical we can use the relations among partial derivatives just proved to write

$$\delta Q_n = \left(\frac{\partial Q_n}{\partial q_m} \frac{\partial P_k}{\partial p_m} - \frac{\partial Q_n}{\partial p_m} \frac{\partial P_k}{\partial q_m} \right) \delta Q_k = (Q_n, P_k) \delta Q_k \quad .$$

From this and analogous arguments we can see that

$$(Q_n, P_k) = \delta_{nk} \quad , \quad (Q_n, Q_k) = (P_n, P_k) = 0 \tag{5.73}$$

as one would expect. These relations can be taken as defining a canonical transformation.

In Sect. 2.6 we saw that the equation of motion (2.61) and the commutation relations (2.62) determine operators in the Heisenberg representation. It is natural to ask if there is a corresponding statement for classical dynamics. Since the structure of classical dynamics is more complicated the answer is not obvious, but a fundamental paper by Pauli [44] shows that if one defines the algebraic structure of the Poisson brackets by (5.65) and assumes the Jacobi identity and an equation of motion in the form (5.64), all of classical dynamics can be derived.

Pathological Lagrangians. Two Lagrangians can differ by a change of scale or by a time derivative and still both give rise to exactly the same equations of motion. Is there any other way in which two Lagrangians can differ without changing anything? There is an extensive literature on this question, which is of particular interest because different Lagrangians can give rise to different versions of quantum mechanics [45]. The alternative Lagrangians, however, are all pathological in some way. For example the Hamiltonian function to which they give rise is not in general an energy, and so they will not fit into quantum mechanics as ordinarily formulated and theory will not agree with experiment [46], [47]. It turns out that in more than one dimension the Lagrangian for any dynamical system that possesses a Lagrangian is unique unless there is some transformation of coordinates that uncouples the various degrees of freedom [48].

Problem 5.41. This problem shows how to transform a hard problem into a simpler one. The Hamiltonian of a nonlinear oscillator is

$$H = \tfrac{1}{2} m (p^2 + \omega^2 x^2) + \alpha x^3 \quad .$$

Find a transformation of the third order in the dynamical variables that for small-amplitude oscillations transforms this system into a simple oscillator. Solve this and return to the original variables, giving the answer $x(t)$ as a Fourier series in ωt.

To begin: Let

$$F_3(p, X) = pX + apX^2 + bp^2 X + cp^3 + dX^3$$

where $a \ldots d$ are constants. This generates the transformation

$$P = p + 2apX + bp^2 + 3dX^2 \quad , \quad x = X + aX^2 + 2bpX + 3cP^2 \quad .$$

To change to the new variables we write these, neglecting terms of higher order, as

$$p = P - 2aPX - bP^2 - 3dX^2 \quad , \quad x = X + aX^2 + 2bPX + 3cP^2 \quad .$$

Now write H in terms of P and X, neglecting terms of order higher than 3, and choose the constants $a \ldots d$ so as to get rid of unwanted terms. The resulting oscillator can then be solved. Note that this procedure is not a panacea, since ω appears in the denominators of terms that are supposedly small. This gives precision to the words "small-amplitude oscillations" in the statement of the problem. They mean "small enough so that ax^3 is small compared with $m\omega^2 x^2/2$".

Problem 5.42. Start with the Lagrangian $L = \dot{x}\dot{y} - \omega^2 xy$. Find and solve the equations of motion. Find the momenta p_x and p_y and the Hamiltonian. Evaluate the Poisson brackets of all the pairs of p's and x's. Taking the commutators in the quantum version of this theory from the Poisson brackets, consider the indeterminacy products $\Delta x \Delta(mv_x)$ etc. Do you like the theory? Does it agree with the rest of physics? Write down the change of variables that makes this an uncoupled system.

Problem 5.43. Show that for any $A(P,Q)$ and $B(P,Q)$ defined as in (5.66), the Poisson bracket is invariant under infinitesimal transformations.

5.5 Generating Finite Transformations from Infinitesimal Ones

Iterating an infinitesimal transformation gives a finite one. How to do this is especially obvious in quantum mechanics, in which

$$f(X) = \left(1 + \frac{i\varepsilon}{\hbar}\hat{G}\right)f(x) \quad .$$

Suppose we wish to make ε cover a finite range γ by iteration. We divide the range into n steps, with n eventually infinite, in each of which $\varepsilon = \gamma/n$. Then

$$f(X) = \lim_{n \to \infty}\left(1 + \frac{i\gamma}{n\hbar}\hat{G}\right)^n f(x) \quad .$$

The limit is the exponential, and even though $\hat{G}$ is an operator we can still write

$$f(X) = \exp\left(\frac{i\gamma}{\hbar}\hat{G}\right)f(x) \quad \text{or} \tag{5.74a}$$

$$f(X) = \left(1 + \frac{i\gamma}{\hbar}\hat{G} - \frac{\gamma^2}{2\hbar^2}\hat{G}^2 + \ldots\right)f(x) \quad . \tag{5.74b}$$

Classical contact transformations are iterated in much the same way, though since the identity from elementary calculus isn't there we have to do a little more work. We have, corresponding to an infinitesimal parameter

$$A_\varepsilon(P, Q) = a(p, q) - \varepsilon(G(p, q), a(p, q)) \quad . \tag{5.75}$$

Let us assume that the finite transformation is of the form

$$A_\gamma = a - \gamma(G, a) + \lambda_2 \gamma^2(G, (G, a)) + \lambda_3 \gamma^3(G, (G, (G, a))) + \ldots \tag{5.76}$$

where the λ's are to be determined. With ε as an infinitesimal increment of γ, we can write (5.59) as

$$\frac{dA_\gamma}{d\gamma} = -(G, A_\gamma) \quad . \tag{5.77}$$

Applying this to (5.76) gives

$$-(G, a) + 2\lambda_2 \gamma(G, (G, a)) + 3\lambda_3 \gamma^2(G, (G, (G, a))) + \ldots$$
$$= -(G, a) + \gamma(G, (G, a)) - \lambda_2 \gamma^2(G, (G, (G, a))) + \ldots$$

and comparison of coefficients gives

$$\lambda_2 = \frac{1}{2} \quad , \quad \lambda_3 = -\frac{1}{6} \quad , \quad \ldots \quad , \quad \lambda_n = \frac{(-1)^n}{n!}$$

so that the finite transformation can be written as

$$A_\gamma = a + \sum_1^\infty \frac{(-\gamma)^n}{n!}(G_n, a) \tag{5.78}$$

where (G_n, a) is the n-times iterated Poisson bracket.

Problem 5.44. Generalize the infinitesimal formula (5.51) to finite transformations, and find an expansion in multiple commutators analogous to (5.78).

Problem 5.45. Equation (5.78) gives the formal solution of equations like (5.64). It is clumsy but it works. Show that it works by using it to construct the solution of a simple harmonic oscillator that starts at $t = 0$ with $x = a$, $p = 0$.

Problem 5.46. What infinitesimal contact transformation is generated by $G = x_n$? By $\hat{G} = x_n$?

Problem 5.47. The infinitesimal rotations (5.48a) can be written in matrix form as

$$\begin{pmatrix} X \\ Y \end{pmatrix} = \begin{pmatrix} 1 & -\varepsilon \\ \varepsilon & 1 \end{pmatrix} \begin{pmatrix} x \\ y \end{pmatrix} = \left[\mathbf{1} + \varepsilon \begin{pmatrix} 0 & -1 \\ 1 & 0 \end{pmatrix} \right] \begin{pmatrix} x \\ y \end{pmatrix} \tag{5.79}$$

where $\mathbf{1}$ is the unit matrix. Iterate this to find the transformation that gives a finite rotation.

The Parameter of the Transformation. The finite transformation (5.78) generates an $A(p, q, \gamma)$ that starts at $A(p_0, q_0, 0)$ and satisfies

$$\frac{dA}{d\gamma} = (A, G) \quad . \tag{5.80}$$

Now suppose that $A(p, q, \gamma)$ (for some suitably chosen A) is solved for $\gamma(p, q)$. If this is done, then setting $A = \gamma$ in (5.80) gives

$$(\gamma, G) = 1 \quad . \tag{5.81}$$

Here γ is a parameter that enters our description of the system, and we see that G is a canonically conjugate parameter. They are not necessarily conjugate coordinate and momentum, though that is what they may sometimes be, as the following example will show.

Example 5.9. We have seen above that $G = xp_y - yp_x$ generates rotations of the axes in the xy plane. If the angle of rotation is γ, we find

$$x = x_0 \cos \gamma - y_0 \sin \gamma \quad , \qquad p_x = p_{x0} \cos \gamma - p_{y0} \sin \gamma$$
$$y = x_0 \sin \gamma + y_0 \cos \gamma \quad , \qquad p_y = p_{x0} \sin \gamma + p_{x0} \cos \gamma \quad . \tag{5.82}$$

These can be solved for γ by writing them as

$$x + iy = e^{i\gamma}(x_0 + iy_0) \quad , \qquad p_x + ip_y = e^{i\gamma}(p_{x0} + ip_{y0})$$

so that

$$\gamma = -i \ln (x + iy) + \text{const} \quad \text{or} \quad \gamma = -i \ln (p_x + ip_y) + \text{const} \quad . \tag{5.83}$$

With

$$G = xp_y - yp_x \tag{5.84}$$

we can now verify with either of the transformations (5.83) that (5.81) is satisfied.

Problem 5.48. Carry out the differentiations that verify (5.81) in the above example.

Now in (5.80), let $A = H$. If G generates a transformation that leaves H invariant, we have

$$\frac{dH}{d\gamma} = (H, G) = 0$$

where the derivative on the left means

$$\frac{dH}{d\gamma} = \frac{\partial H}{\partial P_n} \frac{\partial P_n}{\partial \gamma} + \frac{\partial H}{\partial Q_n} \frac{\partial Q_n}{\partial \gamma} \quad .$$

Thus H is independent of γ, however γ may be expressed in terms of P and Q. In other words, γ is a cyclic parameter – we cannot call it a cyclic coordinate because it is not in general one of the coordinates Q.

We have previously argued that corresponding to any cyclic coordinate there is a conserved momentum canonically conjugate to it. The last few paragraphs turn the argument the other way and show that if there is a conserved dynamical quantity one can construct a canonically conjugate parameter that is cyclic.

These results clarify the relation between H and t. We have already seen that the natural evolution of a system is generated by the Hamiltonian, with time as parameter. Thus t and H are canonically conjugate, as can be seen if one finds a function $t(p, q)$ and puts it into (5.64) to obtain

$$(t, H) = 1 \tag{5.85}$$

as a special case of (5.81). But they are canonically conjugate as parameters and not as coordinate and momentum. Since the ultimate purpose of Hamilton's equations is to find momenta p and coordinates q as functions of t, it makes no sense to take

t as a coordinate. This point should be absorbed carefully since it has caused some confusion in the literature.

Things do not go so well in quantum mechanics, for t in an oscillating system is a multiple-valued function and not easily represented by an expectation value. One might hope to construct an operator $\hat{T}$ corresponding to $t(p, q)$ such that

$$[\hat{T}, \hat{H}] = i\hbar$$

but let us see what happens. Let $|E'\rangle$ be an eigenfunction of $\hat{H}$ belonging to the eigenvalue E', and let

$$|E'\rangle_\omega := e^{i\omega\hat{T}}|E'\rangle$$

with ω arbitrary. Then

$$\hat{H}|E'\rangle_\omega = e^{i\omega\hat{T}}e^{-i\omega\hat{T}}\hat{H}e^{i\omega\hat{T}}|E'\rangle = (E' + \hbar\omega)|E'\rangle_\omega$$

where the last step is taken by applying the result of Problem 5.44 or just by working out a few terms of an expansion in $\omega\hat{T}$. But this is a very peculiar result. Since ω is arbitrary, the eigenvalues of $\hat{H}$ form a continuum that extends to negative infinity. Such a conclusion is inconsistent with the rest of quantum mechanics and contradicts the facts. We must conclude that there is no such operator as $\hat{T}$. The time variable t in ordinary quantum mechanics must be taken as a numerical parameter used to describe the development of the dynamical variables of a physical system and not as a dynamical variable itself. The exception to this statement is the relativistic formulation of Sect. 5.2 in which this role is taken by the proper time, while t is treated on the same footing as the spatial coordinates.

Problem 5.49. Equation (5.85) is not a relation between conjugate coordinate and momentum because t is not a coordinate and H is not a momentum. But according to Problem 5.26, the transformed Hamiltonian is ωP and $X = \omega(t - t_0)$. Clarify the apparent counterexample.

Problem 5.50. A harmonic oscillator moves with $x = A \sin \omega(t - t_0)$. Solve this for $t(x)$ and show that $(t, H) = 1$.

5.6 Deduction of New Integrals

It is trivial in quantum mechanics that if H commutes with dynamical variables a and b it commutes with their commutator, but the consequences are profound. In classical mechanics Jacobi's identity yields

$$\frac{d}{dt}(a, b) = -(H, (a, b)) = (a, (H, b)) - (b, (H, a))$$

so that if a and b are constant, then (a, b) is also. This is a very important result, since it gives a way of finding integrals, and therefore of lowering the order of the equations of motion, without carrying out any integrations.

One can, of course, keep on finding integrals by evaluating Poisson brackets, but after a certain point nothing new will be found. This is because a system with f degrees of freedom has at most $2f$ constants of integration, and for an energy-conserving system one of them is the initial instant t_0. An example will show how this comes about.

The Isotropic Oscillator. A soluble example of the consequences of Jacobi's identity is a simple harmonic oscillator in two dimensions with x and y forces equal, Fig. 5.5. The Hamiltonian (with $m = 1$) is

$$H = \tfrac{1}{2}[p_x^2 + p_y^2 + \omega^2(x^2 + y^2)] \tag{5.86}$$

and from the circular symmetry we know that the angular momentum

$$L = xp_y - yp_x \tag{5.87}$$

is constant. To find another integral, we note that H is the same as the Hamiltonian of two uncoupled oscillators of the same frequency. Since each conserves its energy, the difference in energy is also constant

$$B := \tfrac{1}{2}[p_x^2 - p_y^2 + \omega^2(x^2 - y^2)] \quad . \tag{5.88}$$

It is now a simple matter to find a fourth integral by Jacobi's theorem,

$$C := (L, B) = 2(p_x p_y + \omega^2 xy) \quad . \tag{5.89}$$

Continuing, we find

$$(C, L) = 4B \quad \text{and} \quad (C, B) = -4\omega^2 L \tag{5.90}$$

so that no further new integrals can be found in this way. We have in fact found more than the $2f - 1 = 3$ independent integrals expected, and one can verify that

$$H^2 = B^2 + \tfrac{1}{4}C^2 + \omega^2 L^2 \quad . \tag{5.91}$$

Thus only three of the four integrals are independent.

The integral (5.89) represents a physical property of the oscillator that does not depend only on the energies of its component motions. To see what it is, write the motions in x and y as

$$x = a \, \sin \, [\omega(t - t_0)] \quad , \quad y = b \, \sin \, [\omega(t - t_0) + \beta]$$

where β is a constant phase difference. Then

$$L = abm\omega \, \sin \, \beta \quad , \quad C = 2abm\omega^2 \, \cos \, \beta \quad .$$

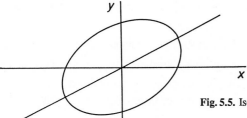

Fig. 5.5. Isotropic oscillator with orbit arbitrarily oriented

The Kepler problem has a vectorial integral that points from the center of force to the aphelion (or perihelion). Since in the present case the center of force is at the center of the orbit, the direction of the vector is ambiguous, and this warns us that some other description will be better. A glance at the integrals H, B, and C shows that they define a constant second-rank tensor

$$A_{ij} = \tfrac{1}{2}(p_i p_j + \omega^2 x_i x_j) \tag{5.92}$$

in terms of which (5.91) becomes

$$|A| := A_{xx} A_{yy} - A_{xy}^2 = \frac{\omega^2}{4} L^2 \quad \text{while} \tag{5.93a}$$

$$\operatorname{tr} A := A_{xx} + A_{yy} = H \quad . \tag{5.93b}$$

To relate the tensor to the orbit, choose the x axis to coincide with the major axis of the ellipse and let $t_0 = 0$. Then $\beta = \tfrac{\pi}{2}$ and the oscillator's motion is given by

$$x = a \cos \omega t \quad , \quad y = b \sin \omega t \quad (a > b)$$

and we readily find that

$$A_{xx} = \tfrac{1}{2}\omega^2 a^2 \quad , \quad A_{yy} = \tfrac{1}{2}\omega^2 b^2 \quad , \quad A_{xy} = 0 \quad .$$

The axes chosen reduce A_{ij} to diagonal form and are called the principal axes of the tensor. Evaluation of H and L gives

$$H = \tfrac{1}{2}\omega^2(a^2 + b^2) \quad , \quad L = -\omega a b$$

whence

$$a^2 = \frac{1}{\omega^2}[H + \sqrt{H^2 - \omega^2 L^2}] \quad , \quad b^2 = \frac{1}{\omega^2}[H - \sqrt{H^2 - \omega^2 L^2}] \quad .$$

When $L = 0$, $b = 0$ and the oscillations are linear. When $|\omega L| = H$ the motion is in a circle, and no larger value for L is possible.

Problem 5.51. Find the eccentricity of the orbit in terms of H and L and in terms of the components of A_{ij}. Show that it reduces to a simple form when principal axes are used.

Problem 5.52. Calculate the infinitesimal transformations of Cartesian coordinates and momenta generated by H, B, and C.

Problem 5.53. Calculate the infinitesimal transformations of coordinates and momenta generated by A_{ij} and L. Note that L generates transformations of spatial coordinates as we have already seen, while H, B, and C are examples of dynamical symmetries. They generate transformations resembling rotations that obviously leave the Hamiltonian invariant but do not represent symmetries in the ordinary geometrical sense.

Problem 5.54. Find the Poisson brackets connecting the various components of A_{ij}.

Problem 5.55. Solve the isotropic harmonic oscillator in quantum mechanics and give the degeneracy of each level.

Problem 5.56. Do the same in polar coordinates and again note the degeneracies.

Problem 5.57. If the oscillator is no longer isotropic and
$$H = H_1 + H_2 \quad , \qquad H_i := \tfrac{1}{2}(p_i^2 + \omega_i^2 x_i^2)$$
then $B := H_1 - H_2$ is still constant but the angular momentum is not. Show that
$$C := \sqrt{H_1 H_2} \cos \Theta \quad ,$$

$$\Theta := \sqrt{\frac{\omega_1}{\omega_2}} \sin^{-1} \frac{\omega_2 x_2}{\sqrt{2H_2}} - \sqrt{\frac{\omega_2}{\omega_1}} \sin^{-1} \frac{\omega_1 x_1}{\sqrt{2H_1}} \tag{5.94}$$

is a new constant and find still another. Now find all the Poisson brackets and show that no more constants result.

Problem 5.58. Except for scale factors, the Poisson bracket relations (5.89) and (5.90) are the same as those of the components of angular momentum. This suggests a way to find the energy levels of the oscillator in quantum mechanics. Transcribe the entire discussion into quantum-mechanical terms, find the commutators, which will again resemble those of angular momentum, and, noting (5.91), use the apparatus of angular-momentum theory to find the energy levels and their degeneracies. Be sure your results are the same as those of Problems 5.55 and 5.56.

Problem 5.59. Show that if the coordinate axes are rotated by (5.82), tr A and the determinant $|A|$ remain the same.

Problem 5.60. For the Kepler problem in the xy plane, the components A_x and A_y of Laplace's vector integral are constant. Does (A_x, A_y) give a new integral?

5.7 Commutators and Poisson Brackets

It is clear from the list of identities shared by commutators and Poisson brackets that the two have a profound formal parallelism, and yet it is not easy to see why they are so much alike. In fact, if the dynamical variables a and b in the Poisson bracket are functions of noncommuting p's and q's, it is not even obvious how derivatives should be taken. This section, following a paper by *F.J. Bloore* [50], will discuss one possible way to take derivatives involving noncommuting quantities and use it to discuss the correspondence (5.61) between commutators in quantum mechanics and Poisson brackets in classical mechanics.

With commuting variables, $\alpha_1, \ldots, \alpha_k$ the ordinary partial derivative of some function $a(\alpha)$ is defined by

$$\frac{\partial a}{\partial \alpha_i} := \lim_{\varepsilon \to 0} \frac{1}{\varepsilon} [a(\alpha_1, \ldots \alpha_i + \varepsilon, \ldots \alpha_k) - a(\alpha)] \quad . \tag{5.95}$$

With noncommuting variables we can define a more general derivative involving an arbitrary function $c(\alpha)$,

$$\frac{\partial a}{\partial \alpha_i}\{c\} := \lim_{\varepsilon \to 0} \frac{1}{\varepsilon} [a(\alpha_1, \ldots \alpha_i + \varepsilon c, \ldots \alpha_k) - a(\alpha)] \quad .$$

Bloore calls this a Fréchet derivative. If c commutes with all the α's, then

$$\frac{\partial a}{\partial \alpha_i}\{c\} = c\frac{\partial a}{\partial \alpha_i} \quad . \tag{5.96}$$

Evidently, the Fréchet derivative satisfies the usual rules that

$$\frac{\partial(a+b)}{\partial \alpha_i}\{c\} = \frac{\partial a}{\partial \alpha_i}\{c\} + \frac{\partial b}{\partial \alpha_i}\{c\} \quad , \quad \frac{\partial(\gamma a)}{\partial \alpha_i}\{c\} = \gamma\frac{\partial a}{\partial \alpha_i}\{c\} \quad .$$

We shall now prove a remarkable theorem involving commutators:

$$\frac{\partial a}{\partial \alpha_i}\{[\alpha_i, b]\} = [a, b] \quad \text{(summed)} \tag{5.97a}$$

where b is another function. The proof is by induction. Write the general function $a(\alpha)$ as the sum of pieces each of which is of the general form $\gamma\alpha_3\alpha_2\alpha_2\alpha_1\alpha_3\alpha_1\alpha_4 \ldots$, and we assume it to have been proved for one such piece, a'. Now add another factor of α_j on the left.

$$\begin{aligned}
\frac{\partial(\alpha_j a')}{\partial \alpha_i}\{[\alpha_i, b]\} &= \lim_{\varepsilon \to 0} \frac{1}{\varepsilon}(\alpha_j + \varepsilon[\alpha_j, b] - \alpha_j)a' + \alpha_j\frac{\partial a'}{\partial \alpha_i}\{[\alpha_i, b]\} \\
&= [\alpha_j, b]a' + \alpha_j[a', b] = [\alpha_j a', b] \quad .
\end{aligned}$$

Thus if (5.97a) holds for a', it holds for $\alpha_j a'$. The theorem holds trivially if $a' = 1$, and therefore the induction is complete. □

To establish the relation with Poisson brackets rewrite (5.97a) replacing i by j, b by α_i, and a by b:

$$\frac{\partial b}{\partial \alpha_j}\{[\alpha_j, \alpha_i]\} = [b, \alpha_i] \quad \text{or}$$

$$\frac{\partial b}{\partial \alpha_j}\{[\alpha_i, \alpha_j]\} = [\alpha_i, b] \quad . \tag{5.97b}$$

Now substitute this into the left side of (5.97a):

$$\frac{\partial a}{\partial \alpha_i}\left\{\frac{\partial b}{\partial \alpha_j}\{[\alpha_i, \alpha_j]\}\right\} = [a, b] \quad . \tag{5.98}$$

This identity will give us a relation between commutators and Poisson brackets. Divide the dynamical variables α_i into two sets, $p_1, \ldots p_f$ and $q_1, \ldots q_f$, with

$$[q_m, p_n] = i\hbar\delta_{mn} \quad , \quad \text{others zero} \quad . \tag{5.99}$$

As α_i and α_j go through all the p's and q's we get

$$\frac{\partial a}{\partial q_n}\left\{\frac{\partial b}{\partial p_n}\{i\hbar\}\right\} + \frac{\partial a}{\partial p_n}\left\{\frac{\partial b}{\partial q_n}\{-i\hbar\}\right\} = [a, b]$$

or by (5.96),

$$[a, b] = i\hbar\left(\frac{\partial a}{\partial q_n}\left\{\frac{\partial b}{\partial p_n}\right\} - \frac{\partial a}{\partial p_n}\left\{\frac{\partial b}{\partial q_n}\right\}\right) \tag{5.100}$$

where the derivatives of b are the usual ones defined in (5.95). (Note that this relation is exact.) The quantity on the right is a version of the Poisson bracket which is appropriate to noncommuting variables and which reduces to the c-number version if the variables commute. This is the origin of the formal similarity between Poisson brackets and commutators.

If we were to overlook all questions of commutation on the right side of (5.100), the Fréchet derivative would become the ordinary derivative by (5.96), which would amount to putting an = sign into (5.61). But there are additional terms left over proportional to $\hbar$, $\hbar^2$, ... from the omitted commutation relations. We have therefore

$$[a, b] = i\hbar(a, b) + O(\hbar^2) \quad . \tag{5.101}$$

From a formal point of view, nothing more can be done to establish the correspondence between classical and quantum theories. Specifically, the correspondence (5.61) can be obtained only by keeping all commutators (5.99) in going from (5.98) to (5.100) and ignoring all commutators in going from (5.100) to (5.101). This is not a consistent procedure. Since there is no universal recipe that tells in what order the numbers of classical physics should be written if they are to become operators, the omitted terms of the order of $\hbar$ must be supplied by other arguments. It is for this reason that although we have taken pains to symmetrize the quantity A in (3.53) and therefore make it hermitian, we recommended in Problem 3.38 that the reader verify that it is in fact the correct form by seeing whether it commutes with the Hamiltonian. Except in the simplest cases the requirement of hermiticity is not enough to make sure that the transcription into operators is correct.

There are two good reasons for taking (5.61) as the best that can be done. The first is that it works in all the simple cases and the second is that both sides obey the Jacobi identity involving double Poisson brackets or double commutators, which deepens the formal correspondence. But this convention does not tell how to write an expression from classical mechanics in terms of quantum variables since Poisson brackets leave the order of factors unspecified. That used to be regarded as an important problem but now one tries to derive quantum physics from considerations more fundamental than classical mechanics and the question rarely arises.

Problem 5.61. Calculate $[p^2, x^2]$ by taking Fréchet derivatives and also by evaluating the commutator directly. Compare the results with $i\hbar(p^2, x^2)$.

Problem 5.62. Calculate the classical Poisson bracket (p^k, q^n). Now calculate $[\hat{p}^k, \hat{q}^n]$ in two ways: by Fréchet derivatives and by setting $\hat{p} = -i\hbar\partial/\partial q$ and performing the differentiations. Are these results the same?

Problem 5.63. In the problems beginning with 5.5, it was required to represent the commutator of a function of p's and q's with a single p or q in terms of a derivative of the function. What is the right way to do this?

Problem 5.64. For practice in taking Fréchet derivatives, evaluate $[A(p, x), px]$ and $-[px, A(p, x)]$, A arbitrary, and show that they are equal.

Problem 5.65. Use the commutator formula (5.100) to tell exactly how to take the gradient in (2.63) when the arguments of V are operators, not numbers.

5.8 Gauge Invariance

It is perhaps surprising that the Hamiltonian equations for a particle in an electro-magnetic field involve the potentials rather than the fields. The fields are defined in terms of forces on charged particles, whereas the potentials do not have any such obvious operational definition (but see [1], Sect. 7.8), and indeed, as we shall see, are not even uniquely defined by the sources that produce them or the forces they exert. When one studies the laws of motion in quantum mechanics one finds that whereas fields govern the orbits of particles, potentials govern not only orbits but phase relations as well, so that an interference pattern, for example, which depends on phases, is affected by the existence of a potential even if the region through which the beams pass is entirely field-free. Thus the occurrence of potentials rather than fields in quantum mechanics is easy to understand, though in classical mechanics it remains somewhat opaque.

The potentials A and ϕ determine the fields according to the relations

$$B = \nabla \times A \quad , \quad E = -\nabla\phi - \frac{\partial A}{\partial t} \quad . \tag{5.102}$$

Now introduce new potentials according to

$$A' = A + \nabla\lambda \quad , \quad \phi' = \phi - \frac{\partial\lambda}{\partial t} \tag{5.103}$$

where $\lambda(x, t)$ is an arbitrary (differentiable) function. Such a change of the potentials is called a gauge transformation. The values of B and E determined by (5.102) are unchanged; this is what is meant by saying that A and ϕ are not uniquely determined by the forces they exert. The latitude of definition is often convenient for computation, but it introduces some complications, as we shall show, for transformation of A and ϕ via (5.103) with arbitrary λ can convert even the potentials of a simple static electric and/or magnetic field, or no field at all, into something time-dependent and elaborate.

The classical equation of motion,

$$m\ddot{x} = e(E + v \times B) \tag{5.104}$$

is gauge-invariant because it depends only on the fields, but in Schrödinger's equation

$$i\hbar\frac{\partial\psi}{\partial t} = \hat{H}\psi \quad , \quad \hat{H} = \frac{1}{2m}(\hat{p} - eA)^2 + e\phi \tag{5.105}$$

the potentials occur explicitly. The equation can, however, be made gauge-invariant like (5.104) if we introduce a change in the (unobservable) phase of the wave function connected with the gauge transformation (5.103):

$$\psi' = e^{i\mu}\psi \quad , \quad \mu := \frac{e}{\hbar}\lambda(x, t) \quad . \tag{5.106}$$

Then, in (5.105),

$$i\hbar\frac{\partial\psi'}{\partial t} - e\phi\psi' = i\hbar\frac{\partial}{\partial t}e^{i\mu}\psi - e\left(\phi - \frac{\partial\lambda}{\partial t}\right)e^{i\mu}\psi$$

$$= e^{i\mu} \left[i\hbar \frac{\partial \psi}{\partial t} - \left(\hbar \frac{\partial \mu}{\partial t} + e\phi - e \frac{\partial \lambda}{\partial t} \right) \psi \right]$$

$$= e^{i\mu} \left[i\hbar \frac{\partial \psi}{\partial t} - e\phi\psi \right] \quad .$$

Similarly (Problem 5.66),

$$(\hat{\boldsymbol{p}} - e\boldsymbol{A}')\psi' = e^{i\mu}(\hat{\boldsymbol{p}} - e\boldsymbol{A})\psi \tag{5.107}$$

from which follows by simple steps the gauge-invariance of (5.105) (Problem 5.67). Pauli has called the transformations (5.106) and (5.107) gauge transformations of the first and second kinds respectively. It is clear that the physical results of the theory should be independent of the choice of potentials, but how this is to be achieved raises some interesting questions.

Problem 5.66. Prove (5.107).

Problem 5.67. Prove that the time-dependent Schrödinger equation (5.105) is gauge invariant.

Problem 5.68. Prove that the eigenvalues of the equation

$$(\hat{H} - e\phi)\psi_n = e_n\psi_n \tag{5.108}$$

are gauge-invariant.

Perturbation Theory. Let us start with a system bound together by its own internal forces and described by a Hamiltonian

$$\hat{H}_0 = \frac{\hat{p}^2}{2m} + eV \tag{5.109}$$

written in one-particle form for simplicity though our considerations are not restricted to one particle. Suppose now that this system is immersed in an external electromagnetic field B, E with potentials A, ϕ, and that we wish to study the effects of this field on the system. The gauge transformations to be considered act on A, ϕ but not on V. The wave equation is now, in Cartesian coordinates,

$$i\hbar \frac{\partial \psi'}{\partial t} = \hat{H}\psi' \quad , \quad \hat{H} = \frac{1}{2m}(\hat{\boldsymbol{p}} - e\boldsymbol{A})^2 + eV + e\phi \tag{5.110}$$

and, as we have seen, it is a gauge-invariant equation.

First of all, a slight digression on the physical meaning of $\hat{H}$. A little thought shows that it is not always an energy. Energy in dynamics is kinetic and potential. The potential energy $\phi(x)$ of a particle located at a certain point is equal to the work done in bringing the particle from some agreed location, often infinity, to the point x. It is understood that the force is path-independent. In the field defined by A, ϕ or B, e, the condition for this is $\nabla \times \boldsymbol{E} = 0$, but in a changing field $\nabla \times \boldsymbol{E} = -\partial \boldsymbol{B}/\partial t$ and this is not in general zero. Thus ϕ is not a potential energy. The unperturbed $\hat{H}_0$ represents an energy but $\hat{H}$ does not, and in fact the energy of the particle cannot even be defined.

Perturbation theory usually starts with an attempt to expand the wave function of a perturbed system in terms of the eigenfunctions ϕ_n belonging to $\hat{H}_0$. If the external potentials are considered as the perturbation, it is natural to put into the perturbing part everything that contains the potentials. If the eigenfunctions of $\hat{H}_0$ are φ_n, one then aims to write the perturbed wave function in the form

$$\psi = \sum c_n(t)\varphi_n \qquad (5.111)$$

where the c_n have their conventional interpretation that $|c_n|^2$ is the probability that the perturbed system would be found in the nth eigenstate of the unperturbed system. But since it is $\hat{p} - eA$ that is gauge-invariant and not $\hat{p}$, this is not a gauge-invariant separation. We could also leave A in the "unperturbed" part and make ϕ the perturbation but this too is not a gauge-invariant procedure and it has the additional difficulty that when A depends on time the functions φ_n may be impossible to find. Lack of gauge invariance does not mean that one will necessarily get wrong results, since nothing has been falsified or omitted, but it is liable to lead to results that are not gauge-invariant unless care is taken in making approximations. For example, suppose that a gauge transformation is carried out on the state ψ, so that it becomes $e^{i\mu}\psi$. The c's in (5.111) will now be entirely different and the conventional interpretation of $|c_n|^2$ will be lost unless the same transformation carries φ_n into $e^{i\mu}\varphi_n$, that is, unless φ_n also satisfies a gauge-invariant equation. Is there any way of making a perturbation theory that is explicitly gauge-invariant?

There are a few situations that offer a way out of the dilemma. The problems at the end of this section aim at systems of general interest in which an atom is immersed in an electromagnetic field whose wavelength is much larger than the atom. They will show that with suitable transformations one can arrive at a gauge in which the numerical value of A is so small that it can reasonably be ignored, and essentially the whole perturbation is in ϕ. The unperturbed Hamiltonian is again $\hat{H}_0$, the $|c_n|^2$ have their conventional significance, and the physical interpretation of the results is once more straightforward [51, 52]. The difficulties encountered here originate in nonrelativistic nature of the theory, in which t appears not as a quantum variable but as a numerical parameter. In a completely relativistic treatment that includes the dynamics of both the matter field and the electromagnetic field it is possible to make a gauge-invariant perturbation theory.

Problem 5.69. Show that when a dynamical variable Ω in quantum mechanics involves potentials which depend explicitly on time, its equation of motion is

$$\frac{d\langle\Omega\rangle}{dt} = \left\langle \frac{i}{\hbar}[H, \Omega] + \frac{\partial\Omega}{\partial t} \right\rangle \quad . \qquad (5.112)$$

Problem 5.70. Show that the corresponding relation in classical mechanics is

$$\frac{d\Omega}{dt} = (\Omega, H) + \frac{\partial\Omega}{\partial t} \quad . \qquad (5.113)$$

Problem 5.71. Show that in classical and quantum mechanics, the complete Hamiltonian $\hat{H}$ defined in (5.110) has a time rate of change which depends on the gauge.

Problem 5.72. Show that the time rate of change of the energy operator in (5.108) is simple, gauge-invariant, and physically reasonable in both classical and quantum mechanics.

Problem 5.73. Show that the operator

$$\hat{\Omega}' := e^{i\mu}\hat{\Omega}e^{-i\mu} \tag{5.114}$$

has the same matrix elements (and expectation values) in the gauge-transformed state (5.106) that the corresponding $\hat{\Omega}$ had in the original state. Show that $\hat{p}' = \hat{p} - e\nabla\lambda$.

Problem 5.74. Just as a canonical transformation to new p's and q's leaves the Poisson brackets unchanged (Poisson's theorem), show that a transformation of the form (5.114) leaves invariant the commutation relations of the transformed operators. (Note that this is still valid if μ is a function of operators as well as coordinates.)

Problem 5.75. Consider in Hamiltonian mechanics the canonical transformation generated by

$$F_3(p, Q, t) = p_n Q_n - e\lambda(Q, t) \quad .$$

Find the new momentum and the new Hamiltonian and comment on the relation between this transformation of particle variables and the gauge transformations (5.103) involving the field variables.

Problem 5.76. Show that if the external field is a plane wave with

$$A = a \sin (k \cdot r - \omega t) \quad , \quad \phi = b \sin (k \cdot r - \omega t) \tag{5.115}$$

the magnetic field B is automatically transverse, while the transversality of E requires that

$$b = \frac{k \cdot a}{k^2}\omega \quad .$$

Show that B is now perpendicular to E.

Problem 5.77. Find the λ that transforms this to

$$A' = a' \sin (k \cdot r - \omega t) \quad , \quad \phi' = 0 \tag{5.116}$$

where the new amplitude a' is perpendicular to k.

Problem 5.78. A weak magnetic field implies that the spatial derivatives of A are small, i.e., in the preceding problem, k is small. Carry out the gauge transformation generated by $\lambda = -A' \cdot r$ and show that the new potentials are

$$A'' = -(a' \cdot r)k \cos (k \cdot r - \omega t) \quad , \quad \phi'' = -E(t) \cdot r \quad .$$

(Note that A'' is longitudinal and ϕ'' is gauge-invariant. Note also that though ϕ'' may look like an electrostatic potential it isn't, since E in general depends on time.) For evaluating matrix elements, r gives the size of the region in which the wave function differs significantly from zero. The magnitude of A'' is therefore smaller than that of a' by a factor of kr, and the "optical approximation", with the wavelength of light a thousand times the radius of an atom, amounts to setting this equal to zero.

Problem 5.79. Show that A'' gives the same magnetic field as the original A. How does it happen that A'' in the situation mentioned above is hundreds of times smaller than A but describes the same field?

6. The Hamilton-Jacobi Theory

The principal object of classical dynamics is to find where everything is at time t; that is, to find a set of $q_n(t)$. The principal object of quantum mechanics is to find a wave function $\psi(q, t)$. From this you cannot calculate where everything is in the sense of classical physics, but you can calculate all there is to know. Lagrangian and Hamiltonian dynamics find $q(t)$ and $\dot{q}(t)$ or $p(t)$ by means of ordinary differential equations. In $\psi(q, t)$, q is in no sense a function of t, since ψ has a value for every q and every t. The equation it solves is a partial differential equation. The differences between the two mathematical descriptions are so wide that they seem to belong to different universes of ideas, but in Ehrenfest's theorems (Sect. 1.3), for example, familiar ordinary differential equations came out of the partial differential equation for $\psi(q, t)$ that are the same as the equations of Newtonian dynamics. It is now our task to show that out of the dynamics of Lagrange and Hamilton comes a partial differential equation that is the same as the eikonal equation (1.31), which was derived from wave optics. In Chap. 7 we will see how to reconstruct quantum mechanics, under suitable assumptions, if the eikonal equation is known. These arguments will help to explain the mathematical relation between the two theories, and a study of the necessary assumptions will do much to illuminate the subtler question of their physical relation.

Further, the wave function is the solution to problems in quantum mechanics in the sense that if a wave function is known, the calculation of expectation values, etc., is then easy. We have a right to expect the same thing in classical theory: when the eikonal equation is solved, the desired answers should be easy to find. There is, however, a catch. Classical mechanics is more specific; the answers are more detailed. Exactly soluble problems are usually about equally hard in classical and quantum theory, but when approximations have to be made, quantum mechanics is much easier. These matters will be discussed in Chap. 8.

6.1 The Hamilton-Jacobi Equation

We are once more in the q, t space of Lagrange, and paths can be sketched as in Fig. 6.1. We have already seen in (5.34) that if the function S is evaluated for a natural path and then compared with the value for a slightly different path running between slightly different end points in a slightly different time, the change in S is given by

$$\delta S = [p_n \delta q_n - H \delta t]_{t_0}^{t} \quad . \tag{5.34}$$

142

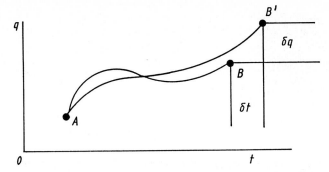

Fig. 6.1. Adjacent paths in qt space leading from a common initial point

If the variations δq and δt are zero at the initial point A, then S is defined by this relation as a function of q and t at the arrival point. Varying the q's and t one at a time gives

$$\frac{\partial S}{\partial q_n} = p_n \quad , \quad \frac{\partial S}{\partial t} = -H(p, q, t) \tag{6.1}$$

and substituting the first of these into the second gives

$$H\left(\frac{\partial S}{\partial q}, q, t\right) + \frac{\partial S}{\partial t} = 0 \quad . \tag{6.2}$$

This is the Hamilton-Jacobi equation.[1] We have seen it before, in Cartesian coordinates for a single particle, in (1.31). Here it occurs in a more general form. The equation is the same but its meaning was entirely different in Chap. 1. There S was the phase of a de Broglie wave; here it is the action integral. (Hamilton called it the *principal function*.)

There is another way to look at the Hamilton-Jacobi equation. Equations (5.38) and (5.39) show how a function $F_1(q, Q, t)$ can be used to change the old variables p, q to a new set, P, Q. For the ensuing discussion interchange the letters: let P, Q, K be the old variables and p, q, H the new ones. The relation (5.39) then becomes

$$K(P, Q) = H(p, q) + \frac{\partial F_1(q, Q, t)}{\partial t} \quad . \tag{6.3}$$

Suppose we are able to choose F_1 so that $K(P, Q) = 0$. Then the canonical equations in the old variables are

$$\dot{P}_n = 0 \quad , \quad \dot{Q}_n = 0 \quad .$$

Integrate them. The P's and Q's are new constants determined by the initial conditions, and provided we know how to transform from these coordinates and momenta

---

[1] This equation was first given by *Hamilton* in 1834 [Ref. 5, p. 103], a year befor the paper containing what are now called Hamilton's equations. Hamilton's discussion was, however, somewhat more involved than that given here, and it was clarified and generalized by Jacobi in his lectures in the winter of 1842–43 [53]. For condensed discussions of the theory's development see [29] and [54].

to the new p's and q's, the calculation is finished. The p's are given at once by $p_n = \partial F_1/\partial q_n$, and the q's can in principle be found by solving $\partial F_1/\partial Q_n = -P_n$.

We have now to find $F_1(q, Q)$. From (6.3), it is the solution of $\partial F_1(q, Q, t)/\partial t = -H(p, q)$. To express H in the same variables as F_1, write

$$\frac{\partial F_1(q, Q, t)}{\partial t} + H\left(\frac{\partial F_1(q, Q, t)}{\partial q}, q, t\right) = 0 \quad . \tag{6.4}$$

If we replace F_1 by S and remember that H is the Hamiltonian in the new variables p, q, this is the same as (6.2).

When everything is solved the q's will be functions of Q and t. Choose $Q_n = q_n(t_0)$. Since q starts at Q and at every moment obeys the canonical equations, this transformation just follows the system's natural motion. In this sense F_1 generates the values of $q(t)$ given that they equal the Q's at t_0. This is therefore the finite form of the infinitesimal transformation from t to $t + dt$ that we saw in (5.68) is generated by the Hamiltonian.

Because a moving system is described by relations of the form $q(t)$ one has to get used to functions like $S(q, t)$ in which q and t, being the coordinates of q, t space, vary independently. (In this way S resembles a wave function.) The following problems are intended to clarify the matter.

Problem 6.1. Show by evaluating $\partial^2 S/\partial q \partial t$ that $\dot{p}_n = -\partial H/\partial q_n$, explaining why $\dot{p}_n$ is a total, not a partial derivative.

Problem 6.2. Evidently,

$$\frac{dS}{dt} = \frac{\partial S}{\partial t} + \frac{\partial S}{\partial q_n}\dot{q}_n = -H + p_n\dot{q}_n = L \tag{6.5}$$

as would be expected. But why is q now regarded as a function of t?

Problem 6.3. Prove that $\dot{q}_n = \partial H/\partial p_n$.

Now we must start integrating. If, as in the examples to be given later, the energy is constant, we can do one integration in (6.1) immediately:

$$S(q, t) = W(q) - (t - t_0)E \tag{6.6}$$

where E is the constant numerical value of the energy, and W, which Hamilton called the *characteristic function*, satisfies

$$H\left(\frac{\partial W}{\partial q}, q\right) = E \quad . \tag{6.7}$$

Equations (6.2) and (6.7) are the classical analogues of the time-dependent and time-independent Schrödinger equations.

The action $S(q, Q, t)$ generates a canonical transformation from p, q at time t_0 to p, q at time t. In the more restricted case considered here, in which H is constant and equal to E, $\partial W/\partial q_n = \partial S/\partial q_n$ and (6.7) has exactly the same content as (6.2). Therefore W generates a transformation of variables from t_0 to t that leaves the energy constant.

Suppose we have solved for $W(q)$. How do we find useful results from it? From (5.34) and (6.6),

$$\delta W = \delta S + \delta[(t - t_0)E] = p_n\delta q_n - E\delta t + E\delta t + (t - t_0)\delta E \quad \text{or}$$

$$\delta W = p_n\delta q_n + (t - t_0)\delta E \tag{6.8}$$

so that W may conveniently be regarded as a function of q and E. (The transformation from $S(q, t)$ to $W(q, E)$ is of a type especially familiar in thermodynamics, called a Legendre transformation. We have already encountered examples in going from $L(q, \dot{q})$ to $H(p, q)$ in Sect. 5.1. and in constructing F_2, F_3, and F_4 in Sect. 5.3.) From (6.8), we have

$$t - t_0 = \frac{\partial W}{\partial E} \quad , \quad p_n = \frac{\partial W}{\partial q_n} \quad . \tag{6,9a, b}$$

The first of these shows the rather curious way in which one deduces the temporal behavior of a system from the timeless description given by solutions of (6.7). The second we have already seen in (1.32). How the procedure works can be seen from a simple example.

Example 6.1. Simple Harmonic Oscillator

$$H = \frac{p^2}{2m} + \frac{1}{2}kx^2 = E \quad \text{so}$$

$$p = \frac{\partial W}{\partial x} = \sqrt{2m\left(E - \frac{1}{2}kx^2\right)} \quad .$$

Thus

$$\begin{aligned}
W &= \sqrt{mk}\int\sqrt{\frac{2E}{k} - x^2}dx \\
&= \frac{1}{2}\sqrt{mk}\left(x\sqrt{\frac{2E}{k} - x^2} + \frac{2E}{k}\sin^{-1}\left[\sqrt{\frac{k}{2E}}x\right]\right) + \text{const} \quad . \tag{6.10}
\end{aligned}$$

This has been written down to show how the solution looks; it is sketched in Fig. 6.2. Actually one is not ordinarily interested in looking at W. Differentiating under the integral sign gives

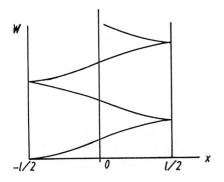

Fig. 6.2. The function $W(x)$ for a harmonic oscillator as given by (6.10). The slope of the curve is the momentum of the particle, and $l/2 = \sqrt{2E/k}$

145

$$\frac{\partial W}{\partial E} = t - t_0 = \sqrt{\frac{m}{2}} \int \frac{dx}{\sqrt{E - kx^2/2}}$$

$$= \sqrt{\frac{m}{k}} \, \sin^{-1} \frac{x}{\sqrt{2E/k}}$$

giving $t(x, E)$. The inverse is

$$x = \sqrt{\frac{2E}{k}} \, \sin \left[\omega(t - t_0) \right] \quad , \qquad \omega^2 = k/m \quad .$$

As promised, the critical point of the solution was finding the integral that gives W. Finding $x(t)$ from a knowledge of W was trivial.

The relation (6.9b) is familiar. We encountered it first in Cartesian form in (1.11), in which $w = \hbar^{-1}W$ is taken to be the spatial part of the phase of a de Broglie wave, and the momentum associated with a quantum is

$$p = \hbar k = \nabla W \quad .$$

The meaning of (6.9a) also becomes clear if we think of $\hbar^{-1}S$ as the entire phase of the wave. S is a function $S(E, q, t)$. In order to make a localized wave packet we superpose waves belonging to a range of values of E as in (1.20) and Fig. 1.2:

$$\int_{-\infty}^{\infty} e^{i\hbar^{-1}S(E,q,t)} A(E) dE$$

where $A(E)$ has its maximum at $\overline{E}$. As in the earlier discussion, the condition for constructive interference is

$$\frac{\partial S}{\partial E} = 0 \quad , \qquad E = \overline{E}$$

and with (6.6) this is the same as (6.9a). Thus as soon as it is realized that $\hbar^{-1}S$ is a phase, the relations (6.9) come out almost trivially. What is remarkable is that the structure of Newtonian mechanics allows these relations to be derived without making the assumption of a wave.

Calculation of $S(q, q_0, t)$. We have seen that $F_1(q, Q, t)$, with q called q_0 and Q called q, is supposed to be the same thing as the principal function S, but comparing (6.6) with (6.10) we do not seem to have succeeded in finding F_1. For one thing, there is no sign of a q_0. Let us start with a much simpler calculation and see how to do it.

For a free particle ($m = 1$) the Lagrangian is $\dot{x}^2/2$. The resulting equation of motion is $\ddot{x} = 0$, with solution $x = x_0 + \beta t$, where x_0 and β are constants of integration. S is given by

$$S = \int \frac{1}{2} \dot{x}^2 dt = \frac{1}{2} \beta^2 t \quad .$$

But this is obviously not the desired answer, since it involves β, a constant of integration that is part of the solution, not the problem, and it lacks both x and x_0. We can fix it up by using the solution to get

$$S = \frac{1}{2t}(x - x_0)^2 \quad . \tag{6.11}$$

This gives the initial momentum,

$$p_0 = -\frac{\partial S}{\partial x_0} = \frac{x - x_0}{t} \quad : \quad x = x_0 + p_0 t$$

and it also gives p,

$$p = \frac{\partial S}{\partial x} = \frac{x - x_0}{t} \quad : \quad x = x_0 + pt : \quad p = p_0 \quad .$$

So given S cleared of undesired constants of integration, the problem is solved at once by differentiation.

S can be found in other ways. For example, for the free particle, we have $\partial S/\partial x = p = -\partial S/\partial x_0 = p_0 = (x - x_0)/t$. Integrating these equations gives the answer at once. To find it by integrating the Hamilton-Jacobi equation (6.2) requires a knowledge of nonlinear first-order partial differential equations that not every reader may have. It is done by the method of characteristics [55, Chap. 2] and leads via Hamilton's equations to a calculation equivalent to the one just given.

Problem 6.4. Verify that the S obtained above solves Equation (6.2.).

Problem 6.5. The Lagrangian corresponding to a particle under a uniform force is $L = \dot{x}^2/2 + ax$. Derive

$$S = \frac{1}{2t}(x - x_0)^2 + \frac{1}{2}a(x + x_0)t - \frac{1}{24}a^2 t^3 \tag{6.12}$$

and show that the complete solution follows as before.

Problem 6.6. Derive S for a harmonic oscillator ($m = 1$). It is

$$S = \frac{\omega}{2\sin\omega t}[(x^2 + x_0^2)\cos\omega t - 2xx_0] \quad . \tag{6.13}$$

Problem 6.7. The relation (6.9a) shows how the time variable is extracted from a knowledge of the characteristic function W. Section 5.2 developed a form of relativistic dynamics in which proper time is the independent variable. What is the characteristic function for this theory? What equation replaces (6.9a)? Earlier, in (5.85) we saw that t and H are conjugate variables. What quantity is conjugate to the proper time?

Problem 6.8. Figure 6.3 shows a pendulum made by wrapping a string around a cylinder. Find the period.

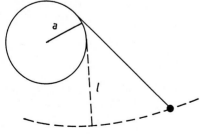

Fig. 6.3. To illustrate Problem 6.8

6.2 Step-by-Step Integration of the Hamilton-Jacobi Equation

The main computational strategy in dynamics is the use of constants of the motion in order to simplify calculations. We will see here how to use constant momenta and their conjugate cyclic coordinates. When E is constant, the integral expression for W is

$$
W = S + E(t - t_0) = \int_{t_0}^{t} (L + E)dt
$$

$$
= \int_{t_0}^{t} (p_n \dot{q}_n - H + E)dt \quad \text{or}
$$

$$
W = \int_{q_0}^{q} p_n dq_n \tag{6.14}
$$

along a natural path on which p and q both depend on t and H has a constant value. If now $q_1, \ldots, q_k$ are cyclic and $p_1, \ldots, p_k$ accordingly constant, we can carry out the corresponding integrations at once,

$$
W(q) = \sum_{n=1}^{k} (q_n - q_{n0})p_n + \int_{q_0}^{q} \sum_{n=k+1}^{f} p_n dq_n \quad .
$$

Call the integral $W_k(q)$. It is

$$
W_k = W - \sum_{n=1}^{k} (q_n - q_{n0})p_n \tag{6.15}
$$

and with (6.8) we have

$$
\delta W_k = \sum_{n=1}^{f} p_n \delta q_n + (t - t_0)\delta E - \sum_{n=1}^{k} p_n \delta q_n - \sum_{n=1}^{k} (q_n - q_{n0})\delta p_n
$$

$$
= \sum_{n=k+1}^{f} p_n \delta q_n + (t - t_0)\delta E - \sum_{n=1}^{k} (q_n - q_{n0})\delta p_n \quad .
$$

Clearly, W_k is a function of the energy and $p_1, \ldots p_k$, all constants, together with the variables $q_{k+1}, \ldots q_f$. It satisfies

$$
\frac{\partial W_k}{\partial p_n} = -(q_n - q_{n0}) \quad n = 1, \ldots k \tag{6.16a}
$$

$$
\frac{\partial W_k}{\partial q_n} = p_n \quad n = k + 1, \ldots f \tag{6.16b}
$$

$$
\frac{\partial W_k}{\partial E} = t - t_0 \quad . \tag{6.16c}
$$

Example 6.2. The Kepler Problem.
With $q_1 = \vartheta$, $q_2 = r$,

$$H = \frac{p_r^2}{2m} + \frac{L^2}{2mr^2} - \frac{\gamma}{r}, \qquad p_\vartheta = L = \text{const}$$

$$p_r = \frac{\partial W_1}{\partial r} = \sqrt{2m\left(E + \frac{\gamma}{r}\right) - \frac{L^2}{r^2}}$$

$$W_1 = \int \sqrt{2m\left(E + \frac{\gamma}{r}\right) - \frac{L^2}{r^2}}\, dr \quad .$$

Everything flows from this integral. To find the orbit, we have

$$\vartheta - \vartheta_0 = -\frac{\partial W_1}{\partial L} = L \int \frac{dr}{r^2\sqrt{2m(E + \gamma/r) - L^2/r^2}} \tag{6.17a}$$

$$t - t_0 = \frac{\partial W_1}{\partial E} = m \int \frac{dr}{\sqrt{2m(E + \gamma/r) - L^2/r^2}} \tag{6.17b}$$

[compare with (3.25) and (3.23)]. Both integrals can be found in the tables. The first does not involve t and is the equation for the orbit,

$$\vartheta - \vartheta_0 = \sin^{-1}\frac{r - L^2/m\gamma}{er} = \sin^{-1}\frac{r - a(1 - e^2)}{er}$$

with e as in (3.28), and this, when solved for r, gives the earlier result. The second equation gives

$$\Omega(t - t_0) = \sin^{-1}\frac{r - a}{ae} - \sqrt{e^2 - \left(\frac{r - a}{a}\right)^2}$$

where Ω is the average orbital angular velocity as in (3.35). These are useful if one wants to read the Solar System like a clock but less so if one wants to predict the positions of the planets. It cannot, of course, be solved to give $r(t)$ in closed form. At any rate, the earlier results are rederived in a few lines.

Example 6.3. Kepler Problem in Special Relativity. Here the relation between energy and momentum is taken from (5.27)

$$[E_T - V(r)]^2 = c^2 p^2 + (mc^2)^2 \tag{6.18}$$

where E_T includes the rest energy mc^2. For comparison with earlier formulas in which E did not include mc^2, write $E_T = E + mc^2$. (For planetary orbits E is again negative.) This gives

$$p^2 = 2m\left[E\left(1 + \frac{E}{2mc^2}\right) - \left(1 + \frac{E}{mc^2}\right)V(r)\right] + \frac{1}{c^2}V(r)^2 \quad . \tag{6.19a}$$

For comparison, the nonrelativistic version is

$$p^2 = 2m[E - V(r)] \tag{6.19b}$$

so that in special relativity it is as if the energy had been increased to

$$E_{sr} = E(1 + E/2mc^2) \tag{6.20}$$

and $V(r)$ had been changed to

$$V_{sr}(r) = \left(1 + \frac{E}{mc^2}\right) V(r) - \frac{1}{2mc^2} V(r)^2 \quad . \tag{6.21}$$

The rest of the work proceeds as in Example 6.2. With $V(r) = -\gamma/r$,

$$p_r^2 = 2m\left[E\left(1 + \frac{E}{2mc^2}\right) + \frac{\gamma'}{r}\right] - \frac{L^2 - \gamma^2/c^2}{r^2} \quad , \quad \gamma' = \left(1 + \frac{E}{mc^2}\right)\gamma \quad . \tag{6.22}$$

The orbit is

$$\vartheta - \vartheta_0 = L \int \frac{dr}{r^2 \sqrt{2m(E_{sr} + \gamma' r^{-1}) - (L^2 - \gamma^2/c^2)r^{-2}}} \quad .$$

The integral is no more difficult than that in (6.17a), and gives

$$r = \frac{(L^2 - \gamma^2 c^{-2})/\gamma(m + E_{sr}c^{-2})}{1 - e_{sr}\sin[\sqrt{1 - \gamma^2/L^2c^2}(\vartheta - \vartheta_0)]} \tag{6.23a}$$

where

$$e_{sr}^2 = 1 + \frac{E_{sr}(L^2 - \gamma^2 c^{-2})}{(m + Ec^{-2})^2\gamma^2} \quad . \tag{6.23b}$$

These are of the same form as (3.27) and (3.28), with only slight differences in the dependence of orbital and dynamical parameters, except for the occurrence of

$$\sin\left[\sqrt{1 - \frac{\gamma^2}{L^2c^2}}(\vartheta - \vartheta_0)\right] \simeq \sin\left[\left(1 - \frac{\gamma^2}{2L^2c^2}\right)(\vartheta - \vartheta_0)\right] \quad .$$

With this, r comes back to a previous value after one cycle when

$$\left(1 - \frac{\gamma^2}{2L^2c^2}\right)(\vartheta - \vartheta_0)$$

has increased by 2π. That is, the increase in ϑ is

$$\Delta\vartheta = 2\pi\left(1 + \frac{\gamma^2}{2L^2c^2}\right)$$

and the orbit precesses forward through an angle

$$\delta\vartheta_{sr} = \frac{\pi\gamma^2}{L^2c^2} \simeq \frac{2\pi}{1 - e^2}\frac{-E}{mc^2} \tag{6.24}$$

in each revolution. In planetary dynamics, with $E \simeq -GMm/2a$, this is

$$\delta\vartheta_{sr} \simeq \frac{\pi GM}{(1 - e^2)ac^2} \quad . \tag{6.25}$$

If the relativistic theory of gravity were put together in this way out of special relativity and Newton's law, this would be the precession of planetary perihelia. The true precession is 6 times as great.

Notes. Equation (6.21) for the effective potential in special relativity has a peculiar feature. Suppose that $V(r)$ corresponds to a repulsive force, say δ/r $(\delta > 0)$. Then for small r, V_{sr} exerts an attractive force. There are no stable bound states for such a potential. If $V(r)$ is less singular, however, there are such bound states, and it is possible that they play some role in physics.

In electrodynamics the electrostatic interaction is represented by the zeroth component of a four-vector. While this is surely correct, there does not seem to be any reason why all velocity-independent interactions of whatever origin must be so represented. There might be some force represented to a good approximation by a scalar quantity $U(r)$, which would occur in the formula

$$c^2 p^2 + [mc^2 + U(r)]^2 = (mc^2 + E)^2$$

where as before, E is the nonrelativistic part of the energy. This gives

$$p^2 = 2m \left[E + \frac{E^2}{2mc^2} - \left(U + \frac{U^2}{2mc^2} \right) \right] \quad . \tag{6.26}$$

If U is repulsive there is now no bound state, and even if it is attractive the effective potential has a repulsive core. Such an interaction is a logical possibility for physics, even if nature has not made use of it.

Problem 6.9. Using the method of Sect. 3.4, investigate the stability of bound orbits determined by the interaction V.

Problem 6.10. Derive (6.26).

Problem 6.11. Suppose the proposal was made that gravity is an attractive scalar interaction. How could one find out whether this is so?

Problem 6.12. Starting from (6.18), write down a relativistic wave equation, derive the energy levels of hydrogen, and exhibit the relativistic correction. Show that there are no stable $1s$ states when the nuclear charge $Z > 137/2$. (These results are at most of methodological interest because they ignore the effects of spin.)

6.3 Interlude on Planetary Motion in General Relativity

The first observable correction to Newtonian dynamics that Einstein calculated from the general theory of relativity was the precession of planetary perihelia. The calculation proceeds in a way roughly analogous to our earlier treatment of light in Sect. 1.6. There we set the speed of light equal to c by setting $ds^2 = 0$ and then determined the path by Fermat's principle. In (5.11) we saw that the variational principle for the motion of a free particle was written in terms of the invariant ds. In general relativity there is no gravitational force introduced as such; a particle in its orbit is considered to be moving freely in a curved space-time. The only invariant law that reduces to (5.11) in flat space-time is again

$$\delta \int ds = 0 \quad .$$

In Chaps. 1 and 4 the separation between two nearby points in the Minkowskian spacetime of special relativity was written as

$$ds^2 = \eta_{ik} dx^i dx^k$$

(summed over $i, j = 0, 1, 2, 3$) where η takes care of some negative signs and factors of c^2. In general relativity spacetime is curved but a similar formula holds,

$$ds^2 = g_{ik} dx^i dx^k$$

where as we have seen in Chap. 1, the g's contain the negative signs and factors of c^2 and also characterize the curvature. The variational principle must be supplemented as in Sect. 5.2 by an auxiliary condition to ensure that

$$g_{ij} u^i u^j = 1 \qquad (6.27)$$

where as before, $u^i := x'^i := dx^i/ds$. We write

$$\delta \int \mathcal{L} \, ds = 0 \quad , \quad \mathcal{L} = 1 + \mu(s)(g_{ij} u^i u^j - 1) \quad .$$

The Euler-Lagrange equations give as before

$$\mu g_{ij,k} u^i u^j - 2\mu g_{ik,j} u^i u^j - 2\mu g_{ik} u'^i - 2\mu' g_{ik} u^i = 0 \quad .$$

(The comma means differentiation with respect to x.) Multiply by u^k and sum:

$$\mu g_{ij,k} u^i u^j u^k - 2\mu g_{ik,j} u^i u^j u^k - 2\mu g_{ik} u^i u'^k - 2\mu' g_{ik} u^i u^k = 0$$

which is the same as

$$-\mu (g_{ik} u^i u^k)' - 2\mu' g_{ik} u^i u^k = 0 \quad .$$

In view of (6.27) this reduces to $\mu' = 0$, and so as before the Lagrangian multiplier is a constant. Its value is unimportant, but to make the later formulas look familiar we can take it to be $mc^2/2$. Dropping additive constants, we have

$$\mathcal{L} = \tfrac{1}{2} mc^2 g_{ij} u^i u^j \quad . \qquad (6.28)$$

This does not look like the earlier Lagrangian, which was essentially 1, but it has the great merit that a solution of equations of motion derived from it will always satisfy (6.27). The canonical momenta are given by

$$\mathfrak{p}_k = \partial \mathcal{L}/\partial u^k = mc^2 g_{kj} u^j$$

and to solve these linear equations for u^i we multiply by the matrix inverse of g_{kj}, which is called g^{ik}:

$$g^{ik} g_{kj} = \delta^i_j$$

where δ is the unit matrix. This gives $mc^2 u^i = g^{ik} \mathfrak{p}_k$, and we can now find the analogue of the Hamiltonian function

$$\mathfrak{H} := \mathfrak{p}_i u^i - \mathcal{L} = \frac{1}{2} mc^2 g_{ij} u^i u^j = \frac{1}{2mc^2} g^{ij} \mathfrak{p}_i \mathfrak{p}_j$$

of which the value is of course a constant, $mc^2/2$. Thus without having calculated anything we know the first integral of the equations of motion,

$$g^{ij}\mathfrak{p}_i\mathfrak{p}_j = m^2c^4 \quad . \tag{6.29}$$

This will lead quickly to a solution of the Kepler problem in general relativity.

The g_{ik} describing the spacetime geometry around the sun are given in one form by (1.56a), but these expressions lead to difficult integrals. We take advantage of the fact that there are many ways to cover a region with a coordinate system and introduce a change in the radial variable. Call the radial variable in (1.56a) r', and define a new one, r, by

$$r' := \left(1 + \frac{\lambda}{2r}\right)^2 r \quad , \qquad \lambda := GM/c^2$$

noting that as soon as $r \gg \lambda$, r' and r are very nearly the same. This transforms (1.56a) into a form that will be easier to handle. In the plane of the orbit we can set the angle ϑ equal to $\pi/2$ and write

$$ds^2 = g_{tt}dt^2 + g_{rr}dr^2 + g_{\varphi\varphi}d\varphi^2$$

and it turns out that

$$g_{tt} = \left(\tfrac{1-\lambda/2r}{1+\lambda/2r}\right)^2 \qquad\qquad g^{tt} = \left(\tfrac{1+\lambda/2r}{1-\lambda/2r}\right)^2$$

$$g_{rr} = -(1 - \lambda/2r)^4/c^2 \qquad g^{rr} = -(1 - \lambda/2r)^{-4}c^2$$

$$g_{\varphi\varphi} = -r^2 g_{rr} \qquad\qquad g^{\varphi\varphi} = -1/(r^2 g_{rr})$$

(the matrix inversion being trivial in this case). Since these contain only r, it follows from Hamilton's or Lagrange's equations that $\mathfrak{p}_t$ and $\mathfrak{p}_\varphi$ are constants. Equation (6.29) is now

$$g^{rr}\mathfrak{p}_r^2 + g^{\varphi\varphi}\mathfrak{p}_\varphi^2 + g^{tt}\mathfrak{p}_t^2 = m^2c^4$$

so that

$$\mathfrak{p}_r^2 = -g_{rr}(g^{\varphi\varphi}\mathfrak{p}_\varphi^2 + g^{tt}\mathfrak{p}_t^2 - m^2c^4) \quad . \tag{6.30}$$

To see that there is nothing strange here, consider the field-free case $\lambda = 0$:

$$\mathfrak{p}_r^2 = -\mathfrak{p}_\varphi^2/r^2 + \mathfrak{p}_t^2/c^2 - m^2c^2 \quad \text{or}$$

$$\mathfrak{p}_t^2 = c^2(\mathfrak{p}_r^2 + \mathfrak{p}_\varphi^2/r^2) + m^2c^4 \quad .$$

The reason for the factor $mc^2/2$ is now clear if it was not before, and once again we see the role of $\mathfrak{p}_t$ as an energy. Actually it is not an energy, for strictly speaking just as general relativity does not speak of force it does not speak of energy either, but we can think of it that way.

Returning to (6.30), write p_φ as $\mathcal{L}$ (this time an angular momentum^2) and p_t as $mc^2 + E$. Let mc^2 be of order 0, E of order 1, and λ/r also of order 1; then carefully expand the g's keeping terms up to order 2. The result is

$$p_r^2 = 2m\left[E\left(1 + \frac{E}{2mc^2}\right) + \left(1 + \frac{4E}{mc^2}\right)\frac{\gamma}{r}\right] - \frac{\mathcal{L}^2 - 6\gamma^2/c^2}{r^2} \quad . \tag{6.31}$$

This can be written in terms of a general relativity potential expressed in the style of (6.21),

$$V_{gr}(r) = \left(1 + \frac{4E}{mc^2}\right)V(r) - \frac{3}{mc^2}V(r)^2 \quad . \tag{6.32}$$

The only significant difference from (6.21) is in the last term, where the correction is 6 times as great as that of special relativity. The predicted perihelion precession is therefore

$$\delta\vartheta_{gr} = \frac{6\pi GM}{(1 - e^2)ac^2} \quad . \tag{6.33}$$

We have already given an estimate of this precession for a nearly circular orbit in (3.42).

The experimental situation is best for Mercury, for which a is small and $e \simeq 0.2$, a relatively large value. Here the predicted value is 43.03 seconds of arc per century. The observed value must be picked out from among a number of much larger effects. In fact, the total precession observed is about 5600″, contributed mostly by the precession of the equinoxes (Sect. 9.3) and the influence of other planets (Sect. 8.7). When Einstein's paper was published, there was a residual discrepancy between observation and Newtonian theory of about 41″,[3] and astronomers were concerned enough about it that they had considered modifying the law of gravity or inventing a small planet (to be called Vulcan) with an orbital radius smaller than Mercury's. The best "observed" value computed from astronomical data is 1.01 ± 0.02 times the theoretical value, computed on the assumption that the sun is a perfect sphere [58]. Since the sun rotates this is certainly not the case, but the amount of oblateness and its gravitational effect are not accurately known [57]. Until they are, a more exact verification of the theory of Mercury's orbit is not possible. The binary pulsar PSR 1913 + 13, though not observed telescopically, provides another test. The precession in this case is much larger, about 4° per year, but the orbital elements are not yet known well enough to provide an exact quantitative test [57].

Problem 6.13. Calculate the precession in another way, by finding how much the Laplace vector A (Sect. 3.5) rotates during one revolution of the orbit. Since A

---

[2] Note that the $\mathcal{L}$ introduced here is not the same as the earlier L, since this one is equal to

$$p_\varphi = mc^2 g_{\varphi\varphi}d\varphi/ds = mr^2(1 - \lambda/2r)^4 d\varphi/ds \quad .$$

The use of s as the time parameter in special relativity was a little strained; we had to imagine a particle carrying a watch, but here it is quite reasonable since almost all our chronometers are attached to the earth.

[3] The generally accepted value was that of *Simon Newcomb* [56], who gave $41.42″ \pm 2.09″$. The effect had first been established as 36″ by U.J.J. Leverrier in 1859.

is constant for an inverse-square force, it is only the added term in the potential, $V'(r) = -3\gamma^2/mc^2r^2$, that changes it. 1) Find $\dot{A}$ by evaluating the appropriate Poisson bracket. $\dot{A}$ is of course a function of the planet's position r. 2) Use the formulas of the appendix to Chap. 8 to calculate $\langle \dot{A} \rangle$, the time average of $\dot{A}$ over a single revolution. 3) The precession will be $\delta A = \langle \dot{A} \rangle T$, where T is the orbital period. Show that $\delta A \perp A$. 4) The angle of precession is then $\delta \vartheta = \delta A/A$.

Problem 6.14. Evaluate the planetary precession by a perturbation calculation starting from the integral for the characteristic function.

Solution. Consider a complete cycle of r, say from aphelion to aphelion. The integral for W over a complete cycle is called the action integral J; it will be further discussed in Chap. 8. The precession comes only from the last term in (6.31), so that

$$J = \oint \sqrt{2m\left(E + \frac{\gamma}{r}\right) - \frac{L^2}{r^2} + \frac{\varepsilon}{r^2}} \, dr \quad , \qquad \varepsilon = \frac{6\gamma^2}{c^2}$$

$$= \oint \sqrt{2m\left(E + \frac{\gamma}{r}\right) - \frac{L^2}{r^2}} \, dr + \frac{\varepsilon}{2} \oint \frac{dr}{r^2 \sqrt{2m(E + \gamma/r) - L^2/r^2}} + \dots \quad .$$

This is to be integrated between the two turning points, all around the orbit. Although the square root vanishes at the turning points and the integrand of the second integral is infinite there, the integral nonetheless converges. The integral can be done by complex methods or taken from tables. Integration over a complete cycle of r gives

$$J = -2\pi \left[\frac{m\gamma}{\sqrt{-2mE}} + L - \frac{\varepsilon}{2L} \right] + \dots$$

and the angle corresponding to the complete radial cycle is not 2π but

$$\vartheta = -\frac{\partial J}{\partial L} = 2\pi \left(1 + \frac{\varepsilon}{2L^2} \right) \quad .$$

Thus the precession is

$$\delta \vartheta = \frac{\pi \varepsilon}{L^2} \tag{6.34}$$

from which follows the previous result.

Problem 6.15. Show that if the perturbing energy is $-C/r^3$, the precession is

$$\delta \vartheta = \frac{6\pi C m^2 \gamma}{L^4} \quad . \tag{6.35}$$

(We will use this result later, in Problem 9.24.)

Fermat's Principle. A ray of light in classical optics follows the path of a photon, i.e., a particle of zero mass. Such a particle travels at the fixed speed c. This is why Fermat's principle of least (or stationary) time is possible, for if the speed were variable one could control the time by throwing the particle a little harder or less hard so as to change its time of flight at will. The next few lines will show that in general relativity as in classical optics, the path of a ray in a static gravitational field is determined by Fermat's principle.

We saw in Sect. 6.3 that orbits in general relativity are determined by the action principle

$$\delta A = 0 \quad , \quad A = \int \Lambda \, ds$$

$$\Lambda = \tfrac{1}{2} mc^2 g_{ij} u^i u^j \quad , \quad u^i := dx^i/ds$$

where s is the proper time. It will be convenient to introduce a new parameter τ defined by

$$ds := mc^2 d\tau \quad , \quad \mathfrak{v} := dx^i/d\tau \tag{6.36}$$

then A becomes

$$A = \int \frac{1}{2} g_{ij} \mathfrak{v}^i \mathfrak{v}^j \, d\tau$$

and m appears only in the scale of proper time. As we shall see, the limit $m \to 0$ produces no problems.

For our purposes, a static gravitational field can be defined as one in which $g_{0i} = 0$ ($i = 1, 2, 3$) and all the g_{ij} are independent of the time x^0. Vary A by varying the dependence of x^0 on τ:

$$\delta A = \int_{\tau_0}^{\tau_1} g_{00} \frac{dx^0}{d\tau} \frac{d\delta x^0}{d\tau} d\tau \quad .$$

Partial integration gives

$$\delta A = - \int_{\tau_0}^{\tau_1} \frac{d}{d\tau} \left(g_{00} \frac{dx^0}{d\tau} \right) \delta x^0 \, d\tau + \left[g_{00} \frac{dx^0}{d\tau} \delta x^0 \right]_{\tau_0}^{\tau_1} \quad . \tag{6.37}$$

First, let δx^0 vanish at the ends of the interval but be arbitrary otherwise and let $\delta A = 0$: these are conditions that define the action principle. It follows that

$$\frac{d}{d\tau} \left(g_{00} \frac{dx^0}{d\tau} \right) = 0 \quad , \quad g_{00} \frac{dx^0}{d\tau} = \text{const} \quad .$$

Now compare two slightly different paths between the same two spatial points, both starting at the same time, while allowing x^0 to be different if it wants to. The remaining term in (6.37) gives

$$\delta A = \text{const} \, \delta x^0(\tau_1) \quad .$$

Now let $m \to 0$. Clearly $A \to 0$ also, and therefore in this case $\delta A = 0$. It follows that $\delta x^0(\tau_1) = 0$, telling us that the two slightly different rays will nevertheless both arrive at the same moment. That is, the time is stationary, and this is Fermat's principle.

Finally, in (6.36), note the fact that helps to define the path of a ray of light: no matter what $d\tau$ may be, ds always vanishes, so that the total proper time along the trajectory is zero. An infinite variety of such paths is possible between any two points of spacetime. The principle of stationary action or time picks out the physical one. This is what was done in Sect. 1.6.

Problem 6.16. Work out the deflection of light in the following way. Write the Schwarzschild line element (1.54a,b) as

$$ds^2 \approx \left(1 + \frac{\alpha}{r}\right) dr^2 + r^2 d\vartheta^2 - \left(1 - \frac{\alpha}{r}\right) c^2 dt^2$$

$$= dx^2 + dy^2 + \frac{\alpha}{r} dr^2 - \left(1 - \frac{\alpha}{r}\right) c^2 dt^2$$

where (Fig. 1.11)

$$dy^2 \approx 0 \quad , \quad r^2 \approx x^2 + a^2 \quad , \quad dr \approx \frac{x}{r} dx$$

so that

$$ds^2 \approx \left(1 + \frac{\alpha x^2}{r^3}\right) dx^2 - \left(1 - \frac{\alpha}{r}\right) c^2 dt^2 \quad .$$

Setting $ds = 0$ gives

$$\frac{dx}{dt} \approx c\left(1 - \frac{\alpha}{2r} - \frac{\alpha x^2}{2r^3}\right)$$

where the first term comes from the "curvature of time" and the second from the "curvature of space". Finish the calculation.

Problem 6.17. In Sect. 2.13 the Maupertuis form of the action principle allowed the endpoint to vary in time. Can that treatment be adapted to calculations in general relativity? Does it lead to Fermat's principle when applied to particles with mass approaching zero?

6.4 Jacobi's Generalization

Section 6.2 leaves the method pretty much where Hamilton left it in 1834. But in solving a partial differential equation there are other ways to introduce constants of integration than as initial values. We need a more general approach that will enable us to use constants of integration however they occur. This was provided by Jacobi in his lectures in the winter of 1842–1843 in Königsberg.

The point is to generate as many constants of the motion as possible.[4] Suppose one can find a contact transformation that transforms p, q into new coordinates and momenta and at the same time transforms H into a number E independent of the new coordinates. Then all the new momenta will be constants and the new coordinates will depend on time in a specially simple way. The trick, of course, is to find the transformation.

Following the pattern of F_2 in (4.34), we wish to find a transformation function which we shall call $-W(q, \alpha)$ that will transform the old p's into new constant α's and the old q's into new coordinates φ with properties to be determined. H is to be

---

[4] It is assumed that $\partial H/\partial t = 0$, since this case is of principal interest to physicists. For an excellent discussion of the general case as well as the one presented here, see *Goldstein* [38], Chap. 10.

transformed to a new E, equal to H but depending only on the constant α's and not on the φ's. The transformation equations are

$$p_n = \frac{\partial W(q, \alpha)}{\partial q_n} \quad , \quad \varphi_n = \frac{\partial W(q, \alpha)}{\partial \alpha_n}, \quad (\varphi_m, \alpha_n) = \delta_{mn} \qquad (6.38\text{a,b,c})$$

and the equations of motion for the new canonical variables are

$$\dot{\alpha}_n = -\frac{\partial E(\alpha)}{\partial \varphi_n} = 0 \quad , \quad \dot{\varphi}_n = \frac{\partial E(\alpha)}{\partial \alpha_n} := \omega_n(\alpha) \quad (\text{const}) \quad . \qquad (6.39)$$

Both equations of (6.39) can be integrated at once: the first states once again that the α's are constant and the second gives

$$\varphi_n = \omega_n t + \beta_n \quad , \quad (\beta_m, \alpha_n) = \delta_{mn} \quad . \qquad (6.40)$$

The constants α and β are known as the first and second integrals of the equations of motion respectively.

What has been done here is very remarkable. A system with f degrees of freedom satisfies f equations of motion that are of second order in t. Solving these equations requires $2f$ integrations, each one producing a constant of motion. Recall how we solved the Kepler problem the first time, performing one integration after another, finding one constant after another. Only one was really trivial: since the Hamiltonian was independent of t, one of the constants was t_0, depending on the moment the clock was started. The rest all referred to conserved quantities and initial conditions. Jacobi's program says: Perform only half the integrations so as to find the α's, and (6.40) gives you the other half.

The α's and the β's. The difference between first and second integrals can be illustrated by the example of a free particle. Its Hamiltonian is $p^2/2m$ and the momentum is constant. This is a first integral. The solved equation of motion gives $x = x_0 + pt/m$, from which the second integral is $x_0 = x - pt/m$. This too is constant, but it is recognized as a second integral because it depends explicitly on t.

The role of the α's is a familiar one. Consider an infinitesimal transformation generated by $\varepsilon_n \alpha_n$: this quantity is constant and so it "commutes" with the Hamiltonian, and therefore the α's generate transformation of p's and q's that leave the Hamiltonian invariant. To see what the β's do, start with

$$\beta_n[q(t), \alpha, t] = \frac{\partial W[q(t), \alpha]}{\partial \alpha_n} - \omega_n(\alpha)t \quad .$$

(In spite of the notation, β_n is constant!) The infinitesimal transformation generated by $\varepsilon_n \beta_n$ gives

$$\delta q_m = \frac{\partial}{\partial p_n} \varepsilon_n \beta_n = 0 \quad \text{and}$$

$$\delta p_m(\alpha, q) = -\frac{\partial}{\partial q_n} \varepsilon_n \beta_n = -\varepsilon_n \frac{\partial^2 W}{\partial \alpha_n \partial q_m} = -\varepsilon_n \frac{\partial p_m}{\partial \alpha_n} \quad .$$

But this is also equal to

$$\delta p_m = \frac{\partial p_m}{\partial \alpha_n} \delta \alpha_n + \frac{\partial p_m}{\partial q_n} \delta q_n$$

158

and since $\delta q_n = 0$, comparison of the last two equations gives

$$\delta \alpha_n = -\varepsilon_n \quad .$$

The α's are constants of motion, their values depending on the initial conditions. The β's change the corresponding α's, and therefore generate changes in the initial conditions.

Completeness and Solubility. There is a point that is usually unimportant but should be mentioned. A point in (p, q) space representing the state of a system is determined by f coordinates and f momenta, and it is assumed that these can be varied independently. Suppose a series of transformations were to produce a momentum p_1 and a second momentum p_2 equal to p_1^2. These both describe the system but they are not independent. A complete solution requires independence, and since the momenta are determined from W, we must express the criterion as a property of W assuring that there is no relation between them of the form $\Phi(p_1, \ldots, p_f) = 0$ that does not involve the α's. If Φ existed, there would be a relation of the form

$$\frac{\partial \Phi}{\partial p_n} \frac{\partial p_n}{\partial \alpha_m} = M_{mn} \frac{\partial \Phi}{\partial p_n} = 0 \quad , \quad M_{mn} := \frac{\partial^2 W}{\partial \alpha_m \partial q_n}$$

and to ensure the nonexistence of such a Φ it is enough to require that the determinant $|M| \neq 0$, that is, that M have an inverse. This condition has further implications for the solubility of the relations connecting the new and old coordinates and momenta, for according to (6.38a) if we vary the α's the p's will change according to

$$\delta p_n = \frac{\partial^2 W}{\partial q_n \partial \alpha_m} \delta \alpha_m = \delta \alpha_m M_{mn}$$

and similarly,

$$\delta \phi_m = M_{mn} \delta q_n \quad .$$

The existence of M^{-1} ensures that these relations are soluble for α and q, that is, that the pairs of quantities uniquely determine each other.

As long as a solution of (6.39) is complete, it makes no difference how the constants α arise and we will see how this procedure is used for computation in Sect. 6.8. By choosing just one of these constants appropriately, however, the procedure can be simplified even further.

A Special Choice of α's. A simple way in which the above conditions can be fulfilled is to choose one of the constants α, say α_1, to be the energy,

$$E(\alpha) = \alpha_1 \quad . \tag{6.41}$$

Then (6.39) gives

$$\omega_1 = 1 \quad , \quad \omega_2 = \omega_3 = \ldots = 0$$

and with $\beta_1 = -t_0$,

$$\varphi_1 = t - t_0 \quad , \quad \varphi_n = \beta_n \quad , \quad n = 2, 3, \ldots f \tag{6.42}$$

159

so that all the coordinates except the first one, like all the momenta, are constants. Now how can one find the transformation function W that does all this? Start from $H(p, q)$ and use (6.38) and (6.41) to get

$$H\left(\frac{\partial W}{\partial q}, q\right) = \alpha_1 \quad . \tag{6.43}$$

This is again the time-independent Hamilton-Jacobi equation for W, but in a more general context, since the constants of integration in the solution are not merely the initial values of the coordinates but any constants of integration however they occur. So far, we have encountered two ways in which constants can occur in W: as conserved quantities and as initial values. In Sect. 6.8 we will encounter a third way: as the constants which arise in the separation of variables of a partial differential equation. However they arise, the constants are $f + 1$ in number: α_1 plus f others that come from integrating the equation. But since W occurs only differentiated in the equation, one of the constants is a trivial additive one that carries no information about dynamics. If we agree to ignore this one, a complete solution depends on f constants $\alpha_1, \ldots \alpha_f$.

Once W has been found, use (6.38) and (6.42) to see that

$$\frac{\partial W(q, \alpha)}{\partial \alpha_1} = t - t_0 \quad , \quad \frac{\partial W(q, \alpha)}{\partial \alpha_n} = \beta_n \quad n = 2, 3, \ldots f \quad . \tag{6.44a,b}$$

The first is already familiar from (6.6). The equations for $n = 2, \ldots f$ do not involve time and describe the shape of the orbits when they are solved to give

$$q = q(\alpha, \beta) \quad . \tag{6.45}$$

Time is introduced only by (6.44a), which is in general a single relation between q's, α's and $t - t_0$. Solving the whole set of equations for $q(\alpha, \beta, t - t_0)$ may be difficult, but it is in principle possible if the α's are the f independent constants of integration that belong to a complete solution of the f equations (6.44).

Integral Surfaces. Let us look at the integration procedure. Suppose for simplicity that there are three degrees of freedom. There will be three constants, α_1 plus two more non-additive ones that arise from the integration. The two equations

$$\frac{\partial W(q, \alpha)}{\partial \alpha_2} = \beta_2 \quad , \quad \frac{\partial W(q, \alpha)}{\partial \alpha_3} = \beta_3 \tag{6.46}$$

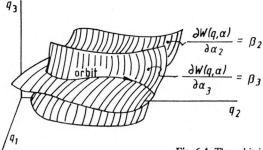

Fig. 6.4. The orbit is the intersection of two integral surfaces

each define an algebraic relation between q_1, q_2, and q_3; in the space whose co-ordinates are q_1, q_2, q_3 they define two surfaces. These surfaces intersect in a line which is the system's orbit Fig. 6.4. Finally, the last equation, (6.44a), defines a third surface which moves as time passes and intersects the orbital line in a moving point. This gives the configuration of the system at time t.

Problem 6.18. Construct the three surfaces mentioned above for a particle moving freely in space.

Problem 6.19. Construct the three surfaces for the path of a projectile in a uniform gravitational field.

6.5 Orbits and Integrals

If one wants to know just what a system will do after having been set in motion in a certain way, there is nothing to do except to solve the equations. But there are other kinds of questions that do not require everything to be solved: is the orbit closed or does it fill a certain region, and in the latter case what region does it fill? What is the character of orbits in phase space? Are there any regions of phase space or configuration space into which a system is forbidden to move? Questions of this kind that require generalizations about classes of solutions are often of more interest than the solutions themselves, and they are sometimes relatively easy to ask and answer in terms of the integral surfaces.

The following paragraphs will illustrate the kind of arguments that can be based on the construction of integral surfaces without making any attempt to solve the Hamilton-Jacobi equations. Because the considerations to follow are geometrical in nature and we are going to illustrate them with pictures, we are obliged to treat systems with only a few degrees of freedom. The methods, however, are general.

Motion Under a Central Force. Suppose we wish to study the general properties of motion under an attractive central potential $V(r)$. The particle's position and motion are given by a point in the phase space whose coordinates are, for example, r, ϑ, p_r, and p_ϑ. Since this is too many for convenience, we ignore the angle ϑ by considering that three-dimensional slice of the four-dimensional space which corresponds to $\vartheta = 0$. In addition, for central forces, p_ϑ is a constant. To see what this says about the orbit, write

$$E = \tfrac{1}{2}(p_r^2 + p_\vartheta^2/r^2) + V(r) \quad \text{and} \tag{6.47}$$

$$p_\vartheta = r\sqrt{2E - 2V(r) - p_r^2} \quad . \tag{6.48}$$

In the space whose coordinates are r, p_r, and p_ϑ this defines a surface; a typical one for $V(r) = r^2/2$ is shown in Fig. 6.5a. The condition $p_\vartheta = \text{const}$ defines another such surface, a plane. The line along which the surfaces intersect, Fig. 6.5b, defines the values of r and p_r compatible with the given values of E and p_ϑ.

If we now represent the system in the space r, p_r with p_ϑ fixed and $\vartheta = 0$, we arrive at the kind of orbital representation shown in Fig. 6.6, in which each

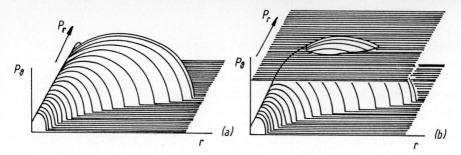

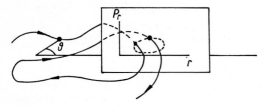

Fig. 6.5. (a) Integral surface corresponding to (6.48). **(b)** Intersection of the integral surface with the plane p_ϑ = const (Only the half of the figure with $p_\vartheta > 0$ is shown)

Fig. 6.6. In this orbit in r, ϑ, p space the locus of points at which the path pierces the plane $\vartheta = 0$ is a closed curve

successive revolution of the particle brings it through the plane $\vartheta = 0$ at a certain point, and these points lie on the contour determined by the construction of Fig. 6.5. We can now distinguish between orbits that are closed and those that are not.

Two-Dimensional Isotropic Harmonic Oscillator. The Hamiltonian of this system is written in polar coordinates (neglecting unimportant constants) as

$$H = \tfrac{1}{2}(p_r^2 + L^2/r^2) + \tfrac{1}{2}r^2 = E \qquad (6.49)$$

with p_ϑ equal to a constant, L. A typical contour in r, p_r for this system is shown in Fig. 6.7a. But there are other integrals of this system (Sect. 5.6). One of them is

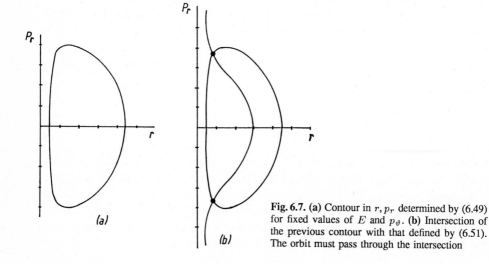

Fig. 6.7. (a) Contour in r, p_r determined by (6.49) for fixed values of E and p_ϑ. **(b)** Intersection of the previous contour with that defined by (6.51). The orbit must pass through the intersection

162

$$B = \tfrac{1}{2}(p_x^2 - p_y^2 + x^2 - y^2)$$

which in polar coordinates is

$$B = \frac{1}{2}\left(p_r^2 + r^2 - \frac{p_\vartheta^2}{r^2}\right)\cos 2\vartheta - \frac{p_r p_\vartheta}{2r^2}\sin 2\vartheta \quad . \tag{6.50}$$

Assume that $B \geq 0$ and look at the plane defined by $\vartheta = 0$. (If $B < 0$, take $\vartheta = \pi/2$.) Then the path in (r, p_r) must satisfy

$$\tfrac{1}{2}(p_r^2 - L^2/r^2) + \tfrac{1}{2}r^2 = B \tag{6.51}$$

at the same time as (6.49). Figure 6.7b shows a pair of typical curves. The allowed values of r and p_r occur where these intersect; the two points correspond to opposite directions around the orbit. There is one of these diagrams for every value of ϑ; thus for each ϑ there is only one possible r and one $|p_r|$, and we now see graphically how the existence of the second integral forces the orbit to be closed.

Problem 6.20. What role does the third integral (5.89) play in fixing the orbit just discussed?

Poincaré Tubes. If an orbit is not closed but angular momentum is conserved, we can still draw conclusions from the r, p_r plot, for each time the path pierces the r, p_r plane with $\vartheta = 0$ it pierces it at a point lying on the closed contour which, as in Fig. 6.5, is determined by the given constant values of E and L. Suppose we change E, leaving L fixed. What will the new contour look like? Only under very special circumstances can any two contours intersect, since if they did there would be two different orbits which at a certain moment had the same values of r, $\dot{r}$, L, and ϑ. With r and L the same, $\dot{\vartheta}$ would be the same and the two orbits would have to coincide throughout. The only way in which two contours can cross is if exactly the same initial conditions at some moment can give rise to two different subsequent motions. This is the situation of a pencil balanced on its point. Contours that do not contain such points of instability cannot intersect. Figure 6.8a shows contours corresponding to three values of E for a central potential. If we continue to decrease E, the contour will narrow to a point. The orbit is now a circle, and no smaller E is possible.

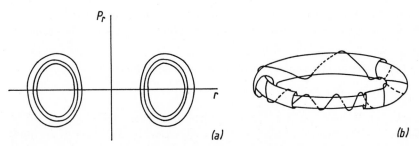

Fig. 6.8. (a) Contours in r, p_r corresponding to different values of E for an attractive central potential. (b) Poincaré tubes whose cross sections are the contours shown in (a)

There is a geometrical construction given by Poincaré that enables the system's dynamical behavior to be visualized: return to Fig. 6.6 and complete the orbit by including the angle ϑ (Fig. 6.8b).

The contour of Fig. 6.8a is now seen as a cross section of a tube in the space r, p_r, ϑ, and the particle's orbit in this space is a path that spirals around the tube but never leaves it. Further, except for the exceptional points at which two orbits intersect, the tubes never touch each other; in this diagram, in which p_ϑ is fixed, the tubes corresponding to smaller E lie entirely inside those corresponding to larger E. The surface illustrated in Fig. 6.8a is known as a *surface of section*.

An Angle-Dependent Potential. The next case is one in which the orbit is not closed. There is no integral corresponding to B above and because ϑ is not a cyclic variable the angular momentum is no longer conserved. The Hamiltonian is

$$H = \frac{1}{2}\left(p_r^2 + \frac{p_\vartheta^2}{r^2}\right) + \frac{1}{2}r^2 + \frac{1}{3}r^3 \cos 3\vartheta \quad . \tag{6.52}$$

The form is chosen here because it has received considerable discussion since it was first studied numerically by *Hénon* and *Heiles* [59] many years ago. The Hamilton-Jacobi equation cannot be solved exactly because it is impossible to separate the variables (Sect. 6.6), and so one must use a computer. See also [60], [61].

The behavior of this simple system shows considerable variety, depending on the value of the energy. Since the potential is not symmetric between $\vartheta = 0$ and $\vartheta = \pi$, set $x = r \sin \vartheta$, $y = r \cos \vartheta$ and plot y and p_y instead of r and p_r, getting a contour corresponding to $\vartheta = \pi$ (negative y) at the same as that corresponding to $\vartheta = 0$ (positive y). Figure 6.9a shows 110 successive passages through the plane. Here again, the points lie on contours. Clearly there is an integral of the motion that is responsible for the contour in the same way as L was in the situations in which it is conserved. In this case the exact analytic form of the conserved quantity is unknown and one must laboriously construct it by approximation [62]. For comparison, Fig. 6.9b shows the corresponding motion when the angular dependence in (6.52) is omitted and the potential is $r^2/2 + r^3/3$, the points lying on a contour determined by the angular momentum integral.

But the system has an unexpected feature. Figure 6.10 represents the same system started at the same point in the y, p_y plane with a slightly greater energy. There is little sign of the effect of an integral here. What has happened is that parts of the integral surfaces are complex in form, with many folds and self-intersections. This produces a strong instability in the observed motion, amplified by the effect of round-off errors in the computer. This will be further explained in the following section.

Problem 6.21. The Hénon-Heiles potential can also be written

$$H_{\mathrm{HH}} = \tfrac{1}{2}(p_x^2 + p_y^2 + x^2 + y^2) + \tfrac{1}{3}y^3 - x^2 y \quad . \tag{6.53}$$

Write a program to make plots in the x, p_x plane like those of Figs. 6.9 and 6.10. To do this let $y = 0$ and let the value of p_y be determined by the value of E. Study the transition from regular to chaotic behavior as a function of E and also of the point in the plane at which motion begins.

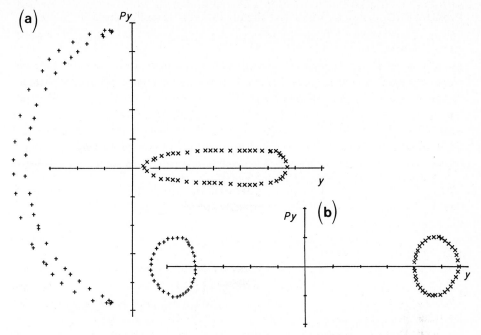

Fig. 6.9. (a) Computed points in a y, p_y plot using the potential (6.52). (b) Same as (a) omitting the factor of $\cos 3\vartheta$ in the potential. $\times$: emerging from the plane; +: entering plane

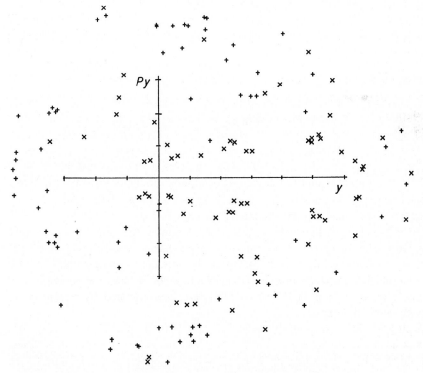

Fig. 6.10. The same plot as in Fig. 6.9a but at a slightly higher energy

165

Problem 6.22. Consider a new and slightly different Hamiltonian [63]

$$H_{CTW} = \tfrac{1}{2}(p_x^2 + p_y^2 + x^2 + y^2) + 2x^3 + xy^2 \quad .$$

Modify the previous program so as to study the x, p_x plane with $y = 0$ and perform some analogous numerical experiments. In what way do the results differ? Unlike H_{HH}, H_{CTW} admits an algebraic integral,

$$I = 4x^2y^2 + y^4 - 4p_y(xp_y - yp_x) + 4xy^2 + 3(y^2 + p_y^2) \quad .$$

How does this new fact relate to the plots your have obtained? Plot some contours of constant I in the x, p_x plane with $y = 0$ and compare them with your results. It may help if you can show that in this plane,

$$p_x^2 = 2E + x^2 + 4x^3 - I/(3 - 4x) \quad .$$

6.6 "Chaos"

Clearly the situation is complicated, but one should not conclude that the Hénon-Heiles model represents some profound pathology. Rather, it is typical of what happens in nonseparable systems. The same phenomena are found in the analysis of systems as clear-cut as the three-body problem and the motion of a star with respect to the rotating mass of a galaxy [59], [64]. The following remarks will be largely qualitative.

The construction of orbits by integral surfaces is straightforward, but it depends on establishing the existence of the integral surfaces. Such a surface is defined by a relation of the form

$$I(p, q, \mu) = 0 \quad , \tag{6.54}$$

where μ stands for one or more constants. A Hamiltonian system is one described by Hamilton's equations with or without an externally applied force, but I will consider only isolated ("autonomous") systems. For such a system the energy is such an integral, and as illustrated above, if the variables are separable the constants of separation are others (see also Sect. 6.9). If the variables are not separable there are of course still integrals: Imagine the equations of motion miraculously solved in terms of a set of initial conditions to give $p(p_0, q_0, t)$ and $q(p_0, q_0, t)$; then miraculously eliminate t between these to give relations of the form (6.54) where μ involves p_0 and q_0. But is this really possible, even in principle? A famous theorem due to Poincaré says no: Except for the energy, a nonseparable system has no integrals that are single-valued and analytic in p and q. Nonseparability is the same as nonintegrability. This fact is of crucial importance in celestial mechanics, for it turns out to imply that perturbation theory (Chap. 8), in which orbits are approximated by expansion in small parameters, diverges.[5]

---

[5] For a straightforward proof of Poincaré's theorem see [64a], which also suggests an escape from its conclusions: there may be integrals that are not analytic. This possibility is illustrated by the KAM theorem, discussed below.

An example of the kind of integral Poincaré was talking about occurs in the separable system of Problem 5.57, where an integral C was defined that cannot be called $C(x, y, p_x, p_y)$ because it is multi-valued and so one cannot know its value without following the system's entire history.

What happens to the integral surfaces if one tries to approximate a calculation by beginning with a separable system and then introducing the nonseparable part of the Hamiltonian as a perturbation? The use of straightforward perturbation theory soon leads to divergent results that cannot be cured by the methods explained in Chap. 8. And yet nature integrates these equations, for a mechanical model of one of these systems would not explode. The mathematical problem is very difficult, but an important theorem by Kolmogorov, Arnold, and Moser, known as the KAM theorem ([65], [66]) explains in a qualitative way how it is possible for the appearance of an integral surface to arise as in Fig. 6.9a, even though the system is not integrable. It states that if the perturbation is small enough the surfaces are not entirely destroyed. The proof requires the perturbation to be exceedingly small, much smaller than in the situation of Fig. 6.9a. Evidently the theorem is of wider validity than the proof, but when the perturbation is large enough (Problem 6.21) the single-valued, analytic integrals no longer exist and "chaos" begins.

To get some idea of what is happening in situations commonly described as chaotic it is useful to make an artificial separation that divides the inquiry into two questions: How much of the apparent disorder in Fig. 6.10 is inherent in the dynamics of this particular system and how much in nonintegrable systems in general? Let us look at the dynamics first.

Singularities in Phase Space. Figure 5.1 illustrated the phase space of a harmonic oscillator. It was only so simple because the oscillator is a confined system: the potential energy rises toward infinity on each side. A pendulum, still simple, is less idealized, and Fig. 6.11 shows orbits in its phase space. An *elliptic point, O*, is a point where the pendulum is stable and at rest. A *hyperbolic point*, X, represents the pendulum in a state of perpetual vertical balance. The phase orbits cross at that point since the pendulum could fall either way, but orbits in the immediate neighborhood

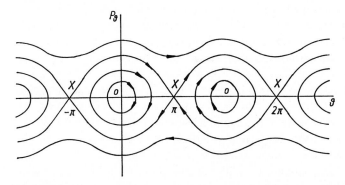

Fig. 6.11. Orbit in phase space of a pendulum, to be compared with Fig. 5.1. In the inner regions it oscillates; in the outer ones it goes round and round. In between are orbits in which, near an X, it comes arbitrarily close to standing vertically

167

of that point do not cross. Note that these *fixed points* represent singularities in the pendulum's time behavior that are not evident in the phase space plot: the closer a phase point comes to an O or an X, the longer it stays there. The succession of O's and X's in the phase plane is characteristic of periodic systems in which there are stable motions corresponding to the O's: one gets from one O to another by way of an X, and examination of the arrows in Fig. 6.11 shows why. Fixed points occur only for special values of the initial conditions: choose them at random and you will not see one.

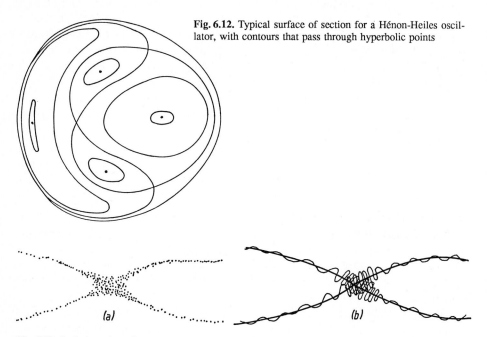

Fig. 6.12. Typical surface of section for a Hénon-Heiles oscillator, with contours that pass through hyperbolic points

Fig. 6.13a,b. Intersection of two contours at a hyperbolic point: (a) as seen in a computer display, (b) as calculated by topological techniques

Figure 6.12 shows some computed orbits in the surface of section for the Hénon-Heiles oscillator. Here also there are O's and X's. At the center of each O is a closed, stable orbit. At each X, if one studies the orbit in detail, one finds that as with the pendulum, the orbit has an unstable equilibrium: if one is lucky enough to get into an orbit that passes close to an X the successive orbits will spend a long time in its vicinity. The remarkable thing is that if one examines closely the annular region between two O's in such a diagram one finds a succession of O's and X's, and within each of those O's the same thing again, ad infinitum. Further, the X's may become very complicated in form and cross at an infinity of points rather than just one. Figure 6.13(a) shows how a typical intersection of two contours at a hyperbolic point looks if one asks the computer for high magnification, while (b) shows what happens theoretically at the intersection. The contours are double, not single, and they intersect more often as the point is approached. In principle this tangle of points is analyzable and the position of each dot in (a) is predictable. In practice, it is not, and *Arnold* [67, App. 8] has shown that the existence of such a structure can enable orbits to escape from the Poincaré tubes that would otherwise have contained them.

No amount of reading and figuring will teach you as much about nonseparable systems as playing with them on a computer. Start with Hénon-Heiles, examine regions of the diagram under different degrees of magnification, then invent your own.

Area-Preserving Mappings. If as in Fig. 6.6 an orbit pierces the surface of section at point P_1 and then goes around and pierces it again at P_2, the position of P_2 is exactly determined by that of P_1 (Problem 6.22). Thus one can think of a complete circuit (or half-circuit as in Fig. 6.9a) as a process that maps P_1 onto its image P_2. The details of the mapping depend on the nature of the system and they may be very complicated, but there is a fact that is easy to prove: these mappings preserve areas. That is, if instead of a point we map a small area, the image of this area remains constant even though the shape may change greatly. The proof needs only the fact that the evolution of a system is given by a canonical transformation (Sect. 6.1), for the preservation of areas is a property of canonical transformations in general.

The mapping of point P_1 in the i-plane with coordinates p_i and q_i onto P_2 with coordinates p_i' and q_i' can be written in any of the standard forms; say that it is generated by an $F_2(x, y, p_x', p_y')$ with

$$x' = -\partial F_2/\partial p_x' \quad , \quad p_x = -\partial F_2/\partial x \tag{6.55}$$

and similarly for the y components. When variables such as x and p_x are transformed to new variables x' and p_x' the element of area becomes $dx' dp_x' = J\, dx\, dp_x$ where J is the Jacobian of the transformation, and areas are preserved if $J = 1$. It is easy to prove that this is so. The Jacobian is

$$J = \begin{vmatrix} \partial p_x'/\partial p_x & \partial p_x'/\partial x \\ \partial x'/\partial p_x & \partial x'/\partial x \end{vmatrix} \tag{6.56}$$

in which, since x and p_x' are independently varying arguments of F_2, $\partial p_x'/\partial x = 0$. In the remaining terms, from (6.55),

$$\partial p_x/\partial p_x' = -\partial^2 F_2/\partial x \partial p_x' \quad , \quad \partial x'/\partial x = -\partial^2 F_2/\partial p_x' \partial x$$

and therefore if one is persuaded that $\partial p_x'/\partial p_x = (\partial p_x/\partial p_x')^{-1}$ it follows that $J = 1$. (Ordinary derivatives can be inverted in this way but with partial derivatives one needs to think, see Problem 6.25). An immediate consequence is that the two closed figures in Fig. 6.9a have equal areas. The value of this result is that it applies to all Hamiltonian systems, including those that are not integrable.

Stability. A striking feature of nonintegrable systems is the emergence, at the right initial conditions, of orbits whose form depends very sensitively on the initial conditions. This instability is explained by the complexity of the integral surfaces, and it has two effects. First, two phase points initially very close together are, after a few complete orbits, imaged very far apart, and second, the round-off errors of a computer are soon so much magnified that the positions of the chaotically distributed dots has little numerical significance, though the qualitative features of the patterns are not generally affected. The criterion for stability is similar to that discussed in Sect. 3.4 but the argument is more subtle.

Instead of a dynamical system I will discuss an area-preserving mapping that *Chirikov* [68] has called the Standard Model, since its qualitative features imitate those of certain dynamical systems. The mapping is

$$x_{n+1} = x_n + y_{n+1}$$
$$y_{n+1} = y_n + K \sin x_n \qquad (6.57)$$

where one can think of x and y as angles and calculate them mod 2π. A simple computer study (please do it yourself) reveals islands of stability surrounded by chaotic regions which become increasingly dominant when K exceeds 2.

Suppose two points start very close together, separated by coordinate intervals δx_0 and δy_0. It is then easy to see that

$$\delta x_{n+1} = (1 + k_n)\delta x_n + \delta y_n$$
$$\delta y_{n+1} = k_n \delta x_n + \delta y_n \quad , \quad k_n := K \cos x_n \quad .$$

This is a linear mapping, but since k_n is different every time it is hard to iterate it to find out how the separation between the two points changes as n increases. First, what happens if k_n is taken to be a constant, k? Express the mapping in matrix form as

$$\delta \boldsymbol{x}_{n+1} = M \delta \boldsymbol{x}_n \quad , \quad M = \begin{pmatrix} 1+k & 1 \\ k & 1 \end{pmatrix}$$

from which $\delta \boldsymbol{x}_n = M^n \delta \boldsymbol{x}_0$. To find M^n, note that

$$M^2 = (2+k)M - 1 \qquad (6.58)$$

so that any power of M can be written in the form

$$M^n = A_n M + B_n \quad . \qquad (6.59)$$

Multiplying both sides by M and using (6.58) leads at once to a pair of linear recurrence relations with constant coefficients,

$$A_{n+1} = (2+k)A_n + B_n \quad , \quad B_{n+1} = -A_n \qquad (6.60)$$

which are solved by setting $A_n = \alpha \mu^n$, $B_n = \beta \mu^n$. There are two solutions for μ,

$$\mu_\pm = \tfrac{1}{2}[2 + k \pm \sqrt{4k + k^2}]$$

and the corresponding values of α's and β's are found by making sure that (6.59) is satisfied for $n = 1$ (Problem 6.28). (The foregoing can also be obtained by diagonalizing M, but for our purposes this way is simpler.) The product $\mu_+ \mu_- = 1$, and after a few iterations the larger of these $|\mu|$ becomes dominant. If $-4 < k < 0$, $\mu_\pm$ is of the form $e^{\pm i\varphi}$. For studying instability this case is not of interest.

Evidently, for large positive k, the growth of both δx and δy is determined by $|2 + k|^n$, but this result, valid for k independent of n, is not at once useful to study the actual behavior of the mapping. It becomes useful when we define a quantity called the KS entropy of the mapping (named after N. Krylov, A. Kolmogorov, and Ya. Sinai) and try to calculate that.

Clearly the separation of the two initially adjacent points increases exponentially with n. Write it as proportional to $e^{n\sigma}$, where σ is the KS entropy. Note that this is a strange term, since σ is actually a rate. The essential point of the argument is this: we know from computer experiments that for large k the phase point wanders all over the surface of section, covering it almost uniformly except for a few small islands of stability. Thus, if we imagine a large number of iterations starting from a given point, the image points will be distributed fairly evenly over the surface of section and σ is effectively given by an average over starting points. That is easy to calculate. Consider a sequence of n iterations. Each one begins at a different point, and assume that these points are distributed at random. For $k > 0$ define σ by

$$e^{n\sigma} := \prod_{i=1}^{n} |\mu_{+i}|$$

where the initial points i cover the surface. This gives

$$\sigma = \frac{1}{n} \sum_{i=1}^{n} \ln |\mu_{+i}| = \langle \ln |\mu_+| \rangle$$

and consideration of negative values of k gives a similar result so that

$$\sigma \approx \langle \ln |2 + K \cos x| \rangle = \frac{1}{2\pi} \oint \ln |2 + K \cos x| dx = \ln (K/2) \quad , \quad K \gtrsim 2 \quad .$$

Chirikov [68] has compared this relation with numerical calculations and for $K = 4$ they agree to within 3 percent. This verifies the assumption that to calculate σ it is not important that k_n is different for every n; one needs only to average over the starting points.

Round-off Errors. One can make the following objection to the foregoing attempt to explain the scattered points that one finds in a computer study of a surface of section or an area-preserving mapping. It is clear from the considerations on instability that any error made in calculating the nth point will be so magnified after (in typical cases) a few dozen further steps that the computed $n + k$th point will bear no relation to the point that would have been calculated had the error not occurred. Of course there are such errors, since the computer rounds off its calculation at every step. Thus there is no reason to think that the computed pattern resembles at all the pattern that would be computed by an error-free computer, and any strange features of the results can be explained as the result of round-off. Experience does not support this argument since if one repeats a calculation using a different degree of precision or a different round-off procedure (say by changing computers) one gets a distribution of dots that looks very much the same as before. Why is this so?

Define an ideal orbit as the one that would be calculated by an error-free computer and a real orbit as one in which a slight error δ_n, with all $|\delta_n| < \Delta$, is introduced in every step. From what has been said, it is clear that if the two orbits start at the same initial point P_0, then typically after a few dozen steps, they will diverge exponentially. But let the ideal orbit begin at a point P_0', slightly different from P_0. There are some remarkable theorems (proved so far only for specific mappings) [69] which show that even for very large N, perhaps 10^7, there exists a P_0' such that for every $n < N$ the separation between the real and ideal orbits remains less than a small assignable quantity. Thus the points one computes from a given P_0 may not be accurate, but that does not mean they do not fall close to an ideal orbit. It is just that the ideal orbit does not start exactly at P_0. And since the investigator usually chooses P_0 at random anyhow, the difference is of small importance. The theorems so far proved refer to specific mappings, but it seems likely that the scattered points one sees in other cases as well do not result from errors of computation.

Determinism? From one point of view, Newtonian mechanics is a perfectly deterministic matter. Solve the equations, put in the initial conditions, and calculate the answer. The reason for studying an area-preserving mapping was to gain some insight into the analysis of dynamical systems that seem to exhibit chaotic behavior. What does it tell us about the Newtonian paradigm? Suppose $n = 10^6$. How accurate must we be in order to compute the position of the nth iteration of the initial phase point with an error of, say, about 50 percent? Take $K = 4$. Then $\sigma = 0.69$ and $\exp(0.69 \times 10^6) \approx 10^{3 \times 10^5}$. This is the factor by which the initial error is magnified. Any attempt to express or compute numbers with the required accuracy obviously belongs to fantasy, even ignoring quantum indeterminacy. (Try calculating the volume of the known universe in terms of the Planck length, 10^{-33} cm.) These remarks show why thoughtful physicists have been reminding us for some time that Newtonian determinism is not what it appears to be on the surface. This is one of those situations in which the answer to a question that expects an unconditional yes or no turns out dependent on numerical criteria, so that what began as a qualitative judgement ends as a quantitative one.

Finally, a note to explain the use of the word entropy to denote the quantity σ. In classical statistical mechanics it is shown that the entropy of an ensemble of systems prepared in a certain way is proportional to the logarithm of the volume of phase space occupied by the points representing the ensemble:

$$S = k_{\mathrm{B}} \ln \Omega + \text{const}$$

where k_{B} is Boltzmann's constant. The coordinates x and y in the standard model are meant to mimic coordinates in phase space. Imagine an ensemble of plane systems with initial phase points very close together; if $n \propto t$ and every dimension of the distribution grows equally, then Ω grows as $\exp(4\nu\sigma t)$, where ν is some constant. Thus

$$S = 4k_{\mathrm{B}}\nu\sigma t + \text{const}$$

and σ measures the rate of growth of the entropy. It is often said that entropy measures a loss of microscopic information. This simple calculation shows how that occurs.[6]

---

[6] Here are some further references. Some assume a knowledge of action-angle variables, which are specially useful for perturbation calculations and are briefly explained in Sect. 8.6. Introductory: [70], [71], [64]. More advanced: [72], [73], [74].

Problem 6.23. By changing the scale of r in the Hénon-Heiles Hamiltonian, show that an increase in the value of the energy is essentially the same as an increase in the size of the cubic terms that make the Hamiltonian nonseparable.

Problem 6.24. Show that the position of a point on a surface of section is determined by that of the preceding or following point.

Problem 6.25. Finish the proof that the mapping from one surface of section to another preserves areas.

Problem 6.26. Start with a two-dimensional oscillator and gradually turn on the cubic terms to produce the Hénon-Heiles potential. Will the cucumber and the banana in Fig. 6.9a have the same area as the two onions in Fig. 6.9b? You may have to invent and prove a generalization of the adiabatic theorem.

Problem 6.27. Show that the standard map (6.57) preserves areas.

Problem 6.28. Solve the recurrence relations (6.60) and show that the values found for μ are the eigenvalues of M. (This is easy if one knows Cayley's theorem that a matrix satisfies its own characteristic equation.)

Problem 6.29. *Hénon* [75] has studied another mapping,

$$x' = x \cos \alpha - (y - x^2) \sin \alpha \quad , \quad y' = x \sin \alpha + (y - x^2) \cos \alpha \quad .$$

Show that it preserves areas and study its properties for different values of α using a computer. Study questions of stability by starting with two points very close together and seeing what happens to them as the mapping is iterated. (You will probably want to clear the screen after each iteration.)

Problem 6.30. The argument developed above can be used to show that if one counts backward from the moment $t = 0$ one also finds entropy increasing. This appears to violate the Second Law. Explain the situation.

Problem 6.31. Invent an area-preserving mapping containing an adjustable parameter and experiment on a computer to see if you can find signs of chaos.

6.7 Coordinate Systems

The power of Jacobi's theory is not evident in simple examples, where almost every constant arises in a fairly obvious way from a cyclic coordinate. These coordinates have been the most familiar ones: straight lines and angles. But sometimes one must use other coordinates, for which the meaning of cyclicity is not intuitively clear. In this section we shall see why certain calculations require the use of special coordinate systems and then go on to discuss some types of coordinates in three-dimensional space in which the Hamilton-Jacobi equation is separable.

The Anisotropic Oscillator. The two-dimensional anisotropic oscillator is defined in Cartesian coordinates by the Hamiltonian

$$H = \frac{1}{2m}(p_x^2 + p_y^2) + \frac{1}{2}(k_1 x^2 + k_2 y^2) \tag{6.61}$$

and the Hamilton-Jacobi equation is

$$\frac{1}{2m}\left[\left(\frac{\partial W}{\partial x}\right)^2 + \left(\frac{\partial W}{\partial y}\right)^2\right] + \frac{1}{2}(k_1 x^2 + k_2 y^2) = \alpha_1 \quad . \tag{6.62}$$

This can be solved by separation of variables. There are two processes that occur by this name: separation into a product of factors each depending on a single variable as with Schrödinger's equation or any other that starts with a laplacian, and separation into a sum, as here. (We shall see later how the two processes are related.) Assuming that

$$W(x, y) = W_1(x) + W_2(y)$$

we find

$$\left[\frac{1}{2m}\left(\frac{dW_1}{dx}\right)^2 + \frac{1}{2}k_1 x^2\right] + \left[\frac{1}{2m}\left(\frac{dW_2}{dy}\right)^2 + \frac{1}{2}k_2 y^2\right] = \alpha_1$$

and the only way this can be satisfied is for each bracketed term to be a constant. This gives

$$\frac{1}{2m}\left(\frac{dW_1}{dx}\right)^2 + \frac{1}{2}k_1 x^2 = \alpha_2 \quad , \quad \frac{1}{2m}\left(\frac{dW_2}{dy}\right)^2 + \frac{1}{2}k_2 y^2 = \alpha_1 - \alpha_2 \quad . \tag{6.63}$$

Since $(dW_1/dx)^2$ and $(dW_2/dy)^2$ cannot become negative, the values of x and y are bounded:

$$-A_x < x < A_x \quad , \quad -A_y < y < A_y \quad \text{where} \tag{6.64a}$$

$$A_x = \sqrt{2\alpha_2/k_1} \quad , \quad A_y = \sqrt{2(\alpha_1 - \alpha_2)/k_2} \quad . \tag{6.64b}$$

How one knows that these limits are actually reached, and that the motion oscillates betweem them, is discussed in Sect. 2.2 under *A Note on Arc-Sines*.

The solutions of (6.63) can be written down from (6.10):

$$W(x, y) = \int \sqrt{2m\left(\alpha_2 - \frac{1}{2}k_1 x^2\right)}\, dx + \int \sqrt{2m\left(\alpha_1 - \alpha_2 - \frac{1}{2}k_2 y^2\right)}\, dy$$

where the integrals may be considered as indefinite because additive constants are ignored. To find the orbits,

$$\frac{\partial W(x, y)}{\partial \alpha_2} = \sqrt{\frac{m}{2}}\left(\int \frac{dx}{\sqrt{\alpha_2 - k_1 x^2/2}} - \int \frac{dy}{\sqrt{\alpha_1 - \alpha_2 - k_2 y^2/2}}\right) = \beta_2$$

or

$$\frac{1}{\omega_1}\sin^{-1}\frac{x}{A_x} - \frac{1}{\omega_2}\sin^{-1}\frac{y}{A_y} = \beta_2 \quad , \quad \omega^2 = k/m \quad . \tag{6.65}$$

What curves this strange formula represents becomes clear if we introduce the time variable as a parameter by (6.44a),

$$\frac{\partial W(x, y)}{\partial \alpha_1} = \frac{1}{\omega_2}\sin^{-1}\frac{y}{A_y} = t - t_0$$

so that

$$\sin^{-1}\frac{y}{A_y} = \omega_2(t - t_0) \quad \text{or} \quad y = A_y \sin\left[\omega_2(t - t_0)\right] \quad . \tag{6.66a,b}$$

Putting the first of these into (6.65) gives

$$x = A_x \sin\left[\omega_1(t - t_0) + \beta_2\right] \quad . \tag{6.66c}$$

Equations (6.66b, c) are the recipe for drawing Lissajous figures; Fig. 6.14 is an example. The curves will ordinarily fill up the entire allowed area defined by (6.64). If these are the orbits, it is clear why Cartesian coordinates were right for separating the variables, since in no other system are the boundaries of this region expressed in terms of one variable at a time.

But suppose the two frequencies happen to be equal. The orbit is now closed (Fig. 6.14b) and it no longer fills the available space. In this case the problem can be separated in polar coordinates also.

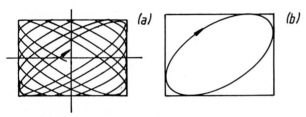

Fig. 6.14a,b. Lissajous figures defined by (6.65) for different frequency ratios. (a) If ω_1 and ω_2 are incommensurable the curve finally fills the allowed region. (b) Here $\omega_1 = \omega_2$ and the allowed region is not filled

Problem 6.32. In Sect. 6.1 it was shown that W can be regarded as the generator of the contact transformation that takes a system from its initial state to its present state. Expressing the constants α_1 and α_2 of the foregoing discussion in terms of the numbers that define the initial state, write down the explicit form of the W that generates this transformation and verify that it does so.

Problem 6.33. Two identical harmonic oscillators are coupled together so that their Lagrangian is

$$L = \tfrac{1}{2}(\dot{x}^2 + \dot{y}^2) - \tfrac{1}{2}\omega^2(x^2 + y^2) - \kappa^2 xy \quad .$$

Separate the variables by introducing new coordinates

$$x = 2^{-1/2}(v - u) \quad , \quad y = 2^{-1/2}(v + u) \quad .$$

Sketch the Lissajous figures in u, v that express the solution of this equation. Sketch them in x, y and note how the boundaries of the figures are expressed in each system.

6.8 Curvilinear Coordinates

The simple examples of the last section suggest how the use of curvilinear coordinates should be regarded: if all the Lissajous figures generated by a certain orbit can be bounded by coordinate lines (or surfaces) in a certain set of coordinates, the variables can be separated. If not, the calculation may be very difficult. The

175

following example in plane parabolic coordinates will serve as an introduction to this approach and will show how one can identify classes of potential functions for which curvilinear coordinates are useful.

Example 6.4. Introduce a new way of covering the xy plane by means of the variables ξ and η defined by

$$x = \sqrt{\xi\eta} \quad , \quad y = \tfrac{1}{2}(\xi - \eta) \quad , \qquad \xi, \eta > 0 \quad . \tag{6.67}$$

Eliminate first ξ and then η, to find

$$y = \frac{x^2 - \eta^2}{2\eta} \quad , \quad y = -\frac{x^2 - \xi^2}{2\xi} \tag{6.68}$$

which, for constant η and ξ, are two sets of confocal parabolas all of whose intersections are at right angles, Fig. 6.15. We can also easily show that

$$\xi = r + y \quad , \quad \eta = r - y \quad , \quad r^2 = x^2 + y^2 \quad . \tag{6.69}$$

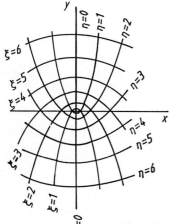

Fig. 6.15. Parabolic coordinates

A routine calculation gives the Lagrangian as

$$L = \frac{1}{8} m(\xi + \eta) \left(\frac{\dot{\xi}^2}{\xi} + \frac{\dot{\eta}^2}{\eta} \right) - V(\xi, \eta) \quad .$$

The momenta are

$$p_\xi = \frac{1}{4} m(\xi + \eta) \frac{\dot{\xi}}{\xi} \quad , \quad p_\eta = \frac{1}{4} m(\xi + \eta) \frac{\dot{\eta}}{\eta} \tag{6.70}$$

and

$$H = \frac{2}{m} \frac{\xi p_\xi^2 + \eta p_\eta^2}{\xi + \eta} + V(\xi, \eta) \quad . \tag{6.71}$$

The Hamilton-Jacobi equation in these variables is

$$\frac{2}{m}\frac{1}{\xi+\eta}\left[\xi\left(\frac{\partial W}{\partial \xi}\right)^2+\eta\left(\frac{\partial W}{\partial \eta}\right)^2\right]+V(\xi,\eta)=\alpha_1 \quad.$$

Suppose now that $V(\xi,\eta)$ is of the form

$$V(\xi,\eta)=\frac{f_1(\xi)+f_2(\eta)}{\xi+\eta} \quad. \tag{6.72}$$

Then

$$\frac{2}{m}\left[\xi\left(\frac{\partial W}{\partial \xi}\right)^2+\eta\left(\frac{\partial W}{\partial \eta}\right)^2\right]+f_1(\xi)+f_2(\eta)-\alpha_1(\xi+\eta)=0$$

and W separates additively into $W_1(\xi)+W_2(\eta)$ where

$$\frac{2}{m}\xi\left(\frac{\partial W}{\partial \xi}\right)^2+f_1(\xi)-\alpha_1\xi=\alpha_2$$

$$\frac{2}{m}\eta\left(\frac{\partial W}{\partial \eta}\right)^2+f_2(\eta)-\alpha_1\eta=-\alpha_2 \quad. \tag{6.73}$$

Note that the constant α_2 has entered not via a cyclic coordinate or the conservation of anything but via the procedure for separating variables in a partial differential equation. Such constants are called separation constants.

Since $(\partial W_1/\partial \xi)^2$ and $(\partial W_2/\partial \xi)^2$ are not negative, we must have

$$\alpha_1\xi+\alpha_2-f_1(\xi)\geq 0 \quad, \quad \alpha_1\eta-\alpha_2-f_2(\eta)\geq 0 \quad. \tag{6.74}$$

Depending on the f's and the values of α_1 and α_2, these may or may not establish limits for ξ and η. Where one of the expressions on the left vanishes is a turning point. If it vanishes twice, the particle oscillates between the two turning points. The role of the constants of integration is clear in these expressions: Even though they may not exactly determine an orbit, they at least restrict it to certain regions of the $\xi\eta$ plane. The equations (6.73) are easy or difficult depending on the forms of f_1 and f_2, but let us see what has already been accomplished. In solving the equations we introduced a separation constant α_2 which is a new constant of the motion, given by either of two forms

$$\alpha_2=\frac{2}{m}\xi p_\xi^2+f_1(\xi)-\alpha_1\xi$$

$$=-\frac{2}{m}\eta p_\eta^2-f_2(\eta)+\alpha_1\eta \quad. \tag{6.75}$$

A special case is Kepler motion, in which

$$V(\xi,\eta)=-\frac{\gamma}{r}=-\frac{2\gamma}{\xi+\eta} \quad.$$

This is of the form (6.72) and we can take

$$f_1=f_2=-\gamma \tag{6.76}$$

Symmetrize the expressions (6.75) by writing α_2 as half their sum

$$\alpha_2=\frac{1}{m}(\xi p_\xi^2-\eta p_\eta^2)-\frac{1}{2}\alpha_1(\xi-\eta) \quad.$$

With α_1 equal to H in (6.71) this is

$$\alpha_2 = \frac{1}{m(\xi + \eta)} [2\xi\eta(p_\xi^2 - p_\eta^2) - m\gamma(\xi - \eta)] \quad . \tag{6.77}$$

To see what it has to do with anything else we know, look at the component of the vector integral A in (3.45) that lies along the axis of the parabolas,

$$A_y = \frac{1}{m^2}(x p_x p_y - y p_x^2) + \frac{\gamma y}{r} \quad . \tag{6.78}$$

Substituting the expressions that transform this into parabolic coordinates gives

$$A_y = -\alpha_2/m \quad . \tag{6.79}$$

(There is no need to ask about A_x, since from (3.48) A has only one independent component.)

Note that the constant which was derived in Chap. 2 by an ingenious device appears here in the course of a routine application of the separation of variables in parabolic coordinates. Of course, the right set of coordinates had to be chosen, but there is not much ingenuity in this, since as we shall see there are only eleven kinds of coordinates in which the separation is possible, and most of these correspond to potentials so esoteric in form that they do not look useful. Further, we have succeeded in generalizing our earlier result, since the functions f_1 and f_2 need not be given by (6.76) but are now arbitrary. Finally, we may if we wish perform an actual calculation. The characteristic function is

$$W(\xi, \eta) = \sqrt{\frac{m}{2}} \int \sqrt{\alpha_1 - \frac{\alpha_2}{\xi} - \frac{f_1(\xi)}{\xi}} d\xi + \sqrt{\frac{m}{2}} \int \sqrt{\alpha_1 + \frac{\alpha_2}{\eta} - \frac{f_2(\eta)}{\eta}} d\eta$$

through which the constants α_1 and α_2 will appear in the solution.

Problem 6.34. Find f_1 and f_2 corresponding to a hydrogen atom in an electric field, and evaluate the integral α_2 in terms of Cartesian coordinates.

Problem 6.35. Carry out the change of variables leading to (6.71).

Problem 6.36. Show that the sets of coordinate lines given by (6.68) all intersect each other at right angles. Why is this fact important for the separation of variables?

Problem 6.37. Set up the equations of motion in 3 dimensions for a particle attracted to a fixed center by an arbitrary potential $V(r)$. Find the physical significance of the two separation constants, and show that when the coordinates are suitably oriented with respect to the initial motion our earlier results are reproduced.

Problem 6.38. Finish solving the Kepler problem in two dimensions in parabolic coordinates.

Problem 6.39. With the major axis of a Keplerian ellipse (Fig. 6.16) along the y axis, show that the formula for the ellipse in parabolic coordinates is

$$\frac{\xi}{1 + e} + \frac{\eta}{1 - e} = 2a \quad . \tag{6.80}$$

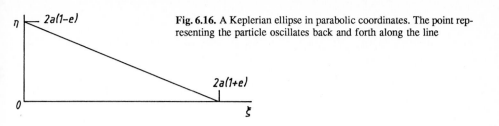

Fig. 6.16. A Keplerian ellipse in parabolic coordinates. The point representing the particle oscillates back and forth along the line

Problem 6.40. Show from (6.74) that $\xi + \eta \le 4a$ in the Kepler problem, and show that this is compatible with (6.80).

Problem 6.41. Discuss the motion of a particle under an inverse-square force in terms of intersecting surfaces in phase space as was done in Sect. 6.5, but here use the ξ, p_ξ plane. You may wish to show what happens if a uniform electric field is also present.

The conditions under which it is possible to separate variables in the Hamilton-Jacobi equation have been much studied, notably by Stäckel in the 1890s. For a particle in three-dimensional Euclidean space it was finally shown in 1934 by *Eisenhart* [76] that there are exactly 11 real coordinate systems (i.e., systems in which the coordinates are real numbers) that satisfy Stäckel's criteria for separability. They are discussed in *Morse* and *Feshbach* [77, Chap. 5], with diagrams in which some readers will see the coordinates in three dimensions. The potentials for which each system allows variables to be separated are the same in classical and quantum mechanics and are given in most economical form in Eisenhart's paper of 1948 [76].

There are only a few calculations for which variables can be separated, though luckily for students and teachers some of them are important. The two-body problem is equivalent to the one-body problem, but when we leave that behind almost nothing is exactly soluble any more except for very special choices of initial conditions. We have to resort to approximate procedures, and here, except for some qualitative considerations, the Hamilton-Jacobi theory has no special advantages. If equations have to be solved approximately, in letters or by machine, ordinary differential equations are easier to deal with, and most practical calculations start from Hamilton's equations. We will discuss a few simple methods in Chap. 8.

6.9 Interlude on Classical Optics

Although it is not possible to create situations in which the index of refraction varies as freely as potentials do in dynamics, the theory developed in this chapter has its close counterpart in optics. Only the derivations are different, since we start from a wave theory.

Let $s(\boldsymbol{r}, t)$ be the phase at a certain point in a monochromatic light ray. If one goes to a neighboring point $\boldsymbol{r} + \delta\boldsymbol{r}$ the phase changes according to (1.10) and (1.11) by $\boldsymbol{K} \cdot \delta\boldsymbol{r}$. If one evaluates it also at a slightly different time the phase changes by $\omega \delta t$. Thus

$$\delta s = \boldsymbol{K} \cdot \delta \boldsymbol{r} - \omega \delta t \tag{6.81}$$

analogously to (5.34). As before, we introduce a time-independent phase by the Lagrange transformation

$$w = s + \omega(t - t_0) \tag{6.82}$$

which satisfies

$$\delta w = \boldsymbol{K} \cdot \delta \boldsymbol{r} + (t - t_0)\delta\omega \tag{6.83}$$

so that

$$\boldsymbol{K} = \nabla w \quad , \quad t - t_0 = \frac{\partial w}{\partial \omega} \quad . \tag{6.84}$$

From the definition of $\boldsymbol{K}$, (1.5) gives

$$(\nabla w)^2 - \left[\frac{\omega}{c} n(\omega, \boldsymbol{r})\right]^2 = 0 \quad . \tag{6.85}$$

The equation is the same as (1.8) but now we know more about how to integrate it.

Consider first the propagation of a signal through a medium of uniform n, taking the x axis parallel to the direction of propagation:

$$\left(\frac{dw}{dx}\right)^2 = \left[\frac{\omega}{c} n(\omega)\right]^2 \quad \text{or} \tag{6.86}$$

$$\frac{dw}{dx} = k \quad , \quad w = kx$$

where $k = \omega n/c$ as usual. From (6.84)

$$t - t_0 = \frac{\partial w}{\partial \omega} = \frac{dk}{d\omega} x \quad \text{or}$$

$$x = \frac{d\omega}{dk}(t - t_0) = v_{\mathrm{g}}(t - t_0) \tag{6.87}$$

where v_{g} is the group velocity. This is in analogy with the particle velocities given by the analogous procedures in dynamics. We shall see in Sect. 7.6 why it comes out this way.

A more interesting case is one treated before in Example 1.2, see Fig. 6.17. with $n(\omega, z)$ we can separate variables and integrate (6.85) to give

$$w(x, y, z) = \alpha_1 x + \alpha_2 y + \int \sqrt{\left(\frac{\omega}{c} n(\omega, z)\right)^2 - \alpha_1^2 - \alpha_2^2} \, dz \quad . \tag{6.88}$$

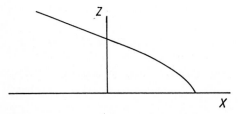

Fig. 6.17. Bending of a ray of light as it enters the atmosphere above a plane earth

The orbit is determined by

$$\frac{\partial w}{\partial \alpha_1} = x - \alpha_1 \int \frac{dz}{\sqrt{k^2(\omega, z) - \alpha_1^2 - \alpha_2^2}} = \beta_1 \quad , \qquad k = \frac{\omega}{c} n(\omega, z)$$

$$\frac{\partial w}{\partial \alpha_2} = y - \alpha_2 \int \frac{dz}{\sqrt{k^2(\omega, z) - \alpha_1^2 - \alpha_2^2}} = \beta_2 \quad \text{whence}$$

$$\frac{x - \beta_1}{\alpha_1} = \frac{y - \beta_2}{\alpha_2} = \int \frac{dz}{\sqrt{k^2(\omega, z) - \alpha_1^2 - \alpha_2^2}} \tag{6.89}$$

and the propagation of a signal along the beam proceeds at a rate given by

$$t - t_0 = \int \frac{k(dk/d\omega)dz}{\sqrt{k^2(\omega, z) - \alpha_1^2 - \alpha_2^2}} \tag{6.90}$$

Problem 6.42. Integrate (6.85) to get (6.88).

Problem 6.43. According to Example 1.1, the propagation of water waves in shallow water of depth $h < g/\omega^2$ (i.e., $kh < 1$) satisfies

$$k(\omega, x, y) = \frac{\omega}{\sqrt{gh}}\left(1 + \frac{\omega^2 h}{6g} + \ldots \right) \tag{6.91}$$

where h is some function of x and y. Assuming a uniformly shelving beach, discuss the motion of a large isolated wave as it approaches a straight beach.

Problem 6.44. Show that (6.89) is equivalent to Snell's law for a horizontally layered medium.

Problem 6.45. Assuming that the atmospheric index of refraction varies according to

$$n(r) = 1 + \nu_0 e^{-\mu(r-R)} \qquad R = \text{earth radius}$$

find the difference between the real and apparent directions of a star whose apparent zenith distance is ϑ_0, Fig. 6.18a.

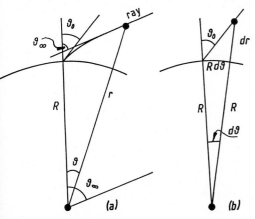

Fig. 6.18. For calculating the bending of a ray of light as it enters the atmosphere above a spherical earth

Solution. The eikonal equation is

$$\left(\frac{\partial W}{\partial r}\right)^2 + \frac{1}{r^2}\left(\frac{\partial W}{\partial \vartheta}\right)^2 = n^2(r)$$

and with $W = W_r + W_\vartheta$ this is

$$\frac{\partial W_\vartheta}{\partial \vartheta} = \alpha \quad , \quad \frac{\partial W_r}{\partial r} = \sqrt{n^2 - \frac{\alpha^2}{r^2}}$$

so that

$$W = \int_R^r \sqrt{n^2 - \frac{\alpha^2}{r^2}}\, dr + \alpha\vartheta \quad . \tag{6.92}$$

The path of the ray is given by $\partial W/\partial\alpha = \beta$, or

$$-\alpha \int_R^r \frac{dr}{r^2\sqrt{n^2 - \alpha^2/r^2}} + \vartheta = \beta \tag{6.93}$$

and setting $r = R$ we see that $\beta = 0$. To introduce ϑ_0, note that from Fig. 6.18b,

$$\tan\vartheta_0 = R\frac{d\vartheta}{dr}\bigg|_{r=R} \quad . \tag{6.94}$$

By (6.93)

$$\frac{d\vartheta}{dr}\bigg|_{r=R} = \frac{\alpha}{R^2(n_0^2 - \alpha^2/R^2)^{1/2}} \quad , \quad n_0 = n(R) = 1 + \nu_0$$

and this, in (6.94), gives

$$\alpha = n_0 R \sin\vartheta_0 \quad . \tag{6.95}$$

To integrate (6.93), write the integral to first order in ν_0 as

$$\int_R^r \frac{dr}{r^2\sqrt{1 - \alpha^2/r^2}} - \int_R^r \frac{\nu_0 e^{-\mu(r-R)}}{r^2(1 - \alpha^2/r^2)^{3/2}}\, dr \quad .$$

Since the atmosphere is only a few kilometers thick we can replace r by R in the denominator of the second integral and find

$$-\frac{1}{\alpha}\sin^{-1}\frac{\alpha}{r} + \frac{1}{\alpha}\sin^{-1}\frac{\alpha}{R} - \frac{\nu_0}{\mu R^2(1 - \alpha^2/R^2)^{3/2}}[1 - e^{-\mu(r-R)}] \quad .$$

In (6.93), this is

$$\sin^{-1}\frac{\alpha}{r} + \frac{\alpha\nu_0}{\mu R^2(1 - \alpha^2/R^2)^{3/2}}[1 - e^{-\mu(r-R)}] + \vartheta = \sin^{-1}\frac{\alpha}{R} \quad .$$

From (6.95) the reader will readily find

$$\sin^{-1}\frac{\alpha}{R} = \vartheta_0 + \nu_0\tan\vartheta_0 + \dots \quad . \tag{6.96}$$

When $r = \infty$, the ray has the asymptotic direction ϑ_∞:

182

$$\frac{\alpha v_0}{\mu R^2 (1 - \alpha^2/R^2)^{3/2}} + \vartheta_\infty = \sin^{-1}\frac{\alpha}{R} \quad .$$

In the first term it is accurate enough to replace (6.95) by $\alpha = R \sin \vartheta_0$, and (6.96) gives

$$\vartheta_\infty - \vartheta_0 = v_0 \tan \vartheta_0 \left(1 - \frac{1}{\mu R \cos^2 \vartheta_0} \right)$$

or, finally,

$$\vartheta_\infty - \vartheta_0 = v_0 \left(1 - \frac{1}{\mu R} \right) \tan \vartheta_0 - \frac{v_0}{\mu R} \tan^3 \vartheta_0 \tag{6.97}$$

if ϑ_0 is not too close to $\pi/2$. The first term comes from Snell's law; the rest is the effect of the curvature of the earth's surface. The angle, though small, is of great importance for astronomers, and very accurate formulas have been developed. The method used here fails when ϑ_0 is very near $\pi/2$, and a more delicate analysis is required.

A value for v_0 under typical conditions is given by the formula of Problem 1.11 as 2.77×10^{-4}, and a corresponding value for μ is $0.14 \, \mathrm{km}^{-1}$. With these

$$\vartheta_\infty - \vartheta_0 = 57.1'' \tan \vartheta_0 - 0.064'' \tan^3 \vartheta_0 \quad .$$

At $\vartheta_0 = 30°$ the deflection is $32.9''$; at $85°$ it has increased to $557''$, or $9.28'$.

Problem 6.46. Find the term in v_0^2 that corrects (6.97).

Problem 6.47. Calculate the apparent change in the shape of the sun's disc as it sinks towards the horizon. (That it often looks different from this can be attributed to inhomogeneities in the atmosphere.) In a clear evening, the apparent height is about 3/4 the width.

7. Action and Phase

It is relatively simple to derive formulas of classical mechanics from those of quantum mechanics. One translates statements about probabilities into statements about measurable quantities, and one finds ways to eliminate $\hbar$, either by considering situations in which phase changes much more rapidly than other parameters (Sect. 1.3) or by taking averages in a suitable way (Sect. 2.1).

To invent quantum mechanics was much more difficult, for ways had to be found in which to introduce $\hbar$ into dynamical law. It was not enough to introduce the $\hbar$ that allows $\hbar^{-1}S$ to be considered as a phase. This is the WKB approximation of quantum mechanics (Sect. 1.1) and turns out finally to produce little new physics: Equation (1.6) for the phase resulting from this approximation is nothing but the Hamilton-Jacobi equation of Newtonian physics, and the only new results furnished by the theory are estimates of energy levels and barrier penetrabilities. It is necessary at some point to change something, to introduce hypotheses that do not belong to Newtonian physics, in order to invent a new dynamics.

Bohr's theory simply grafted a quantum principle involving $\hbar$ onto a Newtonian theory of motion. De Broglie's ideas, though very original and widely underrated, did not point toward new dynamical laws. Heisenberg in 1925 was the first to do that. Starting from an expression of Bohr's theory in terms of Fourier amplitudes (Sect. 7.4) he rewrote it in terms of matrices, making changes as he did so, and arrived at quantum mechanics (Sect. 7.5). Independently of Heisenberg and a year later, Schrödinger united de Broglie's waves with Hamilton-Jacobi dynamics, but making changes as he did so, to invent wave mechanics (Sect. 7.7) which later turned out to be equivalent to Heisenberg's theory. And much later, in 1948 Feynman showed that if one takes $\hbar^{-1}S$ for the phase of a wave and makes careful use of Hamilton's principle and the principle of superposition, one can arrive at quantum mechanics by a path which is particularly free of arbitrary hypotheses (Sect. 7.8).

This chapter will not advance in classical dynamics, but it may deepen understanding of the relation between the old and new theories and, in addition, show something of the ways in which new discoveries are made.

7.1 The Old Quantum Theory

Quantum phenomena were not first noticed at any one instant, but a good place to begin is Wien's observation (1893) that the spectral distribution of black-body radiation has a maximum at a wavelength λ_m inversely proportional to the absolute temperature of the radiation,

$$\lambda_m T = W \quad \text{(const)}$$

and it became clear that the value of the constant of proportionality is about 0.294 cm K, independent of the size, shape, and material of the enclosure: it is a constant of nature. Introducing a factor of Boltzmann's k allows the constant to be expressed in mechanical units. Nowadays one writes Wien's law in terms of the frequency:

$$kT = \frac{kW}{\lambda_m} = \frac{kW}{c}\nu_m \quad .$$

The value of kW/c that could be found from information then available was about 10^{-27} erg s. It was (or, in hindsight, should have been) clear that a successful theory of thermal radiation must contain a universal constant of this size (give or take a dimensionless factor) and units. The constant was introduced into mechanics by Planck in an argument culminating in the conclusion that the energy of an oscillator of frequency ν is quantized in steps separated by an interval $h\nu$.

Although the hypothesis worked, it was hard to justify; as early as 1906 Planck was thinking about expressing it in another way. He had noticed that the orbit in phase space of the point representing an oscillator is an ellipse:

$$\frac{p^2}{2mE} + \frac{q^2}{2E/k} = 1$$

and that the area of the ellipse corresponding to a given energy is

$$J = 2\pi\sqrt{m/k}\,E = E/\nu \quad .$$

As one goes from one energy level to the next, E increases by $h\nu$ and J increases by

$$\Delta J = h \quad .$$

This is a recipe for quantizing the oscillator which puts into evidence the quantity h independently of any physical property of the oscillator, and Planck accordingly felt that it might be of fundamental significance. He and other workers thereupon tried the effect of imposing the same condition on rotators and other systems with one degree of freedom. The area enclosed by a closed orbit in phase space can be written as $\iint dp\,dx$ integrated over the orbit, but a more convenient notation is

$$\oint p\,dx = nh \quad , \qquad n \in \mathbb{Z}_+ \tag{7.1}$$

integrated around the orbit (Fig. 7.1).

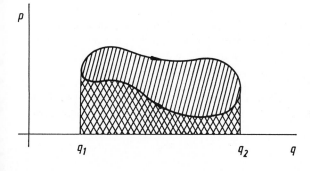

Fig. 7.1. The integral (7.1) around a closed orbit in phase space gives the area of the orbit, since the doubly shaded area is covered twice in opposite directions

The alternative forms of quantization were discussed until 1913, when Bohr's theory appeared. Here the basic quantum hypothesis for circular orbits was that an electron's constant angular momentum satisfied

$$p_\vartheta = nh/2\pi \quad , \quad n \in \mathbf{Z}_+ \quad . \tag{7.2}$$

This does not follow from the quantization of energy in equal steps; in fact, it gives rise to energies

$$E_n = -\frac{2\pi^2 m e^4}{n^2 h^2} \quad , \quad n \in \mathbf{Z}_+ \tag{7.3}$$

but (7.2) does follow from an action principle

$$\oint p_\vartheta \, d\vartheta = 2\pi p_\vartheta = nh \quad . \tag{7.4}$$

The two laws (7.1) and (7.4) appeared to be examples in one dimension of a general principle. In several degrees of freedom it is

$$J = \oint p_m \, dq_m = nh \quad (\text{summed over } m) \tag{7.5a}$$

which was appealing because it can easily be shown that J is invariant under a canonical transformation and that therefore the recipe for quantization has nothing to do with the particular coordinates used. By (5.37), J in two systems of coordinates and momenta satisfies

$$J = \oint p_m \dot{q}_m \, dt = \oint P_m \dot{Q}_m \, dt - \oint \frac{dF_1(q, Q)}{dt} \, dt \quad .$$

But since in a complete cycle both q and Q come back to their initial values, the last term vanishes and the invariance

$$J = \oint p_m \, dq_m = \oint P_m \, dQ_m \quad (\text{summed}) \tag{7.5b}$$

is proved. But it is not correct to quantize Newtonian mechanics by quantizing J, since when applied to systems with more than one degree of freedom such as a hydrogen atom with an elliptical orbit (Problem 7.9) it does not lead to quantized energy levels at all. The right recipe [78] is to separate the terms of (7.5a) and write $J = \Sigma J_v$, where

$$J_v = \oint p_v \, dq_v = n_v h \quad , \quad n_v \in \mathbf{Z}_+ \tag{7.6}$$

and $p_v \, dq_v$ is *not summed* over v. We will use the subscript v in the future to identify products in which the summation convention is suspended. How the procedure works can be seen from a calculation to be carried out in the next section, but the following Example 1 will be instructive to see that in the limit of high quantum numbers, (7.6) is closely connected with arguments based on classical physics.

Example 7.1. A Ball Between Two Walls. The ball moves freely except where it encounters walls at $x = 0$ and $x = l$. The solutions of $(dW/dx)^2 = 2mE$ are of the form

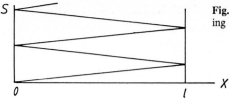

S

0   l X

Fig. 7.2. Characteristic function $S(x)$ for a particle bouncing between two walls

$$W = \pm\sqrt{2mE}(x - x_0)$$

in each segment of the path. Joining segments so as to make S continuous gives the function in Fig. 7.2 (compare Fig. 6.2). Integrating across the box and back gives

$$J = 2\sqrt{2mE}\, l \quad .$$

According to the quantum condition, the ball's energy in its nth state will be

$$E_n = \frac{n^2 h^2}{8ml^2} \quad , \quad n \in \mathbb{Z} \quad .$$

Problem 7.1. What is the quantum-mechanical version of this result? How do they differ?

Problem 7.2. If the quantum conditions are derived from the WKB approximation (see any text on quantum mechanics) n in (7.1) is replaced by $n + 1/2$. The 1/2 corresponds to the zero-point motion. An electron in a box even at zero temperature will not stay still but will bounce up and down. How high will it bounce?

The Correspondence Limit. The following argument is an elementary part of Bohr's theory. Consider circular orbits with quantum number n for convenience; then it is easily shown (Problem 7.6) that the orbital frequency of an electron in the nth orbit is

$$\Omega_n = \frac{me^4}{n^3 \hbar^3} \quad . \tag{7.7}$$

The energy of the state n is

$$E = -\frac{me^4}{2n^2 \hbar^2} \tag{7.8}$$

and the frequency of light emitted in a transition from state n to state $n - \alpha$ is

$$\begin{aligned}
\omega_{n,n-\alpha} &= \hbar^{-1}(E_n - E_{n-\alpha}) \\
&= \frac{me^4}{2\hbar^3}\left(\frac{1}{(n-\alpha)^2} - \frac{1}{n^2}\right) \\
&= \frac{me^4}{2\hbar^3}\left(\frac{1}{n^2} + \frac{2\alpha}{n^3} + \ldots - \frac{1}{n^2}\right) \quad \text{or}
\end{aligned}$$

$$\omega_{n,n-\alpha} = \alpha\Omega_n + \ldots \quad , \quad \alpha \in \mathbb{Z}_+ \quad . \tag{7.9}$$

Thus in the limit of large quantum numbers, the frequencies emitted in transitions starting from state n are just the various Fourier harmonics of the orbital frequency, and the results of quantum theory correspond with what one would expect by classical reasoning. Note that the formula above could have been obtained by writing Taylor's series as

$$\frac{1}{(n-\alpha)^2} = \frac{1}{n^2} - \alpha \frac{\partial}{\partial n} \frac{1}{n^2} + = \frac{1}{n^2} + \frac{2\alpha}{n^3} + \dots$$

but since n varies in integer steps the derivative makes no sense. That when $\alpha \ll n$ the nonsensical procedure leads to the same result as the careful one used above will be important in simplifying the developments to follow.

Derivations of the Quantum Conditions. It is instructive to turn the foregoing argument around and derive the quantum condition (7.6) from the simple notion of a quantum of energy. Suppose an oscillating system with one degree of freedom is in its nth quantum state, where n is a large number and we are in the domain of large quantum numbers. Let the system emit a few $(-\Delta n)$ quanta at its frequency of oscillation, ν, so that the energy $E(n)$ decreases by

$$\Delta E(n) = \Delta n h \nu(n) \quad .$$

Denoting differentiation in the schematic way just explained, we can write this as

$$\frac{dn}{dE} = \frac{1}{h\nu} = \frac{T}{h}$$

where T is the period of oscillation of the oscillator in state n.

From (6.9a) we can write T in terms of J, the value of W on integration around a complete cycle:

$$T = \frac{dJ}{dE} \quad . \tag{7.10}$$

Thus

$$\frac{dJ}{dE} = h \frac{dn}{dE}$$

which integrates to

$$J = nh + \text{const} \quad . \tag{7.11}$$

This gives essentially Planck's result (7.1) for a harmonic oscillator, as we can verify from the value of W given in (6.10). Integration over a complete cycle gives

$$J = 2\pi E \sqrt{m/k} = E/\nu$$

as we have already seen, so that the quantum condition is

$$E = nh\nu + \text{const} \quad .$$

The constant, as we now know, is $h\nu/2$.

The same argument explains why for a multiply periodic system (7.6), and not (7.5), is the correct rule for quantization. Such a system can emit quanta with frequencies and energies corresponding to each of its periodicities. We then have for the vth degree of freedom

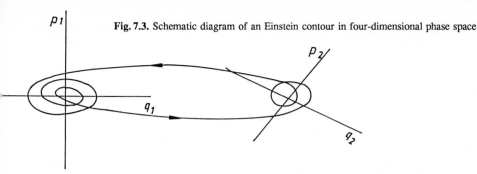

$$\Delta E_v = \Delta n_v h \nu_v(n_v) \quad .$$

If as in all the examples we have given the variables are separable, then J reduces to a sum as in (7.5) and the argument goes through as before for each J_v. For a nonseparable system, which is of course the general case, *Einstein* [79] proposed a new quantum condition: Integrate (7.5a) over any closed path in phase space (Fig. 7.3) that makes an integral number of complete cycles of each of the various periodicities, and set this equal to nh. The anisotropic oscillator discussed in Sect. 6.6 is a convenient (though too elementary) example because most of the calculation has already been done. If we integrate over r cycles of x and s cycles of y, the phase integral is

$$J_{rs} = \frac{r\alpha_2}{\nu_1} + \frac{s(\alpha_1 - \alpha_2)}{\nu_2} = nh \quad , \qquad r, s, n \in \mathbf{Z} \quad . \tag{7.12}$$

(Here α_2 and $\alpha_1 - \alpha_2$ are the energies associated with the motions in x and y.) The only way this relation can be satisfied for all integral values of r and s is if α_2/ν_1 and $(\alpha_1 - \alpha_2)/\nu_2$ are each integral multiples of h. Thus each of the oscillator's degrees of freedom is quantized separately. The trouble with this example is that it is separable, with p_i depending only on q_i; otherwise the situation is much more complicated. Consider an orbit in the plane $q_i q_j$. The point in this plane representing the system will wander around and can cross a given point in the plane moving in various directions. That is, since the momentum is given by $\partial W/\partial q$, W must be a multi-valued function. In complex-variable theory a double-valued function like $\sqrt{x}$ is made single-valued by continuing it onto a second sheet of a Riemann surface; here in the same way W becomes single-valued on a multi-sheeted manifold. In this manifold the momenta corresponding to q_i and q_j are related by

$$\frac{\partial p_i}{\partial q_j} = \frac{\partial^2 W}{\partial q_i \partial q_j} = \frac{\partial p_j}{\partial q_i}$$

so that by Stokes's theorem, as long as one stays on the same sheet and away from singularities, Einstein's integral is independent of the path taken. The trouble is that the behavior of nonseparable systems is so complicated that one does not know much about the sheets or the singularities, and the entire theory is still under development.[1]

---

[1] The literature is immense and often hard to read. An entry is provided by [80–82].

Problem 7.3. Planck's formula for the spectral distribution in blackbody radiation is

$$dI = \frac{8\pi ch\lambda^{-5}}{\exp(hc/\lambda kT) - 1} .$$

Locate the maximum of this curve, make a rough calculation of the constant W in (6.1) and, using a modern value of k, see how good a value of h is implicit in the nineteenth-century measurement.

Problem 7.4. Find how the period of an oscillator obeying the law $V(x) = \kappa x^4/4$ depends on the energy and amplitude of oscillation. The action integral J can be evaluated with the help of the integral formula

$$\int_0^1 \sqrt{1 - x^4} \, dx = \frac{\Gamma^2(1/4)}{6\sqrt{2\pi}} \tag{7.13}$$

which may be derived with profit and enjoyment from the integral for the beta function

$$\int_0^1 t^{m-1}(1 - t)^{n-1} dt = \frac{\Gamma(m)\Gamma(n)}{\Gamma(m + n)} := B(m, n)$$

and some of the identities satisfied by the gamma function $\Gamma(n)$.

Problem 7.5. Find the energy levels of the oscillator in the last problem using the quantum condition in phase space, and show that the calculation is almost exactly equivalent to a solution by the WKB approximation.

Problem 7.6. Derive (7.7) for a circular orbit.

Problem 7.7. The potential $V(r)$ binding two quarks together is usually supposed to increase with distance. Supposing that

$$V(r) = g \ln \frac{r}{r_0}$$

sketch $V(r)$ and show that the quantum condition gives energy levels

$$E_n = g \ln \frac{nh}{2\sqrt{\pi g \mu r_0}}$$

for s states (zero angular momentum) where μ is the reduced mass of the two quarks. The integral is not as difficult as it looks at first. What is the difference in energy between the nth state and the ground state?

7.2 Hydrogen Atom in the Old Quantum Theory

The recipe (7.6) gives the energy levels of hydrogen and their degeneracies. If coordinates are chosen so that the orbit lies in the xy plane, then the Hamiltonian is easily found to be

$$H = \frac{1}{2m}\left(p_r^2 + \frac{p_a^2}{r^2}\right) - \frac{e^2}{r} \qquad (7.14)$$

where p_a is the total angular momentum, which is around the z axis. For an arbitrary orientation of the axes, one finds in spherical coordinates

$$H = \frac{1}{2\mu}\left(p_r^2 + \frac{p_\vartheta^2}{r^2} + \frac{p_\phi^2}{r^2 \sin^2 \vartheta}\right) - \frac{e^2}{r} = E \qquad (7.15)$$

where μ is the reduced mass of electron and proton. Since ϕ is cyclic, p_ϕ is constant and we can write

$$W = W_r(r) + W_\vartheta(\vartheta) + p_\phi \phi \quad . \qquad (7.16)$$

Then

$$\left(\frac{\partial W_\vartheta}{\partial \vartheta}\right)^2 + \frac{p_\phi^2}{\sin^2 \vartheta} = \alpha_\vartheta \;\; (\text{const}) \quad \text{and} \qquad (7.17)$$

$$\left(\frac{\partial W_r}{\partial r}\right)^2 + \frac{\alpha_\vartheta}{r^2} = 2m\left(E + \frac{e^2}{r}\right) \quad . \qquad (7.18)$$

If we write (7.17) as

$$p_\vartheta^2 + \frac{p_\phi^2}{\sin^2 \vartheta} = \alpha_\vartheta$$

and compare (7.15) with (7.14), α_ϑ is recognized as the square of the total angular momentum. Then the quantum conditions are

$$J_\phi = \oint p_\phi d\phi = n_\phi h \qquad (7.19a)$$

$$J_\vartheta = \oint \sqrt{\alpha_\vartheta - \frac{p_\phi^2}{\sin^2 \vartheta}}\, d\vartheta = n_\vartheta h \qquad (7.19b)$$

$$J_r = \oint \sqrt{2\mu\left(E + \frac{e^2}{r}\right) - \frac{\alpha_\vartheta}{r^2}}\, dr = n_r h \quad . \qquad (7.19c)$$

These are (Problem 7.8)

$$J_\phi = 2\pi p_\phi \quad , \quad J_\vartheta = 2\pi(\sqrt{\alpha_\vartheta} - p_\phi)$$

$$J_r = -(J_\vartheta + J_\phi) + \pi e^2 \sqrt{\frac{2\mu}{-E}} \quad . \qquad (7.20)$$

Thus

$$p_\phi = n_\phi h/2\pi \quad , \quad \sqrt{\alpha_\vartheta} = p_\phi + n_\vartheta h/2\pi = (n_\phi + n_\vartheta)h/2\pi \quad .$$

It was customary in Bohr's theory to write $\sqrt{\alpha_\vartheta}$ as $lh/2\pi$ and n_ϕ as m, so that

$$l = m + n_\vartheta \quad .$$

Since m can be positive or negative while $n_\vartheta, n_r > 0$ we have

$$-l \le m \le l \qquad (7.21)$$

as is given also by quantum mechanics, although in the newer theory the square of the orbital angular momentum is written as $l(l+1)\hbar^2$.

From (7.20) the energy is given by

$$E = -\frac{2\pi^2 \mu e^4}{(J_\phi + J_\vartheta + J_r)^2} = -\frac{2\pi^2 \mu e^4}{h^2(n_\phi + n_\vartheta + n_r)^2} = -\frac{2\pi^2 \mu e^4}{h^2(l + n_r)^2} \ . \tag{7.22}$$

Evidently there are a number of sets of integers n_i corresponding to any value of E except the lowest; these levels are said to be degenerate. Problem 7.13 shows that if $n_\phi + n_\vartheta + n_r = n$, the degeneracy of the nth level is n^2.

After the elliptic orbits of hydrogen had been quantized in this way, *Sommerfeld* [83] used the same methods to quantize the orbits of the relativistic theory. His results agreed with experiment, yet we now know that this was a coincidence, for he omitted the important but then unknown effect of electron spin. The same formula was correctly re-derived from quantum mechanics by Dirac in 1927.

Problem 7.8. Carry out the integrations of (7.19b,c).[2] In both, it is necessary to think carefully what the limits of integration are. To start on J_ϑ, take $\cos \vartheta = \mu$ as the variable of integration. The integrand will then involve $(1 - \mu^2)^{-1}$, which can be written as

$$\frac{1}{1 - \mu^2} = \frac{1}{2}\left(\frac{1}{1 + \mu} + \frac{1}{1 - \mu}\right)$$

and the two terms can be integrated separately.

Problem 7.9. Show that if the quantum rule $J = nh$ is adopted, with J given by (7.5a), a continuous spectrum results.

Problem 7.10. Evaluate the energy shift in a hydrogen atom due to an applied electric field F. The action integral is difficult; expand the integrand in powers of F. This is a long and difficult calculation; see *Sommerfeld* [83, Note 11]; *Born* [84, App. 2]; and *Goldstein* [38] for support.

Problem 7.11. Derive (7.15) from the corresponding Lagrangian.

Problem 7.12. Carry out the separation of variables that yields (7.16).

Problem 7.13. Show that if $n_\phi + n_\vartheta + n_r = n$, the degeneracy of the nth level is n^2.

7.3 The Adiabatic Theorem

There is a remarkable argument by which *Bohr* and *Ehrenfest* [85] sought to establish the universality of the quantization procedure (7.6) by showing that if it works for one class of potential in 3 (or fewer) dimensions it works for all. The argument depends on a proposition, originally due to Boltzmann, which Bohr called the principle of mechanical transformability and Ehrenfest called the adiabatic theorem. Suppose

---

[2] The necessary integrals are in Gradshtein and Ryzhik's tables, but they are not hard to evaluate by the use of complex variables. See [83, Note 5], [84], [38, Sect. 10.7].

that in a cyclic system (orbiting or oscillating) we make a change in some of the parameters that describe the system that is slow and does not cause the system to resonate in any way. The theorem states that the change in the action integral J is then *very* slow, in the sense that we can change the parameters as much as we want, without changing J by more than any preassigned amount, provided that we change the parameters slowly enough.

A general proof is rather difficult [39], [85]. The discussion to follow omits the epsilons that would make it a proof, but is intended to show why the theorem is true.

Imagine that the Hamiltonian H contains a parameter (or parameters) $a(t)$ which has been varying slowly since an initial time t_0. Assume that it is possible to define a time interval t_0 to t such that although $a(t)$ changes only by a small amount δa during this time, the system goes through many cycles. As $a(t)$ changes, the Hamiltonian changes at a rate

$$\dot{H}(p, q, a(t)) = \frac{\partial H}{\partial p}\dot{p} + \frac{\partial H}{\partial q}\dot{q} + \frac{\partial H}{\partial a}\dot{a} = \frac{\partial H}{\partial a}\dot{a} \tag{7.23}$$

so that the only change in the numerical value of H is caused by the change in a:

$$\delta H = \frac{\partial H}{\partial a}\delta a + \frac{1}{2}\frac{\partial^2 H}{\partial a^2}(\delta a)^2 + \dots \quad . \tag{7.24}$$

Now calculate

$$\delta J = \oint (p_n \delta \dot{q}_n + \dot{q}_n \delta p_n)dt$$

where δp and δq are changes produced by the change in a. This is

$$\delta J = [p_n \delta q_n] + \oint (-\dot{p}_n \delta q_n + \dot{q}_n \delta p_n)dt \quad .$$

The first term is very small. (For each term of the sum, start the cycle when $p_n = 0$; then when it ends p_n is only of order δa and δq_n is also of order δa.) The rest is

$$\delta J = \oint \left(\frac{\partial H}{\partial q_n}\delta q_n + \frac{\partial H}{\partial p_n}\delta p_n\right)dt$$

$$= \oint \left(\delta H - \frac{\partial H}{\partial a}\delta a\right)dt = \oint \left(\frac{1}{2}\frac{\partial^2 H}{\partial a^2}(\delta a)^2 = \dots\right)dt$$

by (7.24), and so this term too is of second order in the change in a that occurs in one cycle. If one wants to know how J actually changes as a function of a, the analysis is quite difficult [88, 89].

Not every slow change of potential leaves J constant. The change must be small in the time of a complete cycle of the motion. Consider the situation of Fig. 7.4 in which the hump in the potential is slowly raised. As it approaches the level E the particle moves more and more slowly in the vicinity of the hump and the cycle takes longer and longer. Finally, as the period approaches infinity, there is no possible change that is slow compared with that, and the adiabatic theorem ceases to apply. And even if it does apply, the foregoing argument shows only that δJ is small. In Sect. 7.9 a sample calculation will show how small it actually is.

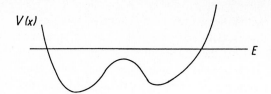

$V(x)$

Fig. 7.4. Potential which violates the adiabatic theorem if the hump is raised

E

The argument of Bohr and Ehrenfest was that if one has a system such as a harmonic oscillator in 1, 2, or 3 dimensions in which the quantization rule (7.6) is known to work, one can imagine it being transformed, by slowly changing the shape of the potential, into an entirely different system, say an atom or even something like H_2^+. But the action integral once equal to nh remains so, if no singular points analogous to that of Fig. 7.4 are encountered, and so the quantization rule is universal.

There is a further argument that makes the adiabatic theorem a plausible basis for a theory of quantization. Suppose a physical system is slowly distorted in the way we have described. Its energy will slowly change as the external forces do their work. This means that there will not be a sudden quantum jump from one value of n to another, for this would involve a discontinuous change in energy or in some other parameter which ought to change slowly and continuously. Thus the quantum numbers should be adiabatic quantities, in accordance with (7.6).

The Quantum Rule For an Oscillator. As a justification for the rule quantizing an oscillator, Lorentz suggested to Einstein that he consider how the period of a pendulum changes if one shortens the string by slowly drawing it up between one's fingers. One can argue as follows.

Suppose the frequency ω of an oscillator is slowly varied. ("Slowly" means that the fractional change in ω over a single period of the oscillator is very small.) The equation of motion is

$$\ddot{x} + \omega^2(t)x = 0 \tag{7.25}$$

of which a solution valid over any short interval of time is

$$x(t) = A \cos\left[\omega(t)t\right] \quad.$$

The energy is

$$E = \tfrac{1}{2}m[\dot{x}^2 + \omega^2(t)x^2] = \tfrac{1}{2}m\omega^2 A^2 \tag{7.26}$$

and it changes at the slow rate

$$\dot{E} = m(\dot{x}\ddot{x} + \omega^2 x\dot{x} + \omega\dot{\omega}x^2) = m(\ddot{x} + \omega^2 x)\dot{x} + m\omega\dot{\omega}x^2 = m\omega\dot{\omega}x^2 \quad.$$

Averaged over a cycle, this is $d\langle E\rangle/dt = m\omega\dot{\omega}\langle x^2\rangle$. In an oscillator the average potential and kinetic energies are equal, so that $m\omega^2\langle x^2\rangle/2 = \langle E\rangle/2$ and

$$\frac{d\langle E\rangle}{dt} = \frac{\dot{\omega}}{\omega}\langle E\rangle : \frac{1}{\omega}\frac{d\langle E\rangle}{dt} - \frac{\dot{\omega}}{\omega^2}\langle E\rangle = \frac{d}{dt}\frac{\langle E\rangle}{\omega} = 0 : \frac{\langle E\rangle}{\omega} = \text{const} \quad.$$

The argument then went: If $\langle E\rangle = n\hbar\omega$ and ω is varied slowly, $\langle E\rangle$ will vary so that no jump in n occurs. It is hard to see how a quantization rule that does not satisfy this simple condition could be true.

Let us do the same calculation via the adiabatically invariant phase integral:

$$J = \oint p\, dx = \oint m\dot{x}^2 dt = \oint m\omega^2 A^2 \cos^2 \omega t\, dt = \oint 2E \cos^2 \omega t\, dt$$

$$J = \langle E \rangle \frac{2\pi}{\omega} \quad . \tag{7.27}$$

The adiabatic invariance of J gives the previous result.

Example 7.2. A ball is set to bouncing on the hard floor of an elevator which slowly changes its acceleration. How does the duration of one bounce vary?

If the ball starts upward at a speed v_0, its trajectory is given by

$$y = v_0 t - \tfrac{1}{2} g t^2$$

where g is the acceleration of gravity plus that of the elevator. One bounce lasts for a time $T = 2v_0/g$. The phase integral is

$$J = \int_0^T m\dot{y}^2 dt = \frac{2}{3}\frac{mv_0^3}{g} = \frac{1}{12} mg^2 T^3 \simeq \text{const} \quad .$$

Thus, $T \propto g^{-2/3}$.

Example 7.3. A particle circles in a uniform magnetic field B which is gradually strengthened. How do the radius and kinetic energy of the orbit change with B?

With B constant, the orbit is determined by the balance of forces,

$$mr\omega^2 = er\omega B \quad , \quad \omega = eB/m \tag{7.28}$$

so that the frequency is determined by B but the radius can have any value. By (5.17), the phase integral is

$$J = \oint \boldsymbol{p} \cdot d\boldsymbol{l} = \oint (m\boldsymbol{v} + e\boldsymbol{A}) \cdot d\boldsymbol{l}$$

where $\boldsymbol{A}$ is the vector potential of the field. In the first term, the velocity $\boldsymbol{v}$ is (nearly) constant in magnitude and parallel to $d\boldsymbol{l}$ The second term is simplified by Stokes's theorem:

$$J = \oint mv\, dl + e \int \nabla \times \boldsymbol{A} \cdot d\boldsymbol{S}$$

$$= 2\pi mvr + \pi r^2 eB \quad .$$

With $v = \omega r = eBr/m$ by (7.28), this is

$$J = 3\pi r^2 eB = 3e\Phi \simeq \text{const}$$

where Φ is the total flux through the orbit. If J is constant, r varies as $B^{-1/2}$, and multiplying the two sides of the equation by $(eB/m)^2 = \omega^2$ shows that the kinetic energy varies as B. One of the standard methods for heating a plasma is to immerse it in a magnetic field which is then slowly increased. Although it might at first be argued that since a magnetic field does no work it cannot possibly change the particle's kinetic energy, remember that a changing magnetic field gives rise to an electric field. It is this that does the work.

Problem 7.14. In the elevator of Example 7.2 is suspended a mass m oscillating at the end of a spring of stiffness k. As the acceleration varies, how do the spring's frequency, amplitude and energy vary? If it was initially in the nth quantum state, how does n change?

Problem 7.15. A magnetic field is slowly established perpendicular to the orbital plane of an electron executing a circular orbit in a Bohr hydrogen atom. How does the electron's energy change?

Problem 7.16. The walls of the box of Example 7.1 are slowly moved. How do the frequency and kinetic energy of the ball vary with the separation l?

Problem 7.17. It has been hypothesized by Dirac and others that certain of the fundamental constants of nature are really changing by about 1 part in 10^{10} per year. Discuss the effect of changes in G on planetary orbits and on the distance to the moon, which can now be measured very accurately.

Problem 7.18. Sketch a graph of the period of an oscillator moving in the potential of Fig. 7.4, showing how it changes as the hump in the center is slowly raised. How will J change?

Problem 7.19. A harmonic oscillator governed as usual by a linear restoring force oscillates at such high speed that relativistic considerations must be used. Find the energy levels. You are invited to ponder the plausibility of the assumption about the force in this regime. Is it possible for a wave to move along a spring at a rate faster than c? Does this matter? What will the spring be doing if the mass moves at such a speed? How would one express the force that it exerts? These questions are not all easy to answer, but they may lead you to some interesting thoughts.

Problem 7.20. As an application of the old quantization rule

$$J = nh$$

calculate the energy levels of the quartic oscillator discussed in Sect. 3.8. Show that $J = 4ET/3$, where T is the period of oscillation, then take T from Problem 7.4.

Problem 7.21. To calculate the speed of sound in Problem 4.29 one had to know that when changed adiabatically, the pressure and volume of a gas are related by $pV^\gamma = \text{const}$, where for a monatomic gas $\gamma = 5/3$. Derive this relation from a simple-minded model which assumes that each molecule of the gas bounces back and forth between a pair of walls of a cube without ever touching the other walls. Having done this, can you now remove the artificial restriction and get the same result?

Problem 7.22.* A pair of linearly coupled oscillators is described by the Hamiltonian

$$N = \tfrac{1}{2}(p_1^2 + p_2^2 + \omega_1^2 x_1^2 + \omega_2^2 x_2^2 + \beta^2 x_1 x_2) \quad .$$

(The linear coupling is special because the two oscillators can be uncoupled by the transformation

$$x_1 = u \cos \vartheta - v \sin \vartheta \quad , \qquad x_2 = u \sin \vartheta + v \cos \vartheta$$

with suitable ϑ.) If now one of the frequencies, say ω_1, is changed slowly, the phase integrals $\oint p_u du$ and $\oint p_v dv$ of the uncoupled oscillators remain nearly constant, but if ω_1 passes through ω_2 the integrals $\oint p_1 dx_1$ and $\oint p_2 dx_2$ may change greatly. Analyze the change and show what happens to these integrals as ω_1 goes through resonance. How is it that a change in ω_1, even very slow, can change them? That is, what violation of the conditions for adiabatic behavior has occurred? You may find you have to explain why the integrals do *not* change when ω_1 is far from resonance. Show qualitatively what happens when the coupling β is very small. Since the algebra in this calculation is tedious, a computer will be helpful.

7.4 Connections with Quantum Mechanics

We have seen in Chap. 1 how the phase integral W arises in the process of approximating the solution of Schrödinger's equation by an expression of the form $A(x) \exp[iw(x)]$. This is the WKB approximation. If the analysis is pursued, one encounters the question what to do if there are the turning points where the approximation fails. Here however an almost exact integration is possible [91, Chap. 7], and on using this to connect solutions in the allowed and forbidden regions one finds that (7.1) is replaced by

$$\oint p\,dx = (n + \tfrac{1}{2})h \qquad\qquad (7.29)$$

which, since p involves E, is an implicit equation for the nth energy level. The wave function between the turning points a and b is approximated by

$$\psi(x) = N k^{-1/2} \cos\left[\int_a^x k(x)dx - \frac{1}{4}\pi \right] , \qquad a \le x \le b \qquad (7.30)$$

and that in the forbidden regions by analogous exponential expressions. This is one of the closest contacts between the old quantum theory and quantum mechanics.

But the WKB theory does more than enable one to derive and correct the equations of the old theory; it helps understand how the new one is related to what went before. Quantum mechanics says that the expectation value of a dynamical variable $a(p, x)$ is given in terms of the corresponding operator $\hat{a}$ by

$$\langle a \rangle = \int \psi^*(x)\hat{a}(p,x)\psi(x)dx \quad . \qquad\qquad (7.31)$$

Let us see how this formula works in the specially simple case in which a is a function of x only. We consider n to be large, but will later return to the question of small n. The wave function in the forbidden regions contributes negligibly to the integral, which we write as

$$\langle a \rangle = \frac{\displaystyle\int_a^b k^{-1}(x)a(x)\cos^2[\int_a^x k(x')dx' - \pi/4]dx}{\displaystyle\int_a^b k^{-1}(x)\cos^2[\int_a^x k(x')dx' - \pi/4]dx} \qquad \text{where} \qquad (7.32a)$$

197

$$\hbar k(x) = p(x) = \sqrt{2m[E - V(x)]} \quad .$$

If n is large, the (cosines)^2 in (7.32a) oscillate many times and can be replaced by their average value of 1/2:

$$\langle a \rangle = \frac{\displaystyle\int_a^b [\hbar k(x)]^{-1} a(x) dx}{\displaystyle\int_a^b [\hbar k(x)]^{-1} dx} \quad .$$

Interpret $\hbar k$ as the classical momentum:

$$\langle a \rangle = \frac{\displaystyle\int_a^b m^{-1}(dt/dx)a(x)dx}{\displaystyle\int_a^b m^{-1}(dt/dx)dx}$$

and finally, change the variable of integration from x to t:

$$\langle a \rangle = \frac{\displaystyle\int_{t_a}^{t_b} a[x(t)]dt}{\displaystyle\int_{t_a}^{t_b} dt} = \overline{a} \tag{7.33a}$$

where $\overline{a}$ is the time average. Thus expectation values in quantum mechanics correspond, in the limit of large quantum numbers, to time averages in classical theory. In particular, the rule which says that the first-order perturbation in energy produced by a perturbing Hamiltonian $\hat{H}_1$ is $\langle H_1 \rangle$ evaluated in the unperturbed state corresponds to a rule that a small perturbation of a classical system produces an energy change given by the time average of the perturbing energy evaluated in the unperturbed system. We will see in Chap. 8 that this rule follows from the adiabatic theorem.

Problem 7.23. Suppose that $\hat{a}$ is a dynamical variable containing $\hat{p}$ and x. Is (7.33a) still true? (Consider first $\hat{p}$ itself, then polynomials containing terms like $x^2 \hat{p} x \hat{p}^3 \ldots$)

Problem 7.24. Using (7.12) and a calculation similar to that carried out above, show that for large m, the dipole matrix elements $\langle n|x|m \rangle$ that define allowed optical transitions are given in classical terms by the Fourier amplitudes

$$\langle n|x|m \rangle = T^{-1} \int_0^T \cos\left(\alpha \Omega_m t\right) x(t) dt \quad , \quad \alpha = m - n \tag{7.33b}$$

where the time T covers a whole number of cycles, Ω_m is the classical frequency in the mth state, and $\alpha \ll m$.

It follows from the result of the last problem that the matrix elements of x correspond to the Fourier components of the classical variable, and we see how to interpret the fact that if $\langle n|x|m \rangle$ vanishes, the quantum transition is not allowed. The Fourier component corresponds to the classical frequency, and if the system is not vibrating at this frequency, one would not expect any radiation in classical physics either. Furthermore, since the intensity of radiation is proportional on the one hand to $|\langle n|x|m \rangle|^2$ and on the other to the square of the amplitude of the electromagnetic wave coupled to the Fourier amplitude, we see that (7.33b) controls intensities as well as selection rules.

The Correspondence Principle. At high quantum numbers, the results of classical and quantum theories are supposed to correspond. This is an obvious requirement of any version of a quantum theory and of course it emerges from the equations of the two theories. What is not trivial is that the WKB approximation is reasonably accurate at low values of n also, and that there is a one-to-one correspondence between states deduced by semi-classical methods and those of the exact theory. One is not led to expect this by the arguments given above, and it furnished a useful though heuristic guide in the early development of quantum theory. It is known as the correspondence principle, although until Heisenberg's theory (Sect. 7.5) explained why it is true Bohr preferred to call it the correspondence *argument.*

How Einstein's rule relates to quantum mechanics is clear if one remembers the connection between action and phase found in Chap. 1: $W = \hbar\phi$. If we circle the configuration space n times in a closed loop, $\Delta W = nJ = nh$, whence

$$\Delta\phi = 2\pi n \quad . \tag{7.34}$$

Evidently the quantum-mechanical "wave" is a wave in configuration space in all its dimensions, however many they are; and (7.34) is the condition of periodicity the wave must satisfy and also the natural generalization of the usual quantum condition to f degrees of freedom. There really isn't any "matter wave" in the same sense that there is a sound wave or an ocean wave. If one studies ocean waves from a moving boat the apparent frequency will change with the boat's speed and direction but the wavelength is a property of the water and nothing else. The wavelength of a "matter wave", on the other hand, is given by the momentum of the matter, and that depends partly on the motion of the observer. The wave function in quantum mechanics is a carrier of phase and amplitude, which are the two crucial parameters in terms of which natural processes are described. The next section will show that though phase and amplitude are essential to quantum mechanics, waves are not.

Improvements in WKB. I mentioned that the WKB approximation, with one foot in classical physics and the other in quantum, is often unexpectedly accurate, even for small quantum numbers. Figure 7.5a shows why: except in the neighborhood of the turning point, the approximation to the wave function is very good. Would there be some way to get the same kind of accuracy there? Beginning in 1961, V. Maslov in the Soviet Union published a number of papers showing how, still in the spirit of the WKB approximation, this could be done. Figure 7.5b shows that an orbit in the phase plane corresponding to an oscillatory motion in space has four turning

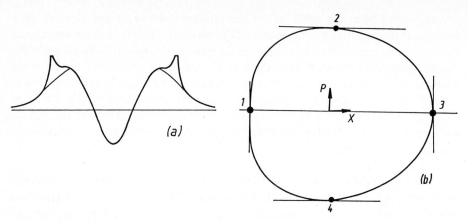

Fig. 7.5. (a) The WKB approximation to a low quantum state of an oscillator is very accurate away from the turning points of the classical orbits. **(b)** The four turning points of a typical oscillator represented in phase space

points, not two, which generally lie quite far apart. Maslov's idea, developed by many authors, is to solve the wave equation in the coordinate representation, which is good at turning points 2 and 4, and then in the momentum representation, where it is good at 1 and 3, and then combine the two to find the Maslov index, an integer α that improves the formula (7.6):

$$J_v = (n_v + \tfrac{1}{4}\alpha)h \quad , \quad n_v \in \mathbb{Z}_+ \quad .$$

For rotating systems in one dimension α comes out to be 0; for oscillating systems it is 2, consistent with the result obtained from the traditional method that integrates through the turning point and then matches with the WKB formulas at each side of it. For nonseparable systems an effort is made to correct Einstein's formula in the same way [92]. A perturbation calculation of the effect of electron-electron interaction in a helium atom agrees with a quantum-mechanical calculation to within a few percent for the lowest states and to within a fraction of a percent for the upper ones [93].

7.5 Heisenberg's Quantum Mechanics

After Hamilton had shown that there is an affinity between the equations of Newtonian mechanics and those of wave motion, almost a century elapsed before evidence was discovered that this formal relationship might have a counterpart in nature. It is not derogating from Schrödinger's discovery to say that if he had not followed his own path to his wave equation (Sect. 7.7) someone else would soon have derived it in another way. The first correct and general formulation of quantum mechanics did not stem from the idea of waves at all, but rather from a series of brilliant guesses about how to correct the foundering quantum mechanics of Bohr and Sommerfeld. The successes of this theory were always more noted than its failures, but by 1925 it was clear that the theory was giving wrong answers – the excited states of helium, for example, should have been easy to calculate by perturbation theory because their energies differ only slightly from those of hydrogen but *Born* and *Heisenberg* [94] were unable to get agreement with experiment. And it should have been possible to quantize the electronic motion in the ion H_2^+ exactly (neglecting the small effects of nuclear motion) and so determine the ion's binding energy [95], but here again, experiment disagreed with the result.

The locus of the discussions that finally gave birth to the "new quantum mechanics" was Bohr's Institute for Theoretical Physics at Copenhagen, though Heisenberg was quite alone when he finally found the right path, in June 1925, on the North Sea island of Helgoland. It is unnecessary to try to follow the full history here – much of it is given by *van der Waerden* [86], full of false starts and good guesses, but it may be instructive to see the formal way in which the old quantum theory was modified into the new.

Heisenberg [96], [97] started with the claim that an atomic theory should involve observable quantities and that it is therefore pointless, for example, to talk about the orbit of an electron around a nucleus. He modified also the most peculiar result of Bohr's theory: that except in the limit of large quantum numbers, the frequency ω_{mn} of light emitted from an atom in a transition from state m to state n is not the same as the frequency of any of the electronic motions inside the atom. The trouble is, of course, that ω_{mn} depends on the final state as well as the initial one:

$$\omega_{mn} = \hbar^{-1}(E_m - E_n) \tag{7.35}$$

and Heisenberg's first step was to assume that *all* dynamical quantities have the same kind of dual dependence: a coordinate, for example, was to occur in the theory as a quantity

$$x_{mn}(t) = X_{mn}e^{i\omega_{mn}t} \tag{7.36a}$$

though its precise relation to measurement remained to be specified. The corresponding momentum is

$$p_{mn} = \mu\dot{x}_{mn} = i\mu\omega_{mn}x_{mn} \tag{7.36b}$$

where μ is the particle's mass.

What these two-index quantities mean was not clear, but fluctuating quantities like the electric dipole moments responsible for radiation manifest themselves during transitions from one state to another $(m \neq n)$, while constant quantities such as energy refer to a single state $(m = n)$. How to calculate with two-index quantities of this kind is suggested, but only suggested, by the fact that in the limit of large quantum numbers, the different spectral frequencies can be understood as Fourier harmonics [see (7.9) and (7.33b)]. Let us examine this relation more closely.

Fourier Amplitudes. Suppose a one-dimensional system is in its mth quantum state, in which, according to Bohr's theory, it oscillates with frequency Ω_m. Then its coordinate can be represented by the Fourier series

$$x_m(t) = \sum_{\alpha=-\infty}^{\infty} X_m^\alpha e^{i\alpha\Omega_m t} \tag{7.37}$$

where the X_m^α are a set of constant Fourier amplitudes appropriate to the state m. They are not all independent because the coordinate $x_m(t)$ is real:

$$x_m^*(t) = x_m(t) \quad \text{or}$$

$$\sum_\alpha X_m^{\alpha*} e^{-i\alpha\Omega_m t} = \sum_\alpha X_m^\alpha e^{i\alpha\Omega_m t} = \sum_\alpha X_m^{-\alpha} e^{-i\alpha\Omega_m t}$$

so that

$$X_m^{\alpha*} = X_m^{-\alpha} \quad . \tag{7.38}$$

The correspondence principle suggests how to relate (7.37) to (7.36). In the limit of large quantum numbers, the radiated frequency as calculated by classical considerations should be the same as that given by (7.35). Classically, a system radiates at the natural frequency Ω_n and its Fourier harmonics:

$$\omega_{\mathrm{cl}} = \alpha\Omega_n \quad , \quad \alpha \in \mathbb{Z}_+$$

whereas according to quantum theory, in a transition from state n to a neighboring state $n - \alpha$, the frequency is

$$\omega_{\mathrm{qu}} = \frac{1}{\hbar}(E_n - E_{n-\alpha}) \simeq \frac{\alpha}{\hbar}\frac{\partial E_n}{\partial n} \quad . \tag{7.39}$$

(Note that here and subsequently, the "derivative" with respect to an index which varies in integer steps refers to a ratio of finite rather than infinitesimal quantities, as already remarked following (7.9).) In the limit of large n, the two frequencies are equal. We learn from this that α in expressions like (7.37) corresponds to $m - n$.

Quantum Amplitudes. Calculations are performed with functions of coordinates and momenta. To see how such functions may be formed, let us see how to represent x_m^2 in a series analogous to (7.37). We write

$$x_m^2(t) = \sum_{\alpha,\beta=-\infty}^{\infty} X_m^\alpha X_m^\beta e^{i(\alpha+\beta)\Omega_m t} \quad .$$

Let $\beta = \gamma - \alpha$. Then

$$x_m^2(t) = \sum_{\alpha,\gamma} X_m^\alpha X_m^{\gamma-\alpha} e^{i\gamma\Omega_m t} \tag{7.40}$$

so that the Fourier amplitude for x_m^2 is

$$\sum_\alpha X_m^\alpha X_m^{\gamma-\alpha} \quad . \tag{7.41}$$

I have suggested above that the Fourier amplitudes correspond to the quantum amplitudes in (7.36), and that α corresponds to $m - n$. Therefore in some sense,

$$X_m^\alpha \quad \text{corresponds to} \quad X_{m,m-\alpha} \quad . \tag{7.42}$$

Now what corresponds to (7.41)? Combining the quantum amplitudes together with their time dependence according to (7.40) gives

$$\sum_\alpha X_{m,m-\alpha} e^{i(E_m - E_{m-\alpha})t/\hbar} X_{m,m-\gamma+\alpha} e^{i(E_m - E_{m-\gamma+\alpha})t/\hbar}$$

corresponding to (7.41). This does not have the required time dependence (7.36a), but if we change it to

$$\sum_\alpha X_{m,m-\alpha} e^{i(E_m - E_{m-\alpha})t/\hbar} X_{m-\alpha,m-\gamma} e^{i(E_{m-\alpha} - E_{m-\gamma})t/\hbar}$$

or, more succinctly,

$$(x^2)_{m,m-\gamma} = \sum_\alpha x_{m,m-\alpha} x_{m-\alpha,m-\gamma} \tag{7.43}$$

then the terms $E_{m-\alpha}$ cancel and $(x^2)_{m,m-\gamma}$ has a time dependence in agreement with (7.36a) as well as with the corresponding quantity (7.40). The notation can be simplified: let $m - \alpha = k$ and $m - \gamma = n$. Then (7.43) is

$$(x^2)_{mn} = \sum_k x_{mk} x_{kn} \tag{7.44}$$

where k runs over all quantum states of the system. This is Heisenberg's first invention. Note that at this point he had stopped merely transcribing the old quantum theory into a new notation; by changing an equation so as to make it agree with the frequency condition (7.39) he was introducing new physics. The reader will recognize the rule for matrix multiplication but Heisenberg did not recognize it. Apparently he had never heard of matrices. He noted that according to this recipe for forming products, $(xy)_{mn} \neq (yx)_{mn}$ and went on to draw momentous conclusions.

The Quantum Condition. In the old quantum theory, Planck's constant made its appearance via the action integral,

$$\oint p_v dq_v = m_v h \quad, \qquad v = 1, 2, \ldots f \tag{7.45}$$

(Sect. 7.1), where the m_v are integers. Heisenberg's next task was to find how h occurs in a calculation based on the two-index quantities. Again, the calculation by Fourier amplitudes furnished a guide. Let

$$p_m(t) = \sum_\alpha P_m^\alpha e^{i\alpha \Omega_m t} \quad.$$

Then in the mth state of the one-dimensional system,

$$J_m = \oint p_m \dot{x}_m dt = \sum_{\alpha,\beta} \int_0^{2\pi \Omega_m^{-1}} P_m^\alpha e^{i\alpha \Omega_m t} i\beta \Omega_m X_m^\beta e^{i\beta \Omega_m t} dt \quad.$$

The integral is zero except for the terms in which $\beta = -\alpha$:

$$J_m = \frac{2\pi}{\Omega_m} \sum_\alpha (-i\alpha \Omega_m) P_m^\alpha X_m^{-\alpha} \quad \text{or}$$

$$J_m = -2\pi i \sum_\alpha \alpha P_m^\alpha X_m^{-\alpha} \quad. \tag{7.46}$$

The basic assumption of Bohr's method of quantization is that J_m increases in steps of h, so that

$$J_m = -2\pi i \sum_\alpha \alpha P_m^\alpha X_m^{-\alpha} = mh + \text{const} \quad .$$

Transcribe it in terms of quantum amplitudes just as (7.41) was transcribed into (7.43),

$$-2\pi i \sum_\alpha \alpha p_{m,m-\alpha} x_{m-\alpha,m} = mh + \text{const} \tag{7.47}$$

in which the exponential time factors now cancel. Now differentiate with respect to m!

$$-2\pi i \sum_\alpha \alpha \frac{\partial}{\partial m}(p_{m,m-\alpha} x_{m-\alpha,m}) = h \tag{7.48}$$

with "differentiation" as in (7.39). Heisenberg constructs the integer relation of which this is the quantum limit. He writes

$$\alpha \frac{\partial}{\partial m} f(m) = f(m+\alpha) - f(m)$$

so that in the limit of small α, (7.48) becomes

$$\sum_\alpha (p_{m+\alpha,m} x_{m,m+\alpha} - p_{m,m-\alpha} x_{m-\alpha,m}) = i\hbar \tag{7.49}$$

with $\hbar = h/2\pi$. This is Heisenberg's second invention. Let $m + \alpha = k$ in the first term and $m - \alpha = k$ in the second. Then

$$\sum_k (x_{mk} p_{km} - p_{mk} x_{km}) = i\hbar$$

or in matrix notation,

$$(xp - px)_{mm} = i\hbar \quad . \tag{7.50}$$

This is the first time that either the imaginary i or non-commuting dynamical quantities, what Dirac later called q-numbers, had ever occurred in physics in any essential way.

A Simple Application. The application of these ideas to a harmonic oscillator is so simple that it is not an adequate test of the theory, and even in his first paper Heisenberg went on to consider the more difficult problem of the energy levels of an anharmonic oscillator, but the simpler problem contains some, at least, of the essential ideas. With $\mu = $ mass and $p_{mn} = \mu \dot{x}_{mn}$, (7.50) is

$$i\hbar = i\mu \sum_k (x_{nk}\omega_{kn} x_{kn} - \omega_{nk} x_{nk} x_{kn}) \quad \text{or}$$

$$\hbar = 2\mu \sum_k \omega_{kn} x_{nk} x_{kn}$$

because $\omega_{kn} = -\omega_{nk}$. The great simplification here is that the oscillator is harmonic; its motion is at the frequency $\omega_{n,n-1} = \omega$ and there are no higher harmonics. Thus the sum has only two nonvanishing terms:

$$\hbar = 2\mu\omega(x_{n,n+1}x_{n+1,n} - x_{n,n-1}x_{n-1,n}) \quad . \tag{7.51}$$

Heisenberg assumes, in analogy with (7.38), that

$$x_{mn} = x_{nm}^*$$

so that (7.51) is

$$|X_{m,m+1}|^2 = |X_{m,m-1}|^2 + \frac{\hbar}{2\mu\omega} \quad . \tag{7.52}$$

In this relation let $m = 0$, corresponding to the ground state. Since there is no lower state, $X_{m,m-1} = 0$ and

$$|X_{0,1}|^2 = \frac{\hbar}{2\mu\omega} \quad .$$

We can now use (7.52) to find all the $|X_{m,m+1}|^2$:

$$|X_{m,m+1}|^2 = (m+1)\frac{\hbar}{2\mu\omega} \quad . \tag{7.53}$$

Note that $X_{mm} = 0$: the oscillator's displacement x has, of course, no zero-frequency Fourier component, and the time-dependence of x is manifest only during a radiative transition, in which X_{mn} $(m \neq n)$ plays a part.

Now calculate the energy. Assume the formula from classical physics

$$H = \tfrac{1}{2}\mu(\dot{x}^2 + \omega^2 x^2)$$

and transcribe it as

$$H_{mn} = \tfrac{1}{2}\mu[(\dot{x}^2)_{mn} + \omega^2(x^2)_{mn}] \quad .$$

The nonvanishing H_{mn} can have $n = m \pm 2$ and $n = m$. We find

$$\begin{aligned}
H_{m,m+2} &= \tfrac{1}{2}\mu(\dot{x}_{m,m+1}\dot{x}_{m+1,m+2} + \omega^2 x_{m,m+1}x_{m+1,m+2}) \\
&= \tfrac{1}{2}\mu[(-i\omega)x_{m,m+1}(-i\omega)x_{m+1,m+2} + \omega^2 x_{m,m+1}x_{m+1,m+2}] = 0
\end{aligned}$$

and similarly, $H_{m,m-2} = 0$. The diagonal matrix element is

$$\begin{aligned}
H_{mm} = \tfrac{1}{2}\mu[&\dot{x}_{m,m+1}\dot{x}_{m+1,m} + \dot{x}_{m,m-1}\dot{x}_{m-1,m} \\
&+ \omega^2(x_{m,m+1}x_{m+1,m} + x_{m,m-1}x_{m-1,m})]
\end{aligned}$$

or

$$H_{mm} = \omega^2\mu(|x_{m,m+1}|^2 + |x_{m,m-1}|^2) \quad . \tag{7.54}$$

By (7.53), this is

$$H_{mm} = \omega^2\mu(m+1+m)\frac{\hbar}{2\mu\omega} \quad \text{or}$$

$$H_{mm} = (m + \tfrac{1}{2})\hbar\omega \quad . \tag{7.55}$$

Thus the energy is represented by the elements of a time-independent diagonal matrix. The extra half in the energy was not entirely new. It had been conjectured earlier as a necessary correction to the Planck formula but this was the first time it had emerged from a mathematical scheme.

The unfamiliar quantity in all this is the quantum amplitude X_{mn}. Clearly, it has something to do with the amplitude of oscillations. Let the energy of state m be $(m + 1/2)\hbar\omega$. Then the amplitude A_m as one would calculate it classically is given by

$$E_m = \frac{1}{2}(2m + 1)\hbar\omega = \frac{1}{2}\mu\omega^2 A_m^2 \quad , \qquad A_m^2 = (2m + 1)\frac{\hbar}{\mu\omega} \quad .$$

This is something like (7.53) but not the same. And the remarkable thing about the calculation of (7.55) is that nowhere, even implicitly, has it been assumed that the oscillator is executing sinusoidal motion. The idea of a trajectory does not occur.

Equation of Motion. Suppose that some dynamical variable f is represented by a matrix

$$f_{mn} = F_{mn}e^{i\omega_{mn}t} \quad .$$

Then its time derivative is

$$\dot{f}_{mn} = i\omega_{mn}F_{mn}e^{i\omega_{mn}t}$$

$$= \frac{i}{\hbar}(H_{mm} - H_{nn})f_{mn}$$

$$= \frac{i}{\hbar}(H_{mm}f_{mn} - f_{mn}H_{nn}) \quad .$$

If H is a diagonal matrix, this is

$$\dot{f}_{mn} = \frac{i}{\hbar}(Hf - fH)_{mn} \quad . \tag{7.56}$$

Heisenberg's theory was put together out of a belief that the dynamical variables of atomic physics represent something other than coordinates and momenta in the classical sense, and out of an understanding of the correspondence relations. He was also able to assume that, transcribed into his new quantum amplitudes, the Hamiltonian functions from classical physics would still be of use. It is the miraculous effectiveness of classical Hamiltonians clothed in new mathematics that has led to the triumphs of atomic theory in the last half-century. The difficulties of the future theory that will explain the structure and properties of particles may be much greater, for there is no reason to think that classical physics will suggest what to write down.

In Heisenberg's original paper it is not at all clear what the quantum amplitudes mean or how a dynamical variable can refer to two states at once. That was clarified gradually in the next few years. And the theory was ill-adapted to computation. Heisenberg tried and failed to derive the Balmer formula; Pauli accomplished it after hard work. In the next year Schrödinger invented wave mechanics and shortly afterward showed that it is mathematically equivalent with matrix mechanics where the two overlap. They overlap for physical systems that can be described by coordinates and momenta having a continuous range of variables – the r, ϑ, ϕ of atomic orbitals for example – but situations involving spin and isospin, for example, are usually treated by matrix methods, which are in principle more general.

Problem 7.25. Assuming that Heisenberg's quantum amplitude X_{mn} is the matrix element $\langle m|x|n \rangle$, verify that the equations involving this quantity derived above are in fact correct.[3]

Problem 7.26. Show that the nondiagonal elements $(xp - px)_{mn}$ $(m \neq n)$ are zero for the harmonic oscillator.

Problem 7.27. Heisenberg constructed the commutation relation (7.50) as the expression of the old quantum condition (7.45) in terms of his new quantum amplitudes. In order to appreciate the subtlety of the argument, start with the commutation relation, plus anything else from quantum mechanics that you need, and derive the old quantum condition from it.

Shortly after Heisenberg's paper, *Born* and *Jordan* [97] showed how to generalize (7.50) to include the off-diagonal elements as well,

$$(xp - px)_{mn} = i\hbar \delta_{mn} \quad . \tag{7.57}$$

The theory was then complete in that if one worked hard enough one could solve or approximate any dynamical system with a finite number of degrees of freedom.

7.6 Matter Waves

In 1923 *de Broglie* [98], [99] proposed the existence of matter waves, derived the relation between wavelength and momentum by a relativistic argument, and proposed experiments by which the hypothesis could be checked. We will not follow his arguments here, but will instead show some connections with what we have already done.

The propagation constant k of a matter wave is given by

$$k = \hbar^{-1}p = \sqrt{2m\hbar^{-2}[E - V(r)]} \tag{7.58}$$

and the phase change along an element of path dr and a time dt is

$$d\phi = \hbar^{-1}(p \cdot dr - E\,dt) \quad , \qquad E = \hbar\omega$$

so that the phase change in going from a point r_0 at t_0 to a point r at t is

$$\Delta\phi = \hbar^{-1} \int_{r_0,t_0}^{r,t} (p \cdot \dot{r} - E)dt$$

integrated along the natural path connecting r_0 and r. Since it is a natural path, we can replace E by $H(p, r)$ so that

$$\Delta\phi = \hbar^{-1} \int_{r_0,t_0}^{r,t} L(r, \dot{r})dt = \hbar^{-1}S(r, t) \tag{7.59}$$

---

[3] Having established this, we can verify that X_m^α and $X_{m,m-\alpha}$ are actually the same in the correspondence limit. See (7.42).

where the notation reflects the fact that S is (for a given r_0 and t_0) a function only of the end point of the integration. Thus, in this approximation, a de Broglie wave originating at r_0, t_0 is given by a "wave function"

$$\psi(r, t) = a(r, t) \exp\left[\frac{i}{\hbar} S(r, t)\right] = a(r, t) \exp\left[\frac{i}{\hbar} [W(r) - Et]\right] \tag{7.60}$$

where a is some suitably chosen amplitude. This is nothing but the phase-integral approximation derived from the wave equation in Sect. 1.1, but here derived from classical mechanics plus de Broglie's hypothesis. Classical mechanics provides no $\hbar$, though there must be such a constant if S is to be considered as a phase.

In (7.5a) we have noted that the phase integral J is invariant under contact transformations. Thus the advance in phase corresponding to a complete oscillation of the system is independent of the coordinates used. In (7.34) we have the statement of a hypothesis more general still: that the same applies to a system of arbitrary complexity described by any number of variables, not just the three that occur in (7.60). In its full generality, (7.35) describes a wave in any number of canonical coordinates. The wave is a wave in q space, which only for a single point particle coincides with the x, y, z of ordinary space.

Stationary Phase. The phase W in (7.60) was written as a function of spatial co-ordinates because the derivation went that way. But we know that W is invariant under canonical transformations. We can write it as $W(q)$, and for systems more complicated than a single particle, q may represent any set of generalized coordinates.

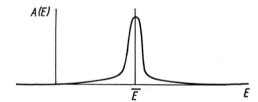

Fig. 7.6. Distribution in $A(E)$ that forms a wave packet

Superpose waves of the form (7.60) with phases $\phi = \hbar^{-1}[W(q) - Et]$ and initial amplitudes distributed in E as in Fig. 7.6. According to the principle of stationary phase (Sect. 1.2) the amplitude will subsequently be large only where $\partial\phi/\partial E = 0$, or $\partial W/\partial E = t$, with E evaluated at $\overline{E}$, as in (6.6a). The wave in configuration space contains Hamiltonian mechanics in generalized coordinates as a limiting case.

One can go further. Construct a new form of characteristic function which will be called the *orbital function* U, defined as

$$U := W(q, \alpha) - \alpha_n \beta_n$$

and construct as before a wave packet by superposing waves with different α's:

$$\psi(q) = \int A(\alpha) e^{i\hbar^{-1}U} d^f\alpha \quad . \tag{7.61}$$

The condition for constructive interference now says that

$$\left. \frac{\partial U}{\partial \alpha_n} \right|_{\alpha_n = \bar{\alpha}_n} = 0 \quad n = 1 \ldots f \quad \text{or}$$

$$\left. \frac{\partial W}{\partial \alpha_n} \right|_{\alpha_n = \bar{\alpha}_n} = \beta_n$$

as in (6.44). We see that (7.61) represents a wave packet whose center satisfies Jacobi's equations.

In classical dynamics, with (6.44) satisfied, a variation in U is given by

$$\delta U = \frac{\partial W}{\partial q_n} \delta q_n + \frac{\partial W}{\partial \alpha_n} \delta \alpha_n - \beta_n \delta \alpha_n - \alpha_n \delta \beta_n \quad \text{or}$$

$$\delta U = \frac{\partial W}{\partial q_n} \delta q_n - \alpha_n \delta \beta_n \quad .$$

Thus U is conveniently regarded as a function of q and β with

$$\frac{\partial U}{\partial q_n} = \frac{\partial W}{\partial q_n} = p_n \quad , \quad \frac{\partial U}{\partial \beta_n} = -\alpha_n \quad .$$

The difference in meaning between U and W is clear when we remember that the α's correspond to conserved dynamical variables and the β's to initial positions in configuration space. Thus W, with given values of the α's, allows one to calculate families of orbits corresponding to different initial positions, whereas U, with given initial positions, allows one to calculate families of orbits corresponding to different constant values of the α's. It is the interference between different members of these α-families that creates the wave packets.

It is now clear why the speed of a light signal in (6.87) came out to be the group velocity. But note that what has been done here is in no sense a mere extension of the Hamilton-Jacobi theory or a new way of expressing the argument, for we have introduced an idea that is entirely new to mechanics, the possibility of superposing waves to form a wave packet. Though this is a fundamental feature of classical wave theories it makes no sense at all within the framework of classical mechanics. The assimilation of Hamilton-Jacobi mechanics into a wave theory depends on two entirely new ideas, superposition of wave amplitudes and introduction of a universal constant having the same units as W. From the publication of Hamilton's first paper "On a general method of expressing the paths of light, and of the planets, by the coefficients of a characteristic function" (1833) to de Broglie's tentative hypothesis, rather useless because it still lacked a wave equation, took 90 years. From the Wright brothers to the moon landing took 66 years.

Problem 7.28. Show, by adapting standard methods and definitions, that it is impossible to form a wave function of the form (7.61) that corresponds simultaneously to arbitrarily narrow distributions of α_n and β_n. What "indeterminacy principle" results?

Problem 7.29. How does one change (7.61) so as to produce a time-dependent wave function?

7.7 Schrödinger's "Derivation"

De Broglie's idea of matter waves was original and entirely unexpected; in the establishment of a dynamical theory it went about as far as Huygens' principle went in the direction of a wave theory of light. Heisenberg was the first to write an equation of motion for quantum mechanics, but the matrix theory was clumsy from a computational point of view and had no visible connection with de Broglie's waves. Erwin Schrödinger, in a series of brilliant papers published in 1926, found the wave equation that is the proper dynamical basis for de Broglie's construction and applied it successfully to calculations that only a virtuoso could carry through with matrices.

In his first paper (see Sect. 10.7) Schrödinger justifies his time-independent wave equation with some mathematical legerdemain (see Sect. 10.7) that leaves the reader no wiser than before. The second paper, a month later, presents a careful line of reasoning that leads from specific hypotheses to the desired result. The hypotheses are not, of course, justified by any considerations more fundamental, but they are clear and simple. The version to be given here does not do justice to the original. It duplicates the line of thought but omits the art, for Schrödinger works in generalized coordinates describing a system of arbitrary complexity, while I will treat a single particle in terms of x, y, and z.

Make four assumptions:

1. Mechanical quantities are correctly analyzed by the theory of Hamilton and Jacobi.
2. The particle is represented by a wave of which the wave fronts are surfaces of constant S.
3. The frequency of the wave is $\nu = E/h$, where E is the particle's total energy.
4. Matter waves obey d'Alembert's equation.

Since $S = W - Et$, a surface of constant S advances at a rate given by $\Delta W = E\Delta t$. To find ΔW, integrate $\nabla W = p$ over an infinitesimal element of path perpendicular to the wave front to give

$$\Delta W = \int p \cdot ds \quad .$$

Since as we have seen in Sect. 1.4 such a path is parallel to the momentum of the particle, we can write the integral as $p(x)\Delta s$ where

$$p(x) = \sqrt{2m[E - V(x)]} \quad .$$

Thus the surface of constant S advances at a rate $v_S = E/p(x)$, and this is assumed to be the phase velocity of the matter waves. Assuming that $E = h\nu$ now gives

$$v_S = \frac{h\nu}{p(x)} \tag{7.62}$$

from which the wavelength is $\lambda = h/p(x)$, de Broglie's relation, and it is easy to show (Problem 7.30) that the group velocity of such a wave equals the particle velocity.

A short step leads to Schrödinger's equation. D'Alembert's equation, which we have assumed, reads

$$\nabla^2\psi = \frac{1}{v_S^2}\frac{\partial^2\psi}{\partial t^2} = -\frac{(2\pi\nu)^2}{v_S^2}\psi$$

and by (7.62) this is

$$\nabla^2\psi = -\frac{2m[E - V(x)]}{\hbar^2}(2\pi)^2\psi$$

or with $\hbar = h/2\pi$,

$$-\frac{\hbar^2}{2m}\nabla^2\psi + V(x)\psi = E\psi \quad .$$

The time-dependent equation was written down three months later after what was evidently a struggle over introducing imaginary quantities, in Paper IV [100], [23]. Shortly before this *Schrödinger* had used the theory of eigenfunction expansions to show that his wave theory, which looked so different from Heisenberg's matrices, was in fact equivalent to it [101], [23]. We will not need this development.

Seldom in the history of science has the author of a great advance communicated it to the public with such care, style, and even charm as Schrödinger, and these epochal papers are still worth reading to see the speed and skill with which their author overcomes the conceptual difficulties that stood in his way when he began. Nevertheless the "derivation" is not satisfactory. It mixes particle variables with wave variables in a way that is not clear, it assumes that surfaces of constant phase coincide with surfaces of constant S without saying what is the relation between them, it assumes that matter waves satisfy d'Alembert's equation, which they don't. One may ask whether there is really any use in pretending to derive quantum mechanics from classical mechanics. It is impossible of course – the theories answer entirely different questions – but nevertheless there is one strong connection between them, the correspondence principle. In the limit of large quantum numbers, when one has decided how statements from one theory can be translated into the language of the other, the results of quantum and classical dynamics should ordinarily agree. There are ambiguities in the theory of fields, especially gauge fields, which can be removed by considering their classical counterparts, and for this reason it is important to have arguments that lead from the classical to the quantum domain in a way that makes the assumptions as clear as possible. Such an argument will be sketched in the next section.

Problem 7.30. Show that the group velocity that corresponds to the phase velocity v_S is the particle velocity in mechanics.

Problem 7.31. Contrary to Schrödinger's hypothesis, his matter waves do not in fact obey d'Alembert's equation. Does this fact invalidate the "derivation"?

Remarks on the Nature of Dynamics. When the quantum mechanics of Heisenberg and Schrödinger came along, most people, specially those in the habit of thinking deeply about physics, were profoundly puzzled. How could it be, in the coordinate picture for example, that momentum appears as an operator (involving $\sqrt{-1}$!) and not as a number? In 1927 Heisenberg's paper on indeterminacy showed why: exact position and exact momentum are idealizations, and something in nature prevents us from specifying both at once. How to give this vague statement an exact meaning was the subject of long debates in the subsequent years and even now that it is settled to everyone's satisfaction one finds that no two people seem to have settled it in the same way. What surprised people the most was being told that ever since the Middle Ages, in basing theories of dynamics on a knowledge of a body's position ($\Delta x = 0$) as it moves around, dynamicists had been operating in a regime in which $\Delta p = \infty$ and they knew nothing whatever about its momentum. Yet still, they had been getting answers that agreed with observation. This is the approximation of classical dynamics: one assumes one knows $\langle x \rangle$ with arbitrary precision and then defines velocity as $d\langle x \rangle/dt$ and momentum as $md\langle x \rangle/dt$. It is as if one thought one could learn all about motion by looking at successive frames of a motion picture film. One could learn much about motion – everything, perhaps, except what motion actually is. The inventors of quantum mechanics found that change had always been imagined in terms of a sequence of static situations, and nobody until then had ever seen real dynamics at all. The difference becomes clear when one compares the quantum dynamical variable $\hat{x}$ with $\langle x \rangle$ and $\hat{p}$ with $\langle p \rangle$.

The situation has reminded many authors of Zeno's paradox known as The Arrow: If you claim that an arrow is in a certain place and moving in a certain way you are guilty of a contradiction, for if it is really in that place it cannot be in motion. The inventors of calculus grappled with this problem in the 17th and 18th centuries and it was finally settled with appropriate arguments involving limits and continuity in the 19th, but Zeno was talking about arrows, not mathematics, and if you apply to arrows the lessons learned from calculus you arrive at a dynamical theory which assigns to the arrow coordinates and momenta simultaneously and is therefore false. Quantum mechanics cannot be said to be the resolution of Zeno's paradox since we are allowed to assign values to x and p_y at the same time, but it may resolve some more sophisticated statement of it.

7.8 Construction of a Wave Function

The difficulty with a wave function like (7.61) is that it refers only to the natural paths of the particles described, whereas we know that waves diffract and do not follow the natural paths exactly. The following argument, due to *Feynman* [102], shows how one can use the principle of superposition to construct true quantum-mechanical wave functions out of action integrals evaluated along paths that are not natural. Begin with a special case.

To find the wave amplitude at r in Fig. 7.7, Huygens's principle states that one must superpose the rays that have travelled from r_0 to r via the different slits, each with its proper phase. Assuming that the wave function to the left of the grating is known, construct the amplitude at r at time t as the sum of amplitudes contributed by the slits, the ith one being

$$a_i w b_i(\boldsymbol{r}) e^{iS_i(\boldsymbol{r},t)/\hbar} \ . \tag{7.63}$$

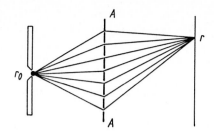

Fig. 7.7. The amplitude at r is found by coherent superposition of contributions from the slits

Here a_i is the amplitude at slit i from the source at r_0, w is the slit width, and

$$wb_i(r)e^{iS_i(r,t)/\hbar}$$

is the amplitude, together with its proper phase, produced at r at time t by a wave of unit amplitude arriving at the slit. The factor b_i involves the inverse-square law and some obliquity factor, but we do not know it a priori. It must come out of the theory. $S_i(r, t)$ is, of course, Hamilton's principal function evaluated along the natural path from the slit to r. It is also a function of the initial point r_0 and is therefore, in each straight section, the free-particle action given in (6.11).

The role of the barrier was to allow us to discuss the contribution to the amplitude at r made by Huygens wavelets originating at different points along the plane of the grating. If we take away the grating, all points of the plane contribute, and we have the situation of Fig. 7.8. The amplitude at r is not just that contributed by the natural path from r_0 to r, which is a straight line.

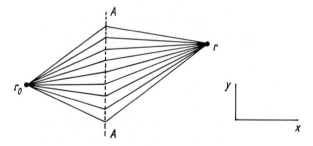

Fig. 7.8. Every point of the plane can be regarded as a slit

Finally there is nothing special about the plane A-A or this particular set of broken-line paths. We therefore consider all imaginable paths from r_0 and r that occupy the same time interval. Figure 7.9 shows the natural path together with some of the varied paths. We have introduced a force field, so that the natural path is curved. We must evaluate S along each path, calculate $e^{iS/\hbar}$, and find some way of summing over the different paths. The values of S differ for different paths; because $\hbar$ is small, the $S/\hbar$ differ widely, and the terms of the sum tend to cancel each other.

Now however we make use of the stationary property of S: for slight variations away from the natural path, it changes very little. The situation is the same as in Sect. 1.2 and Sect. 7.6, where we found the motion of a wave packet by the principle

213

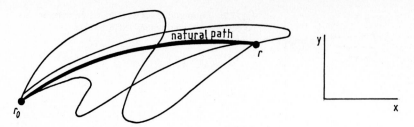

Fig. 7.9. Every path contributes to the amplitude at r

of stationary phase. Only varied paths in the immediate neighborhood of the natural path contribute appreciably to the sum.

We can generalize (7.63) as follows. Let $\psi(x',0)$ be the amplitude of the wave function at time 0. (This is analogous to a_i in (7.63).) Consider a particular trajectory leading from x' to x. It is characterized not only by a curve along which the particle travels from one point to the other but also by its rate of travel from point to point along the curve; I will call it a history, H. Let the Lagrangian belonging to this particular history be L_H, and let

$$S_H(x, x', t) = \int_0^t L_H \, dt \quad .$$

Then, still looking at (7.63), the amplitude at x at time t contributed by this history is

$$A_H(x, t) = e^{i\hbar^{-1}S_H} \psi(x', 0)$$

and the hypothesis is that the wave function at x, t is given by the sum of all such terms, contributed by different paths from x' to x and from the amplitudes at different points x',[4]

$$\psi(x, t) = \int \sum_H e^{i\hbar^{-1}S_H} \psi(x', 0) dx' \tag{7.64}$$

where dx' is analogous to the slit width w in (7.63). The sum in this expression is called a propagator, since it takes a wave function from one point to another in spacetime. How the sum is to be carried out is not specified and indeed there is no general procedure, but the idea is that it may not matter very much. The values of S for different histories will produce contributions to the amplitude that tend to cancel except for histories very close to the natural one. Here S is stationary, and the contributions of these neighboring histories will tend to add. Thus we can hope that the propagator will have the form

$$\sum_H e^{i\hbar^{-1}S_H(x, x', t)} = b e^{i\hbar^{-1}S_N(x, x', t)} \tag{7.65}$$

where S_N is evaluated for the natural path and b is some normalizing factor analogous

---

[4] That wave functions might be constructed in this way was suggested long ago by *Dirac* [104].

214

to b_i in (7.63), contributed by paths close to the natural one. The phases for different initial points x' are given by S_N and so b may depend on t but not on x or x'. Its value is determined when the sum over histories is carried out. For simple cases this can be done [103], but it will be convenient here to proceed differently.

If S_H is Infinitesimal. The argument sketched above is global, seeking to construct a wave function over a finite (or infinite) interval of spacetime. It is perhaps analogous to Gauss's theorem in electrostatics. In electrostatics one can also take an infinitesimal approach, studying the field contributed by an infinitesimal charge element: this leads to the differential equation $\nabla \cdot D = \varrho$. In his first paper [102] *Feynman* showed how the infinitesimal version of his argument also leads to a differential equation.

Consider the simplest case, motion in one dimension, drawn in xt space in Fig. 7.10(a). The last infinitesimal bit of the history is shown in (b), where the trajectories have drawn close together and only a time ε remains before they all arrive at x, t. Now $\psi(x, t)$ becomes

$$\psi(x, t) = b \int e^{i\hbar^{-1} S_N(x, x', \varepsilon)} \psi(x', t - \varepsilon) dx'$$

where dx' corresponds to w in (7.63). Since only paths close to the natural one contribute, b is taken outside the integral.

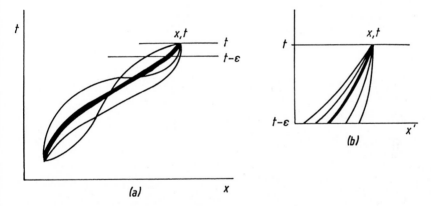

Fig. 7.10. (a) Path in one dimension represented in Lagrangian qt space. (b) Magnified view of the end of the path

The intervals are now all infinitesimal. We write $\dot{x}$ as $(x - x')\varepsilon^{-1}$ so that

$$L \simeq \frac{1}{2} m \left(\frac{x - x'}{\varepsilon} \right)^2 - V(x) \quad \text{and}$$

$$S \simeq L\varepsilon = \frac{1}{2} m \frac{(x - x')^2}{\varepsilon} - \varepsilon V(x) \quad . \tag{7.66}$$

It is convenient to introduce

$$\xi = x - x' \quad , \quad dx' = -d\xi$$

215

so that

$$\psi(x,t) = b \int_{-\infty}^{\infty} \exp\left(\frac{i}{\hbar}\left[\frac{m\xi^2}{2\varepsilon} - \varepsilon V(x)\right]\right) \psi(x - \xi, t - \varepsilon)d\xi \quad .$$

Since only values of x' near x matter, we expand in powers of ξ,

$$\psi(x,t) = be^{-i\varepsilon\hbar^{-1}V(x)} \int_{-\infty}^{\infty} \exp\left(\frac{i}{\hbar}\frac{m}{2\varepsilon}\xi^2\right) \left[\psi(x, t - \varepsilon) - \xi\frac{\partial\psi(x, t - \varepsilon)}{\partial x}\right.$$

$$\left. + \frac{\xi^2}{2}\frac{\partial^2\psi(x, t - \varepsilon)}{\partial x^2} + \ldots\right]d\xi \quad .$$

Since

$$\int_{-\infty}^{\infty} e^{-ax^2} dx = \sqrt{\pi a^{-1}} \quad , \qquad \text{Re } a \geq 0 \tag{7.67}$$

we have

$$\int_{-\infty}^{\infty} e^{iax^2} dx = \sqrt{i\pi a^{-1}} \qquad \text{while} \tag{7.68}$$

$$\int_{-\infty}^{\infty} xe^{iax^2} dx = 0 \quad , \qquad \int_{-\infty}^{\infty} x^2 e^{iax^2} dx = \frac{i}{2a}\sqrt{i\pi a^{-1}} \quad . \tag{7.69}$$

Thus

$$\psi(x,t) = b\sqrt{\frac{2\pi i\hbar\varepsilon}{m}}e^{-i\varepsilon\hbar^{-1}V(x)}\left[\psi(x, t - \varepsilon) + \frac{i\varepsilon\hbar}{2m}\frac{\partial^2\psi(x, t - \varepsilon)}{\partial x^2} + \ldots\right] \quad .$$

Expand the right side in powers of ε, keeping only ε^0 and ε:

$$\psi(x,t) = b\sqrt{\frac{2\pi i\hbar\varepsilon}{m}}$$

$$\times \left[\psi(x,t) - \varepsilon\frac{\partial\psi(x,t)}{\partial t} - \frac{i\varepsilon}{\hbar}V(x)\psi(x,t) + \frac{i\hbar\varepsilon}{2m}\frac{\partial^2\psi(x,t)}{\partial x^2}\right] \quad .$$

We get agreement in the limit $\varepsilon \to 0$ if

$$b = \sqrt{\frac{m}{2\pi i\hbar\varepsilon}} \quad . \tag{7.70}$$

The first power of ε then gives

$$i\hbar\frac{\partial\psi}{\partial t} = -\frac{\hbar^2}{2m}\frac{\partial^2\psi}{\partial x^2} + V(x)\psi$$

which may be familiar.

The mathematical method is not familiar. "One feels," wrote Feynman, "as Cavalieri must have felt calculating the volume of a pyramid before the invention of calculus." The derivation just given is Feynman's. Shortly afterwards Pauli, wishing to see what it has to do with quantum mechanics and what the approximations really are, did it over in a completely different way [106, Chap. 7]. Pauli's derivation deserves careful study. The method's possibilities are explored in [103]; see also [107–110].

Finding the Momentum. As an example of elementary calculations involving sums over histories, I will calculate $\langle p \rangle$, the expectation value of the momentum of a particle in a given state. Consider first a trajectory from x_0 to x_1, Fig. 7.11. The trajectory is divided into two parts with histories H and H', on each side of the point x, t. The procedure will be to evaluate $p_H(x)$, the momentum at x contributed by a path with a certain history, and then sum over histories, using the phase factors as before to weight contributions from the different paths. Finally, since x can be any point, integrate over x. This gives

$$\langle p \rangle \propto \int \sum_{H,H'} \left\{ p_H(x) \exp\left[\frac{i}{\hbar} \int_{t_0}^{t} L_H dt\right] \cdot \exp\left[\frac{i}{\hbar} \int_{t}^{t_1} L_{H'} dt\right] \right\} dx \quad .$$

But

$$p_H(x) \exp\left[\frac{i}{\hbar} \int_{t_0}^{t} L_H dt\right] = p_H(x) \exp\left[\frac{i}{\hbar} S_H(x,t)\right] = -i\hbar \nabla \exp\left[\frac{i}{\hbar} S_H(x,t)\right]$$

so that

$$\langle p \rangle \propto \int \sum_{H,H'} \left\{ -i\hbar \nabla \exp\left[\frac{i}{\hbar} \int_{t_0}^{t} L_H dt\right] \cdot \exp\left[-\frac{i}{\hbar} \int_{t_1}^{t} L_{H'} dt\right] \right\} dx \quad .$$

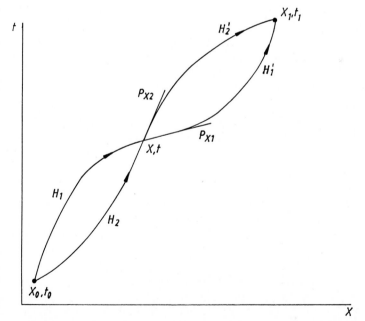

Fig. 7.11. Contributions to the x component of the particle's momentum at point x, t from two possible histories. The momenta are determined by the slopes of the tangents marked p_{x1} and p_{x2}

217

This is for given initial and final points x_0 and x_1. Equation (7.64) shows how to construct the wave function at (x, t): multiply the first factor by $\psi(x_0, 0)$, integrate over x_0, and sum, and similarly for the second factor, taking account of the difference in sign. This gives

$$\langle \boldsymbol{p} \rangle \propto \int -i\hbar \nabla \psi(x, t) \cdot \psi^*(x, t) dx \quad .$$

Finally, since nothing has been said about normalization, divide by $\int \psi \psi^* dx$ to get $\langle \boldsymbol{p} \rangle$.

This derivation is given in order to show as clearly as possible what is mathematically new in the Feynman construction: the use of $\exp(i\hbar^{-1} S_{\mathrm{H}})$ as a complex weighting factor in computing an average.

The Propagator For a Free Particle. Constructing a propagator is equivalent to finding the complete wave function, and so this must normally be done by approximations. A few are easy, and I will illustrate with the easiest, the propagator for a free particle. Since the action evaluated along a natural path is given by (6.11) (including the mass m) as

$$S_{\mathrm{N}} = \frac{m}{2t}(x - x')^2$$

we have

$$\psi(x, t) = b \int \exp\left[\frac{i}{\hbar} \frac{m}{2t}(x - x')^2\right] \psi(x', 0) dx' \quad . \tag{7.71}$$

This is not as hard as it looks. Represent $\psi(x', 0)$ as a Fourier integral,

$$\psi(x', 0) = \int \tilde{\psi}(k) e^{ikx'} dk \quad .$$

Then

$$\psi(x, t) = b \int dk \tilde{\psi}(k) e^{ikx} \int e^{ik\xi} e^{im\xi^2/2\hbar t} d\xi \quad , \qquad \xi = x' - x$$

$$= b \sqrt{\frac{2\pi i\hbar t}{m}} \int \tilde{\psi}(k) e^{i(kx - \hbar k^2 t/2m)} dk \quad .$$

Let $t \rightarrow 0$. The expression then reduces to an identity if we choose

$$b = \sqrt{\frac{m}{2\pi i\hbar t}}$$

(compare (7.70)). Further, it satisfies Schrödinger's equation, which was not assumed in deriving it. We have thus constructed the free-particle propagator

$$K_0(x, x', t) = \sqrt{\frac{m}{2\pi i\hbar t}} e^{im(x-x')^2/2\hbar t} \tag{7.72}$$

from the action S_{N} as calculated by classical mechanics and shown that it has two important properties: it satisfies Schrödinger's equation and also

$$\lim_{t \rightarrow 0} K_0(x, x', t) = \delta(x - x') \quad . \tag{7.73}$$

These properties will be useful when we use K_0 to construct approximate solutions in the next chapter.

Problem 7.32. Write the Fourier transform of a free-particle wave function as $\tilde{\psi}(k,t)$. How do you express (7.71) as a relation between Fourier transforms?

Insights Gained from Huygens's Principle. What has been done here is very important, for it furnishes by far the most direct route from the equations of classical physics to the Schrödinger equation. There are certain cases in which this direct-ness is a great advantage, notably in cases in which the usual prescriptions for "quantizing" a classical theory do not yield unambiguous results. Examples are the derivation of unambiguous computational rules for Yang-Mills gauge theories [111] and for quantum mechanics in curved spacetime [112].

An equation of motion, however, is not a theory. One needs an interpretive apparatus, and one needs to be able to generalize the dynamical laws to new situations. The interpretation of $|\psi|^2$ as a probability density and other parts of the interpretive apparatus are explained by *Feynman* and *Hibbs* [103]. The mathematics is done somewhat more generally and the commutation relations are derived by *Yourgrau* and *Mandelstam* [113]. Feynman's theory is not readily generalized to include the dynamics of spin and isospin, any more than Schrödinger's is, so that physics still needs Heisenberg's royal road of Hamiltonian mechanics. But Feynman's theory leads in a straight line from the Hamilton-Jacobi theory to the wave equation and shows where and why $\hbar$ comes in, and with it the argument of Sect. 1.1, in which classical theory was derived from the quantum theory, has been reversed.

Finally, there is an interesting question of physical content. In Sect. 3.1 the equations of motion were expressed in the form of Hamilton's principle, which states that of all paths connecting given initial and final states in qt space, the one actually followed is the one that minimizes the action integral. This can of course be taken as a purely mathematical prescription for finding equations of motion, but if one tries to give it a physical sense one encounters a question of the general form "How does the system know which path gives the least action if it tries only one of them?" Classical dynamics leaves the question unanswered, but in the light of this discussion we can say that it tries all of them, and that the natural path emerges as a result of the property of stationary phase.

Problem 7.33. Show by considering the area under successive loops, that the integral of (7.68) converges. What about (7.69)? How can the derivation be justified?

Problem 7.34. Derive by Huygens' principle the wave function of a free particle moving in a unit circle starting with an arbitrary $\psi(x,t_0)$.

Solution [112]. There is a trick here, for the sum over all possible natural paths leading to x includes not only paths in which it moves a distance from x' to x but also those in which it travels several times around the circle in going from x' to x. The action corresponding to n revolutions is

$$S_n = \frac{m}{2t}(x + 2\pi n - x')^2 \quad \text{and}$$

$$\psi(x,t) = b \sum_{n=-\infty}^{\infty} \int_0^{2\pi} e^{i\hbar^{-1} S'_n(x,x',t)} \psi(x',0) dx' \quad .$$

Now show, assuming that $\psi(x',0)$ is single-valued on the unit circle, that

$$\psi(x,t) = b \int_{-\infty}^{\infty} \exp\left[\frac{im}{2\hbar t}(x-x')^2\right] \psi(x',0) dx' \tag{7.74}$$

and complete the calculation by showing that $\psi(x,t)$ is also single-valued. What goes wrong if one forgets to include the S_n for $n \neq 0$?

Problem 7.35. The uncertainty principle is usually quoted as saying that if a particle's momentum is known one knows nothing about its position. But if a particle with definite momentum moves around a loop, one knows it is somewhere on the loop, which may be quite small. State the uncertainty principle more exactly so as to answer this objection.

7.9 Phase Shifts in Dynamics

This section will discuss some examples, from classical and quantum theory, of changes in phase that arise from changing parameters. To start, the following question: My wristwatch contains a crystal oscillator that vibrates 4096 times per second. How much time does it gain or lose when I raise my hand to block a yawn (a displacement of 1 meter parallel to the direction of vibration, taking 1 second)? Assume that everything happens in one dimension; then the equation of motion is

$$\ddot{x} + \omega^2[x - R(t)] = 0 \tag{7.75}$$

where R is first stationary, then changes by ΔR in a time τ, and finally levels off. For ease in calculation, assume that the motion is given by

$$\dot{R}(t) = \frac{\tau \Delta R}{\pi(t^2 + \tau^2)} \quad . \tag{7.76}$$

Write the equation of motion in Hamiltonian form,

$$\dot{x} = p \quad , \quad \dot{p} = -\omega^2(x - R)$$

and let $\Pi := p + i\omega(x - R)$ denote position in the complex phase plane. Now the equation of motion is of first order,

$$\dot{\Pi} = i\omega\Pi - i\omega\dot{R}(t)$$

and therefore

$$\Pi(t) = (\Pi_0 - i\omega \int \dot{R} e^{-i\omega t} dt) e^{i\omega t} \quad .$$

With (7.76), and integrated to a time $\gg \tau$, this gives

$$\Pi(t) \sim (\Pi_0 - i\omega \Delta R e^{-\omega\tau}) e^{i\omega t} \approx \Pi_0 \exp\left[i\omega\left(t - \frac{\Delta R}{\Pi_0} e^{-\omega\tau}\right)\right] \quad .$$

Here Π_0 is complex, say $|\Pi_0|e^{i\delta}$, and contains the initial phase, so that the correction to t can be positive or negative. Suppose $\delta = 0$. Then if the amplitude of the oscillation is $A := |\Pi_0|/\omega$, the watch is slow by

$$\Delta t = \frac{\Delta R}{A\omega}e^{-\omega\tau}$$

and if $A = 1\,\mu$ this amounts to about 10^{-1770} s. The phase change is small but it happens. The motion satisfies the criterion for adiabaticity, $\omega\tau \gg 1$, and in the approximation studied here the amplitude of oscillation does not change. Note that A occurs in the denominator: approaches via a small-amplitude approximation will not succeed.

Phase Shifts in Quantum Mechanics. Of course phase in quantum mechanics does not mean the same as in classical mechanics, where it corresponds to the overall phase of a wave packet or, more generally, to a coherent state. Nevertheless, similar conclusions can be drawn concerning adiabatic behavior.

Suppose the Hamiltonian operator contains a slowly varying real numerical parameter that will again be called $R(t)$, even though it can be of any nature. Then

$$i\hbar\frac{\partial\psi}{\partial t} = H[R(t)]\psi \quad . \tag{7.77}$$

At any moment, since H is changing slowly, it makes sense to specify a set of normalized eigenfunctions satisfying

$$H(R)\varphi_n(R) = E_n(R)\varphi_n(R) \tag{7.78}$$

where the variables depend on time only through their dependence on R, and we assume there are no degeneracies or near degeneracies. The phase of φ_n is arbitrary, but it cannot depend on $R(t)$, for then (7.78) would be inconsistent with (7.77). Let it be some ϕ_n that need not be further specified. Now write $\psi(x,t)$ as

$$\psi(t) = \sum c_n(t)\varphi_n(R)\exp\left[-\frac{i}{\hbar}\int E_n(R)dt\right] \tag{7.79}$$

where the time phase is the obvious extension of the usual one. Put this into (7.77), use (7.78), multiply by $\varphi_m^*(R)$, and integrate over the spatial coordinates:

$$\dot{c}_m \exp\left[-\frac{i}{\hbar}\int E_m(R)dt\right] = -\sum c_n\dot{R}\int\varphi_m^*(R)\nabla_R\varphi_n(R)(dr)$$
$$\times\exp\left[-\frac{i}{\hbar}\int E_n(R)dt\right]$$

where ∇_R means $\partial/\partial R$. Suppose that initially the system is in state a, so that c_a is much larger than the others. Then

$$c_m \approx -\int\left\{\dot{R}\int\varphi_m^*(R)\nabla_R\varphi_a(R)(dr)\right.$$
$$\left.\times\exp\left[-\frac{i}{\hbar}\int[E_a(R) - E_m(R)]dt\right]\right\}c_a(t)dt \quad . \tag{7.80}$$

Since we have assumed adiabatic change in R and no resonances or near resonances, the exponential oscillates many times, unless $m = a$, during a small change in the rest of the integrand, and it is shown at the end of this section that in the adiabatic limit the integral approaches zero. From this two conclusions follow: First, if a quantum system starts in a certain state a and the parameter R changes very slowly compared with the frequencies given by the energy differences, the system will end up in the corresponding state of the altered system, no matter how different it is. This is one version of the *adiabatic theorem* in quantum mechanics; another will be given in Sect. 8.3. Second, if $m = a$, we can write c_a as $\exp[i\gamma_a(t)]$, where

$$\gamma_a(t) = i \int_0^t \left(\dot{R}(t) \int \varphi_a^*(R) \nabla_R \varphi_a(R)(dr) \right) dt \tag{7.81}$$

and since

$$\nabla_R \int \varphi^*(R)\varphi(R)(dr) = 2\,\mathrm{Re}\left[\int \varphi^*(R)\nabla_R \varphi(R)(dr) \right] = 0 \tag{7.82}$$

the γ so defined is a real number called Berry's phase [114]. Its existence was first pointed out in 1985 and recently it has been much discussed. It is independent of the phase ϕ attached to the stationary states φ and we see further that since a bound state of a one-dimensional system can be described by a real φ, it is zero in that case. Most examples quoted are three-dimensional.

The most important examples of Berry's phase are those in which R makes an excursion returning to its initial value. The system is then unchanged except that sometimes the process may give its wave function a change of phase. It is found that if the change is in only a single parameter R the phase change is in general zero; there must be more than one R and the excursion takes the form of a loop in the parameter space defined by these Rs. For this purpose the integrand in (7.81) is changed, in a convenient notation, to $\dot{R} \cdot \langle a | \nabla_R | a \rangle$. In evaluating γ one soon gets involved in computational details some of which are dealt with in Berry's paper. Here is an example, from the same paper, that is simple and at the same time illustrates an important physical principle.

Example 7.4. The Aharonov-Bohm Effect [114]. Figure 7.12a shows a two-slit interference experiment in which a very small iron whisker carrying an intense magnetic field has been introduced in a place between the two beams where neither of them touches it. The rest of the region is magnetically shielded. One would at first think that the presence of the whisker would not affect the interference pattern, but analysis, as well as experiment [115], show that this is not so. Figure 7.12b shows a box containing an electron that is carried once around the whisker without changing its orientation. The Hamiltonian is of the form $\hat{H}(\hat{p} - eA, \hat{r} - R)$ where A is the vector potential produced by the magnetic field.

It is easy to construct the wave function of a particle in a potential where there is no field (see Sect. 5.8 for ideas):

$$\varphi(r - R, A) = \exp\left[\frac{ie}{\hbar} \int_R^r A(r') \cdot dr' \right] \varphi(r - R, 0) \tag{7.83}$$

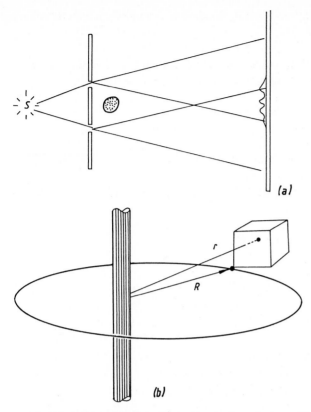

Fig. 7.12. (a) Illustrating the Aharonov-Bohm effect. An iron whisker containing an intense magnetic field is placed between the beams in a diffraction experiment. Though the beams do not touch it, the whisker affects the diffraction pattern. **(b)** A box containing one or more electrons is taken around the magnetized whisker without change of orientation

where R is an arbitrary point fixed in the box and the vanishing of $\nabla \times A$ assures that the integral depends on the points r and R but not on the path followed. Since

$$\nabla_R \varphi(r - R, A) = \left[-\frac{ie}{\hbar} A(R) \varphi(r - R, 0) + \nabla_R \varphi(r - R, 0) \right]$$

$$\times \exp \left[\frac{ie}{\hbar} \int\limits_R^r A(r') \cdot dr' \right]$$

we have

$$\int \varphi^*(r - R, A) \nabla_R \varphi(r - R, A) dr = -\frac{ie}{\hbar} A(R)$$

where the second integral vanishes by (7.82) because a particle in a box can be described by a real wave function. Now carry the box once around the whisker (in this case the change need not be adiabatic). The phase shift is

$$\gamma = \frac{e}{\hbar} \oint A(R) \cdot dR = \frac{e}{\hbar} \int \nabla \times A \cdot dS = \frac{e}{\hbar} \Phi \qquad (7.84)$$

223

where Φ is the total magnetic flux inside the loop, and it can be seen experimentally [115].

The Integral in (7.80). The integrand is the product of a rapidly changing exponential and another function that changes adiabatically. Write it in the form

$$I = \int_0^t e^{i\lambda \int f(t)dt} g(t)dt$$

where f and g change on the scale of minutes or hours, say, and λ is therefore a very large number. (If for example the frequencies are optical, λ is of the order of 10^{16}.) Change the variable of integration to $s := \int f(t)dt$ and solve for $t(s)$; then

$$I = \int_0^s e^{i\lambda s} G(s)ds \quad , \quad G(s) := \frac{g[t(s)]}{f[t(s)]}$$

$$= \left[\frac{1}{i\lambda} e^{i\lambda s} G(s)\right]_0^s - \frac{1}{i\lambda}\int_0^s e^{i\lambda s} G'(s)ds \quad .$$

The partial integrations can be continued and thus, for large λ, I is arbitrarily small. This argument is a trivial extension of the Riemann-Lebesgue lemma which occurs in the theory of Fourier integrals.

Problem 7.36. Choose some other bell-shaped distribution, perhaps a Gaussian one, for $\dot{R}$ in (7.76) and see what difference it makes in the result.

Problem 7.37. Show that (7.83) satisfies the time-independent Schrödinger equation for a particle in a region containing a magnetic vector potential but no magnetic field.

Problem 7.38. A system starts in a state φ_a as an eigenstate of $\hat{H}_0$ and a perturbing V is gradually turned on. Using the quantum adiabatic theorem, find the first- and second-order approximations to the system's perturbed energy.

Problem 7.39. Analyze the interference experiment shown in Fig. 7.12a to show what phase shift is predicted.

Problem 7.40. To illustrate departures from adiabatic behavior, show that in the example of the wristwatch at the beginning of this Section (assuming $\delta = 0$ and inserting a factor of m), the change in J is

$$\Delta J = \pi m \omega (\Delta R)^2 e^{-2\omega\tau} \quad .$$

8. Theory of Perturbations

The number of calculations that can be carried out exactly, either in classical or in quantum dynamics, is comparatively small and the number of useful results so obtained is even smaller. The greater part of physics, engineering, and dynamical astronomy involves approximations, and the worst difficulties arise when the variables of a problem are not separable. What one hopes to do is to find a corresponding problem in which not only are the variables separable but the resulting differential equations can be solved exactly, and then move from the solved problem to the unsolved one by a systematic procedure of approximation, usually involving expansion in series. The procedures in quantum and in classical mechanics can be carried out in somewhat analogous fashions, but there is no use in pursuing the comparison very far since what is observable differs in the two theories. An astronomer, for example, wishes to calculate positions in the sky and has little interest in knowing energies, whereas in atomic physics an electron's position is not an observable while its energy is.

Because astronomers need great precision, the classical theory of perturbed orbits has been carried very far, and many of the best mathematicians in history have had a hand in it. Because it is a very specialized subject we will mostly leave it alone, and our discussion of the classical theory of perturbations will mostly be confined to examples which throw some light on analogous situations in quantum mechanics.

8.1 Secular and Periodic Perturbations

To someone who has encountered perturbation theory in quantum mechanics, it might seem quite obvious how to proceed in classical mechanics. Let us say that we wish to calculate the perturbation of the Earth's orbit by the action of Mars. Assume that the perturbing force or energy carries a factor ε and write the Earth's position as

$$r(t) = r_0(t) + \varepsilon r_1(t) + \varepsilon^2 r_2(t) + \ldots \quad .$$

Then put it back into the equations of motion, and gather together terms of the same order in ε. In this procedure it is assumed that the perturbation terms are small, that $|\varepsilon r_1| \ll |r_0|$, but this is not so, since even a small perturbation, if allowed to run long enough, will finally bring the Earth to a point very far from where it would otherwise have been.

To see what happens if one disregards this warning and proceeds, let us use perturbation theory on a trivial problem, the perturbation of a harmonic oscillator by the addition of another spring that changes the stiffness k to $k + \varepsilon$ and the frequency to $\omega + \varepsilon \omega_1$. Then if the oscillator starts from rest its displacement becomes

$$x(t) = a \cos (\omega + \varepsilon \omega_1)t \tag{8.1}$$
$$= a \cos \omega t + \varepsilon \omega_1 \frac{d}{d\omega} a \cos \omega t + \ldots$$
$$= x_0(t) - \varepsilon a \omega_1 t \sin \omega t + \ldots \quad . \tag{8.2}$$

This result is called a secular perturbation. It is spurious because it has $|x - x_0|$ increase without limit, whereas of course it remains bounded. To get a finite result we must effectively sum an infinite number of terms of the perturbation series in ε. Let us see how this can be done in another simple example.

Anharmonic Oscillator. The example is a harmonic oscillator perturbed by an anharmonic force so that the equation of motion is

$$\ddot{x} + \omega_0^2 x = \varepsilon x^3 \quad . \tag{8.3}$$

If one tries to solve this by letting

$$x = x_0 + \varepsilon x_1 + \varepsilon^2 x_2 + \ldots \tag{8.4}$$

a secular perturbation arises:

$$\ddot{x}_0 + \omega_0^2 x_0 + \varepsilon(\ddot{x}_1 + \omega_0^2 x_1) + \varepsilon^2(\ddot{x}_2 + \omega_0^2 x_2) = \varepsilon(x_0^3 + 3\varepsilon x_0^2 x_1 + \ldots)$$

whence

$$\ddot{x}_0 + \omega_0^2 x_0 = 0$$
$$\ddot{x}_1 + \omega_0^2 x_1 = x_0^3$$
$$\ddot{x}_2 + \omega_0^2 x_2 = 3x_0^2 x_1 \tag{8.5}$$

etc. The first equation has a solution

$$x_0 = a \cos \omega_0 t \quad .$$

Then the second equation is

$$\ddot{x}_1 + \omega_0^2 x_1 = a^3 \cos^3 \omega_0 t = \tfrac{1}{4} a^3 (3 \cos \omega_0 t + \cos 3\omega_0 t) \quad .$$

If one encountered this equation without knowing its context one would judge it to be the equation of a harmonic oscillator driven by a force having a Fourier component in resonance with the oscillator. The resonance would transfer an ultimately divergent amount of energy to the oscillator and its amplitude would ultimately increase proportionally to t (Problem 8.1). This is a secular perturbation.

To get out of the difficulty we suppose [as in the primitive example (8.1)] that the frequency changes from ω_0 to ω, where

$$\omega_0^2 = \omega^2 + \varepsilon \lambda_1 + \varepsilon^2 \lambda_2 + \ldots \tag{8.6}$$

and choose λ_1 so that the secular perturbation vanishes. The terms through ε^2 are given by

$$\ddot{x}_0 + \varepsilon \ddot{x}_1 + \varepsilon^2 \ddot{x}_2 + (\omega^2 + \varepsilon \lambda_1 + \varepsilon^2 \lambda_2)(x_0 + \varepsilon x_1 + \varepsilon^2 x_2) = \varepsilon(x_0 + \varepsilon x_1)^3 \quad . \quad (8.7)$$

Suppose we require that

$$x(0) = a \quad , \quad \dot{x}(0) = 0 \quad . \tag{8.8}$$

The zeroth-order equation from (8.7) is

$$\ddot{x}_0 + \omega^2 x_0 = 0$$

of which the solution that satisfies (8.8) is

$$x_0 = a \cos \omega t \quad .$$

The next equation is

$$\begin{aligned} \ddot{x}_1 + \omega^2 x_1 &= -\lambda_1 x_0 + x_0^3 \\ &= -\lambda_1 a \cos \omega t + \tfrac{1}{4} a^3 (3 \cos \omega t + \cos 3\omega t) \end{aligned}$$

and the resonant secular term cancels if

$$\lambda_1 = \tfrac{3}{4} a^2 \quad . \tag{8.9}$$

The remaining equation

$$\ddot{x}_1 + \omega^2 x_1 = \tfrac{1}{4} a^3 \cos 3\omega t \tag{8.10}$$

has a solution

$$x_1 = \frac{a^3}{32\omega^2}(\cos \omega t - \cos 3\omega t) \tag{8.11}$$

which when combined with x_0 is consistent with the boundary condition (8.8). The frequency found thus far is

$$\omega = (\omega_0^2 - \varepsilon \lambda_1)^{1/2} \simeq \omega_0 \left(1 - \frac{\varepsilon \lambda_1}{2\omega_0^2} \right) = \omega_0 \left(1 - \frac{3\varepsilon a^2}{8\omega_0^2} \right) \tag{8.12}$$

and the solution thus far is

$$x(t) = a \left(1 + \frac{\varepsilon a^2}{32\omega^2} \right) \cos \omega t - \frac{\varepsilon a^3}{32\omega^2} \cos 3\omega t \tag{8.13}$$

[compare (2.29]. The secular term is gone and only periodic perturbations remain. The oscillator is called anharmonic because its frequency depends on the amplitude of swing. In one sense we have stopped after the first term of the series expansion in ε, but in fact if we were to expand $x(t)$ in powers of ε it would be an infinite series because ω contains ε. Difficulties of this kind occur in quantum mechanics; notable examples are in the theory of many-body systems and in the theory of interacting fields. The method used here is there called "summing an infinite number of diagrams," and is necessary if finite results are to be obtained.

Problem 8.1. A series L-C circuit (no resistance) with a natural frequency $\omega_r = (LC)^{-1/2}$ is driven by a sinusoidal voltage of frequency ω. Find the current $i(t)$ when $\omega < \omega_r$, $\omega = \omega_r$, and $\omega > \omega_r$.

Problem 8.2. Find the general solution of (8.10) and verify that (8.11) is the right solution to choose.

Problem 8.3. Look again at (2.27) and verify that the results of Sect. 2.3 agree with those obtained here.

Problem 8.4. Find the next terms in ε in the example above.

Problem 8.5. Discuss the motion of the anharmonic oscillator if it is very slightly damped with a force proportional to $\dot{x}$.

Problem 8.6. Show that an exact form of (8.2) is

$$x(t) - x_0(t) = -2a \, \sin(\tfrac{1}{2}\varepsilon\omega_1 t) \, \sin \, [(\omega + \tfrac{1}{2}\varepsilon\omega_1)t]$$

so that for small ε there are really two series to be summed: one for the slowly varying amplitude and the other for the motion with perturbed frequency.

8.2 Perturbations in Quantum Mechanics

In quantum mechanics the calculations are easier because the questions we ask are easier: though the oscillator in classical physics may end up far from where it would have been with no perturbation, a bound-state wave function, in a nondegenerate state at least, is little changed by a small perturbation. We expect an expansion in powers of ε to go through, and it does, if we handle the secular perturbation correctly. It is easy to arrange things so that the danger of secular perturbations is not in evidence. This is what is usually done in expositions of perturbation theory, but frequency is energy, and I will find the change in energy by finding the change of frequency produced by a perturbation as in the oscillator calculations just performed.

The problem to be solved is

$$i\hbar \frac{\partial \psi}{\partial t} = (H_0 + \varepsilon V)\psi \tag{8.14}$$

where, in general H_0 and V are both operators, and assume for the moment that V is independent of time. Assume also that it has been possible to solve the unperturbed problem to find the stationary states,

$$i\hbar \frac{\partial \psi_n^{(0)}}{\partial t} = H_0 \psi_n^{(0)} = E_n \psi_n^{(0)} \tag{8.15}$$

$$\psi_n^{(0)} = \varphi_n e^{-iE_n t/\hbar}$$

where φ_n is independent of t. The problem is to find that state which in the absence of the perturbation would coincide with one of the unperturbed stationary states, say $\psi_a^{(0)}$.

Assume that the spatial part of ψ can be represented by a sum

$$\varphi_a + \sum_{n \neq a} c_n \varphi_n \tag{8.16}$$

where c_n is of order ε, but that the time dependence, instead of being that of φ_a, is changed by the addition of an energy ΔE:

$$\psi = (\varphi_a + \sum{}' c_n \varphi_n) e^{-i(E_a + \Delta E)t/\hbar} \tag{8.17}$$

where the prime indicates that the term $n = a$ is omitted. Putting this into (8.14) gives

$$(E_a + \Delta E)(\varphi_a + \sum{}' c_n \varphi_n) = E_a \varphi_a + \sum{}' c_n E_n \varphi_n + \varepsilon V \varphi_a + \varepsilon \sum{}' c_n V \varphi_n$$

or

$$\Delta E \varphi_a + \sum{}'(E_a + \Delta E - E_n) c_n \varphi_n = \varepsilon V \varphi_a + \varepsilon \sum{}' c_n V \varphi_n \quad . \tag{8.18}$$

Now multiply by φ_m^* and integrate. If $m = a$,

$$\Delta E = \varepsilon \langle a|V|a \rangle + \varepsilon \sum{}' c_n \langle a|V|n \rangle \tag{8.19}$$

where the matrix element $\langle a|V|n \rangle$ is defined as

$$\langle a|V|n \rangle := \int \varphi_a^* V \varphi_n dx \quad . \tag{8.20}$$

To order ε, (8.19) gives the familiar formula for the first-order perturbation of energy,

$$\Delta E = \varepsilon \langle a|V|a \rangle + O(\varepsilon^2) \quad . \tag{8.21}$$

To continue, return to (8.18), this time multiplying by φ_m^* and integrating with $m \neq a$:

$$(E_a + \Delta E - E_m) c_m = \varepsilon \langle m|V|a \rangle + \varepsilon \sum{}' c_n \langle m|V|n \rangle$$

$$c_m = \frac{\varepsilon}{E_a + \Delta E - E_m} (\langle m|V|a \rangle + \sum{}' c_n \langle m|V|n \rangle) \quad , \quad m \neq a \quad . \tag{8.22}$$

Equations (8.19) and (8.22) are exact, and they form the basis for an iterative evaluation of ΔE and c_m. For example, using the first term of (8.22) in (8.19) gives

$$\Delta E = \varepsilon \langle a|V|a \rangle + \varepsilon^2 \sum{}'_n \frac{\langle a|V|n \rangle \langle n|V|a \rangle}{E_a + \Delta E - E_n} + \dots \tag{8.23}$$

and further terms may be written by inspection. The only difficulty with the application of this formula is that it contains ΔE implicitly and must be solved for it, but this expansion is simpler to write down than the usual expansion in ε that contains ΔE only on the left-hand side. It was first given by Léon Brillouin in 1933.

As in the discussion of classical perturbation theory, this is not really an expansion in ε, since the expressions (8.17) for ψ and (8.23) for ΔE contain all powers of ε, but it achieves a formal simplicity beyond that of the more conventional Schrödinger perturbation theory precisely because it goes to the root of the problem of secular perturbations. Although the problem solved in this way can be dealt with in other forms of perturbation theory, most forms of the quantum theory of field require the correction to the frequency to be made explicitly as is done here. The process by which this is done is known as renormalization.

Calculation of Perturbed Wave Functions. Using the first term of (8.22) to calculate a "first-order" perturbation to the wave function gives

$$c_m = \frac{\varepsilon}{E_a + \Delta E - E_m} \langle m|V|a \rangle + O(\varepsilon^2) \quad \text{and} \tag{8.24a}$$

$$\varphi = \varphi_a + \Delta\varphi \quad , \quad \Delta\varphi = \varepsilon \sum_m {}' \frac{\langle m|V|a\rangle}{E_a + \Delta E - E_m} \varphi_m \quad . \tag{8.24b}$$

Sometimes only a few of the matrix elements $\langle m|V|a\rangle$ are different from zero and the sum is easy, but sometimes it has many terms and also it may involve a contribution in the form of an integral from a range of energies E_m in which the spectrum is continuous. For this reason it is often convenient to reduce the determination of $\Delta\varphi$ to the solution of a differential equation. Multiplying (8.24b) by $\hat{H}_0 - E_a - \Delta E$ gives

$$(\hat{H}_0 - E_a - \Delta E)\Delta\varphi = -\varepsilon \sum_m {}' \langle m|V|a\rangle \varphi_m$$

$$= -\varepsilon \left(\sum_m \langle m|V|a\rangle \varphi_m - \langle a|V|a\rangle \varphi_a \right) \quad .$$

The sum is just the expansion of $V\varphi_a$ in terms of φ_m, so with $\Delta E \simeq \varepsilon\langle a|V|a\rangle$ as in (8.23),

$$(\hat{H}_0 - E_a - \varepsilon\langle a|V|a\rangle)\Delta\varphi = -\varepsilon(V - \langle a|V|a\rangle)\varphi_a \quad . \tag{8.25}$$

The quantity on the right involves only the unperturbed wave function. In the ordinary one-particle Schrödinger equation, where

$$\hat{H}_0 = -\frac{\hbar^2}{2m}\nabla^2 + V_0(r)$$

it is useful to write

$$\Delta\varphi = \varepsilon\chi(r)\varphi_a(r) \quad .$$

A short calculation then shows that χ satisfies

$$-\frac{\hbar^2}{2m}\left[\nabla^2\chi + 2\frac{\nabla\phi_a \cdot \nabla\chi}{\phi_a}\right] - \langle a|V|a\rangle\chi = -V + \langle a|V|a\rangle \quad . \tag{8.26}$$

This may be reasonably easy to solve.

Problem 8.7. Show that (8.19) is, in an obvious notation,

$$\Delta E = \varepsilon(\langle a|V|a\rangle + \langle a|V|\Delta\varphi\rangle)$$

Problem 8.8. Find the first-order perturbed wave function and the first- and second-order perturbed energies of a simple harmonic oscillator perturbed by an interaction εx^4.

Problem 8.9. Using the method just explained, find the lowest-order correction to the ground-state energy and wave function of a hydrogen atom in an electric field F.

Problem 8.10. Show that if ΔE is smaller than all the energy differences $E_n - E_a$, the implicit equation (8.23) can be solved for ΔE to give, to third order, an expansion for ΔE:

$$\Delta E = \varepsilon\langle a|V|a\rangle - \varepsilon^2 \sum_n {}' \frac{\langle a|V|n\rangle\langle n|V|a\rangle}{E_n - E_a}$$

$$+ \varepsilon^3 \left[\sum_m{}' \sum_n{}' \frac{\langle a|V|m\rangle\langle m|V|n\rangle\langle n|V|a\rangle}{(E_m - E_a)(E_n - E_a)} \right.$$

$$\left. - \langle a|V|a\rangle \sum_m{}' \frac{\langle a|V|m\rangle\langle m|V|a\rangle}{(E_m - E_a)^2} \right] + \dots \quad . \qquad (8.27)$$

This series in ε is the original perturbation expansion given by Schrödinger in his earliest papers. The second part of the third-order term is called a wave function renormalization term. Such terms appear in increasing profusion when the series is carried further (though, in practice, nobody does).

8.3 Adiabatic Perturbations

In this section I will compare the classical and quantum theories of a perturbed bound state, in physical content and quantitative result. The physical content of the quantum perturbation theory is clear, especially if the unperturbed state is nondegenerate: a bound state is determined by a potential energy and boundary conditions. One tries to alter the known solution of a problem similar to the one in hand in such a way that the eigenvalue equation is satisfied approximately and the boundary conditions exactly. If the unperturbed state is degenerate, one determines a set of stabilized eigenfunctions in a way that is explained in texts on elementary quantum mechanics and it is these that are altered by the perturbing potential. The physical sense of the classical theory is not so clear, for there is no criterion for a stationary state to be met, and any number of states of roughly the same energy could be taken as the effect of perturbing a given unperturbed state. We need a criterion by which "the" perturbed state corresponding to a given unperturbed one may be defined.

The criterion is found in the adiabatic principle: consider the calculation not as a certain expansion in powers of ε but as the attempt to find out what happens if by some means, physical or miraculous, the perturbation is established very slowly. The adiabatic theorem (Sect. 7.3) then states that

$$J := \oint p_n dq_n = \oint p_n \dot{q}_n dt$$

remains almost constant in value. It follows from the definition of energy in generalized coordinates that this is

$$J = 2 \oint [E(t) - V(t)] dt \quad .$$

Suppose that V is initially constant, then changes slowly to $V + \Delta V$, then stays constant at its new value. Then E will change to $E + \Delta E$ in the same manner and the orbit will also change from O_1 to O_2. But according to Hamilton's principle, if the change in V is not too large, the integral around O_2 will be nearly equal to the integral around O_1:

$$\oint_{O_1} (E - V) dt \approx \oint_{O_2} (E + \Delta E - V - \Delta V) dt$$

$$\approx \oint_{O_1} (E + \Delta E - V - \Delta V) dt \quad .$$

If T is the orbital period and time averages are denoted by angular brackets then

$$\oint_{O_1} (E - V)dt = T(E - \langle V \rangle_{O_1}) \quad,$$

and similarly for the other integral; therefore

$$\Delta E \approx \langle \Delta V \rangle_{O_1} \quad. \tag{8.28}$$

In quantum mechanics ΔE is approximated by the expectation value of ΔV in the unperturbed state; and remember that as shown in Sect. 7.4, expectation values correspond to time averages in classical mechanics. Is the perturbed energy in quantum mechanics also the result of adiabatically establishing the perturbation? This can be answered when we have found the quantum-mechanical description of adiabatic behavior.

Adiabatic Transitions in Quantum Mechanics. If the perturbing energy V depends on t, we shall have to slightly modify the earlier treatment. Since ΔE now depends on t, replace (8.17) by

$$\psi = (\varphi_a + \sum{}' c_n(t)\varphi_n) \exp\left[-i\hbar^{-1} \int (E_a + \Delta E)dt\right] \quad.$$

Substitution into the wave equation gives

$$(E_a + \Delta E)(\varphi_a + \sum{}' c_n \varphi_n) + i\hbar \sum{}' \dot{c}_n \varphi_n$$
$$= E_a \varphi_a + \sum{}' c_n E_n \varphi_n + \varepsilon V \varphi_a + \varepsilon \sum{}' c_n V \varphi_n \quad \text{or}$$

$$\Delta E \varphi_a + \sum{}'(E_a + \Delta E - E_n)c_n \varphi_n + i\hbar \sum{}' \dot{c}_n \varphi_n = \varepsilon V \varphi_a + \varepsilon \sum{}' c_n V \varphi_n \quad.$$

From this, as before,

$$\Delta E(t) = \varepsilon \langle a|V(t)|a \rangle + \varepsilon \sum{}' c_n(t)\langle a|V(t)|n \rangle \tag{8.29}$$

while if $m \neq a$,

$$i\hbar \dot{c}_m + (E_a + \Delta E(t) - E_m)c_m = \varepsilon \langle m|V(t)|a \rangle + \sum{}' c_n \langle m|V(t)|n \rangle \quad. \tag{8.30}$$

If we calculate $\dot{c}_m$ only to the first order, (8.30) becomes

$$i\hbar \dot{c}_m + (E_a - E_m)c_m = \varepsilon \langle m|V(t)|a \rangle \quad. \tag{8.31}$$

Now suppose that $V(t)$ has been turned on continuously from $t = -\infty$; $V(t) = e^{\gamma t}V_0$, where γ is eventually to become very small. The solution of (8.31) is

$$c_m(t) = -\varepsilon \frac{\langle m|V(t)|a \rangle}{E_m - E_a - i\hbar\gamma} \tag{8.32}$$

which with $\gamma \to 0$ is exactly (8.24a) to first order. The energy shift becomes

$$\Delta E(t) = \varepsilon \langle a|V(t)|a \rangle - \varepsilon^2 \sum{}' \frac{\langle a|V(t)|m \rangle \langle m|V(t)|a \rangle}{E_m - E_a} \quad. \tag{8.33}$$

That is, nothing (to second order) is changed in our earlier result except that ΔE

has become a slowly-varying function of time. The adiabatic and stationary state perturbation theories give the same answer. This is another version of the quantum adiabatic theorem. The first, in Sect. 7.9, stated that when a system is perturbed adiabatically in such a way that two levels never cross, if it starts in the ath eigenstate of the unperturbed Hamiltonian it will end in the corresponding eigenstate of the perturbed one. While this is good to know, it does not help much in computation, since the perturbed eigenfunctions are in general unknown. The formulas of this section approximate the adiabatically perturbed eigenfunctions in terms of the unperturbed ones. The same distinction is seen if one compares (8.28) with the results just found, since the expectation value $\langle a|V(t)|a \rangle$ is taken with respect to unperturbed wave functions and the time average in (8.28) is with respect to the unperturbed motion of the system. The analogy is pursued in Problem 8.14; let us see how good it is from a numerical point of view.

Example 8.1. Anharmonic Oscillator. The Hamiltonian corresponding to (8.3) is

$$H = \frac{p^2}{2m} + \frac{1}{2}m\omega_0^2 x^2 + V \quad , \qquad V = \frac{1}{4}\varepsilon m x^4 \quad .$$

Classical mechanics. An unperturbed motion has

$$x_0 = a\cos\omega_0 t \quad , \qquad E_c^{(0)} = \tfrac{1}{2}m\omega_0^2 a^2 \quad .$$

The first-order perturbation of energy, which is the time average of V over a cycle of the unperturbed motion, is

$$E_c^{(1)} = \tfrac{1}{4}\varepsilon m \overline{x^4} = \tfrac{3}{32}\varepsilon m a^4 \quad . \tag{8.34}$$

In terms of $E_c^{(0)}$ this is

$$E_c^{(1)} = \frac{3}{8}\frac{\varepsilon}{m\omega_0^4}E_c^{(0)2} \quad . \tag{8.35}$$

Quantum mechanics

$$E_{qn}^{(1)} = \langle n|V|n \rangle = 3\varepsilon m\left(\frac{\hbar}{4m\omega_0}\right)^2(2n^2 + 2n + 1) \qquad \text{or} \tag{8.36}$$

$$E_{qn}^{(1)} = \frac{3}{8}\frac{\varepsilon}{m}\frac{\hbar^2}{\omega_0^2}\left[\left(n + \frac{1}{2}\right)^2 + \frac{1}{4}\right] \quad .$$

Since $E_{qn}^{(0)} = (n + \tfrac{1}{2})\hbar\omega_0$, this is

$$E_{qn}^{(1)} = \frac{3}{8}\frac{\varepsilon}{m\omega_0^4}[E_{qn}^{(0)2} + E_{q0}^{(0)2}] \quad . \tag{8.37}$$

The classical and quantum results correspond at high energies where the zero-point energy $E_{q0}^{(0)}$ can be ignored.

Problem 8.11. Calculate (8.34) and (8.36).

Problem 8.12. Calculate and compare the perturbed energies that result in classical and quantum theory when a harmonic oscillator is placed in a uniform field of force. (This is a second-order calculation.)

Problem 8.13. Solve the preceding problem exactly by introducing a new coordinate $\xi = x + \alpha$ and choosing α appropriately.

Problem 8.14. Let ψ be the exact solution of

$$\hat{H}\psi = E\psi \quad \hat{H} = \hat{H}_0 + V$$

and $\psi^{(0)}$ be an eigenfunction of $\hat{H}_0$

$$\hat{H}_0\psi^{(0)} = E^{(0)}\psi^{(0)}$$

normalized so that $\int \psi^*\psi(dx) = 1$. Show that

$$E = E^{(0)} + \int \psi^{(0)*}V\psi(dx) \tag{8.38}$$

exactly. If ψ is the state that results from $\psi^{(0)}$ when the perturbation is established adiabatically, this is the closest (but not very close) quantum analogue of (8.28).

8.4 Degenerate States

Everybody knows that in quantum mechanics, when a perturbation is applied to a degenerate state the degenerate energy level usually splits. (We will see below what the criterion for "usually" is.) In classical mechanics the situation is essentially the same, if for energy we read frequency. In the old quantum theory an extensive apparatus was developed (see for example [40]) for dealing with this situation, but it lies beyond the cutoff of this book and moreover is no longer current. Since the subject is a commonplace of elementary quantum mechanics, while in classical mechanics it is of narrow interest, my discussion will be very short.

A Classical Oscillator. Start with a nondegenerate oscillator in two dimensions, for which

$$H_0 = \frac{p_1^2 + p_2^2}{2m} + \frac{1}{2}m(\omega_1^2 x_1^2 + \omega_2^2 x_2^2) \quad . \tag{8.39a}$$

To this precessing system (Fig. 8.1) add a symmetrical but nonlinear restoring force

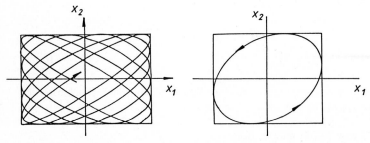

Fig. 8.1. Orbit of a harmonic oscillator in 2 dimensions. (a) $\omega_1 \neq \omega_2$. (b) Degenerate case, $\omega_1 = \omega_2$

$$V = \tfrac{1}{4}\varepsilon(x^2 + y^2)^2 \quad . \tag{8.39b}$$

Writing the unperturbed oscillation as in Sect. 5.6

$$x = a \sin \omega_1 t \quad , \quad y = b \sin (\omega_2 t + \phi)$$

note that

$$E^{(0)} = \tfrac{1}{2}m(\omega_1^2 a^2 + \omega_2^2 b^2) \tag{8.40}$$

independent of ϕ, but that $E^{(1)}$ involves ϕ:

$$E^{(1)} = \tfrac{1}{4}\varepsilon[a^4\overline{\sin^4 \omega_1 t} + b^4\overline{\sin^4 (\omega_2 t + \phi)} + 2a^2 b^2\overline{\sin^2 \omega_1 t \, \sin^2 (\omega_2 t + \phi)}] \quad .$$

In the first two terms, $\overline{\sin^4 \omega t} = 3/8$, but the last average must be done carefully. We have to average

$$\tfrac{1}{4}\{ \cos^2 [(\omega_1 + \omega_2)t + \phi] - 2 \cos [(\omega_1 + \omega_2)t + \phi] \cos [(\omega_1 - \omega_2)t - \phi]$$
$$+ \cos^2 [(\omega_1 - \omega_2)t - \phi]\} \quad .$$

If $\omega_1 \neq \omega_2$, the first and last terms give $\tfrac{1}{2}$ and the middle term gives zero. If $\omega_1 = \omega_2$ the last term gives $\cos^2 \phi$, while the other two are unchanged. Thus

$$E^{(1)} = \frac{1}{4}\varepsilon\left[\frac{3}{8}(a^4 + b^4) + \frac{1}{2}a^2 b^2\left(1 + \frac{1}{2}\delta \cos 2\phi\right)\right] \quad \text{where} \tag{8.41a}$$

$$\delta = \begin{cases} 0 & \omega_1 \neq \omega_2 \\ 1 & \omega_1 = \omega_2 \end{cases} \quad . \tag{8.41b}$$

These formal expressions reveal that the degenerate case is essentially different from the nondegenerate one. If the unperturbed state is nondegenerate, the perturbed energy depends only on the amplitudes of the unperturbed oscillations and hence on the energies associated with the x_1 and x_2 directions. In the degenerate case, in which the unperturbed motion is a stationary ellipse, $E^{(1)}$ depends on ϕ but not on the spatial orientation of the ellipse (Problem 8.15). For the space-filling curve of the nondegenerate motion there is no such thing as orientation. The existence of different perturbed energies corresponding to the same unperturbed energy is the classical analogue of the splitting of degenerate levels by a perturbation in quantum mechanics.

We may ask how the discontinuity in (8.41) arises out of a continuous analysis. It arises, of course, only because we have made an idealization; we are willing to wait an infinite time in order to take the time average. In fact, if $\omega_2 - \omega_1 = 2\pi/(1$ year), say, it is only a matter of convenience whether we say that the system is degenerate. In this case the ellipse precesses very slowly in space, and whether we say it is stationary depends on what we wish to emphasize. Whether we say that two energy levels are degenerate depends not on the exact vanishing of the difference between them but on the practical consequences of assuming it.

Problem 8.15. Show that in the above example if the frequencies are equal and the semidiameters of the resulting ellipse are A and B, the shift in energy is

$$E^{(1)} = \tfrac{1}{32}\varepsilon(3A^4 + 2A^2 B^2 + 3B^4) \quad .$$

Degeneracy in Quantum Mechanics. If the unperturbed system has several states corresponding to the same energy, we must choose the right linear combination of them as the unperturbed state from which to start the perturbation expansion. That is, if the orthogonal time-independent degenerate unperturbed states are $\varphi_1 \ldots \varphi_d$, where d is the degree of degeneracy, we form the linear combination

$$\Phi^{(0)} = \sum_{i=1}^{d} c_i \varphi_i \qquad (8.42)$$

corresponding to the energy $E^{(0)}$. Then from

$$(H_0 + V)\Phi = (E^{(0)} + E^{(1)})(\Phi^{(0)} + \Phi^{(1)})$$

we find, to first order,

$$(H_0 - E^{(0)})\Phi^{(1)} + (V - E^{(1)}) \sum c_i \varphi_i = 0 \quad .$$

(The ε previously used to distinguish orders of perturbation is omitted here.) Multiply this equation by any φ_j^* and integrate. This gives

$$\sum c_i (\langle j|V|i\rangle - E^{(1)}\delta_{ij}) = 0 \quad , \qquad j = 1, \ldots d$$

or, written out,

$$(\langle 1|V|1\rangle - E^{(1)})c_1 + \langle 1|V|2\rangle c_2 + \langle 1|V|3\rangle c_3 + \ldots = 0$$

$$\langle 2|V|1\rangle c_1 + (\langle 2|V|2\rangle - E^{(1)})c_2 + \langle 2|V|3\rangle c_3 + \ldots = 0$$

etc. The condition that there exist a nontrivial solution for the c's is that the determinant of the coefficients vanish,

$$|\langle j|V|i\rangle - E^{(1)}\delta_{ij}| = 0 \qquad (8.43)$$

(recall (4.28)) and this, with given $\langle j|V|i\rangle$ and unknown $E^{(1)}$, is an algebraic equation of dth order which in general produces d different values of $E^{(1)}$ corresponding to d different linear combinations of unperturbed states. These combinations are known as stabilized eigenfunctions.

Examples of this procedure are found in all books on quantum mechanics – it is usually the same example, the linear Stark effect in hydrogen – so these remarks will suffice. In the perturbed oscillator discussed above, the ground state and the first excited state are still degenerate under the perturbation, but higher energy levels are split. (Problem 8.18).

Almost Degenerate Levels. It is interesting to note what happens in quantum mechanics as the separation of the unperturbed levels becomes very small. A two-level example shows that the transition from the nondegenerate to the degenerate case is a continuous one.

Suppose a system has a pair of levels close together:

$$H_0 \varphi_1 = E_1^{(0)} \varphi_1 \quad , \qquad H_0 \varphi_2 = E_2^{(0)} \varphi_2$$

and by "close" is meant that the difference $E_2^{(0)} - E_1^{(0)}$ is comparable with the energy shift produced by the perturbation. Assume that other states of the system have energies far away from these. We will see that under the perturbation V the two states, even though not degenerate, become mixed. In the equation

$$(H_0 + V)\Phi = E\Phi$$

approximate the solution by a linear combination of the two unperturbed functions,

$$\Phi \simeq c_1\varphi_1 + c_2\varphi_2 \quad . \tag{8.44}$$

Substituting into the equation and multiplying first by φ_1^* and then by φ_2^* and integrating gives a pair of equations

$$(V_{11} - E + E_1^{(0)})c_1 + V_{12}c_2 = 0$$
$$V_{21}c_1 + (V_{22} - E + E_2^{(0)})c_2 = 0 \tag{8.45}$$

where V_{ij} abbreviates $\langle i|V|j\rangle$. The resulting quadratic equation for E has the solutions

$$E^{(1)} = \tfrac{1}{2}(E_1^{(0)} + E_2^{(0)}) +$$
$$+ \tfrac{1}{2}\{V_{11} + V_{22} \pm \sqrt{(V_{11} - V_{22} + E_1^{(0)} - E_2^{(0)})^2 + 4|V_{12}|^2}\} \tag{8.46}$$

with equally long formulas for c_1 and c_2.

In Fig. 8.2 we have assigned numerical values to $E_1^{(0)}$ and $E_2^{(0)}$, and have taken V as proportional to a parameter λ which is allowed to vary from 0 to 1. The energy levels do not cross, and the plot of the coefficients $|c_1|$ and $|c_2|$ shows that when the matrix elements are comparable in magnitude to the splitting of the unperturbed levels there is considerable mixing of states.

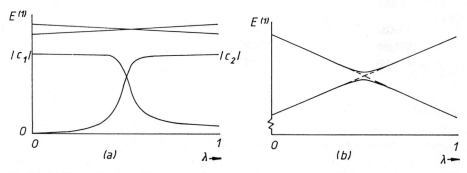

Fig. 8.2. (a) Energy levels of (8.46) and coefficients of (8.44) as functions of the strength of the perturbation; c_1 and c_2 are drawn using the upper sign in (8.46). (b) Plot of the energy levels in (a) on an expanded scale

Thus in classical theory from a practical point of view and in quantum theory by calculation, the situation when there are degenerate levels is only a point on a continuum.

If one looks at Fig. 8.2a the energy levels appear to cross; only in the enlarged view is it seen that they do not. Do they ever actually cross? Problem 8.17 shows that they do if $V_{12} = 0$. The vanishing of this quantity is determined by a simple theorem:

Let $\hat{T}$ be a hermitian operator representing a symmetry of H, so that $[\hat{T}, V] = 0$. Further, let φ_1 and φ_2 be eigenfunctions of $\hat{T}$ belonging to different eigenvalues, t_1 and t_2. Then $V_{12} = 0$.

Proof:

$$\int \varphi_1^*(\hat{T}V - V\hat{T})\varphi_2 dx = 0 \quad .$$

Therefore

$$(t_1 - t_2)V_{12} = 0 \quad .$$

But $t_1 - t_2 \neq 0$, therefore $V_{12} = 0$. □

If φ_1 and φ_2 have different symmetries, the levels cross. This is often the case. Two atomic states with different magnetic quantum numbers but otherwise identical correspond to different eigenvalues of $\hat{L}$, and there are many examples in molecular physics.

Problem 8.16. Show that if $E_n^{(1)}$ are the eigenvalues of the determinantal equation (8.43), then

$$\sum_{i=1}^d E_i^{(1)} = \sum_{i=1}^d \langle i|V|i \rangle \quad .$$

According to this useful rule, if the diagonal elements $\langle i|V|i \rangle$ are all zero, the center of gravity of the perturbed levels is also zero.

Problem 8.17. Find a formula for the minimum separation of the two values of $E^{(1)}$ and show that it vanishes when $|V_{12}| = 0$.

Problem 8.18. Study the quantum-mechanical degeneracies of the unperturbed and perturbed states of the oscillator defined by (8.39).

8.5 Quantum Perturbation Theory for Positive-Energy States

If a particle is unbound, as it is when aimed at a target, its energy is determined by the experimenter and is not a quantity to be calculated. Books are written about scattering theory and there is no reason here to go into details, but Feynman's perturbation technique that uses a diagrammatic representation of the scattering process is easily accessible from the theory of propagators developed in Chap. 7.

Figure 8.3 represents the mental picture. In this picture a particle starts at point x_0 at time t_0. It propagates freely to x_1, t_1 where it interacts with the external potential V, after which it advances to x_2, t_2 and repeats the process, finally arriving at x, t. Since the x_n, t_n are not definite points we integrate over them to find the total amplitude, understanding that only points close to the natural path from x_0, t_0 to x, t will contribute appreciably. The resulting amplitude is represented as the sum of contributions from $0, 1, 2, \ldots$ of these interactions. Look at the first two terms:

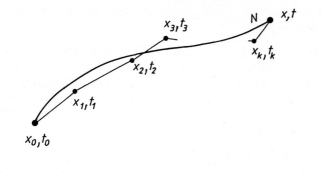

Fig. 8.3. The action of a force that perturbs the straight-line path from x_0, t_0 to x, t into a natural (classical) path N is approximated in quantum mechanics by a series of interactions with the perturbing potential at $x_1, t_1, \ldots$

$$\psi(x, t) = \int dx_0 K_0(x - x_0, t - t_0)\psi_0(x_0, t_0)$$

$$-\frac{i}{\hbar} \int_0^t dt_1 \int dx_1 K_0(x - x_1, t - t_1)V(x_1, t_1)$$

$$\times \int dx_0 K_0(x_1 - x_0, t_1 - t_0)\psi_0(x_0, t_0) + \ldots \quad . \tag{8.47}$$

(Note there is no integration over t_0). The others are formed in the same way, with a factor of $-i/\hbar$ for each interaction whose origin is clear in (7.66). The first term, by (7.71), is just $\psi_0(x, t)$. Now form the quantity $\mathcal{D}\psi(x, t)$, where $\mathcal{D}$ represents the differential operator

$$\mathcal{D} := i\hbar \frac{\partial}{\partial t} + \frac{\hbar^2}{2m} \frac{\partial^2}{\partial x^2} \quad .$$

Since K_0 satisfies the free-particle Schrödinger equation, the first term gives zero. The same happens in the integrand of the second term but there is another term that comes from differentiating with respect to the limit t of the t_1 integral:

$$\mathcal{D}\psi(x, t) = \int dx_1 K_0(x - x_1, 0)V(x_1, t)$$

$$\times \int dx_0 K_0(x_1 - x_0, t - t_0)\psi(x_0, t_0) + \ldots \quad .$$

We know from (7.13) that $K_0(\xi, 0) = \delta(\xi)$ and so the integration over x_1 can be carried out. The result is

$$\mathcal{D}\psi(x, t) = V(x, t) \int dx_0 K_0(x - x_0, t - t_0)\psi(x_0, t_0) + \ldots \quad .$$

If the series (8.47) is continued in the same way, we see that $\mathcal{D}\psi = V\psi$, and so the series, constructed using only the free-particle propagator, generates a wave function that satisfies the equation with interaction. Of course the integrals get increasingly difficult, but the first one is trivial, the second may be easy, and the third has been done. This picture of interactions as represented by a series of increasingly complicated diagrams is at the heart of most perturbation calculations in the theory of fields, with the difference that the interactions are with other fields and normally result in the creation or absorption of field quanta, i.e., particles. Then the lines have branches and it gets interesting.

Problem 8.19. In the first approximation (called the first Born approximation), find the wave function and the differential and total cross sections for scattering by a potential $V(r) = -ge^{-\kappa r}$. Compare the total current flowing outward with that flowing inward. Comments.

8.6 Action and Angle Variables

There is a special set of canonical variables that is particularly useful in the analysis of systems that are periodic (or nearly so) but nonseparable. Such systems include the mutual perturbations of planetary orbits and the establishment of chaotic behavior in oscillators with nonlinear coupling such as the Hénon-Heiles oscillator discussed in Sect. 6.5. To provide access to the literature on such matters this section will provide a brief introduction to action-angle variables, illustrating it with a calculation relating to coupled oscillators. The next section will discuss another version of perturbation theory and apply it to a problem in celestial mechanics.

Action-angle variables were introduced by the astronomer Charles Delauney in 1846. We will start by considering a separable periodic system and later we will introduce nonseparable terms as a perturbation. Assume that in the unperturbed system the entire motion is periodic, so that for each degree of freedom the system's path in phase space is either a closed orbit, as with a pendulum, or an infinite line, as with a rotating one; Fig. 6.12 shows both. The action integrals J_i are constant and the idea is to take them as the constants α_i in Jacobi's dynamics. Start with the characteristic function $W(q, \alpha)$ which since the variables are separable is the sum of terms $W_i(q_i, \alpha_i)$. Any J_v is the integral W_v evaluated over a single cycle, $\oint p_v dq_v$ (not summed). Calculate the J's in terms of the α's, invert to find $\alpha(J)$, and write (for each degree of freedom) $W(q, J)$. The energy is $H(J)$. Now consider W as a function of the type $-F_2$ that generates a canonical transformation from variables p, q to new variables w, J:

$$p(q, J) = \frac{\partial W(q, J)}{\partial q} \quad , \quad w(q, J) = \frac{\partial W(q, J)}{\partial J} \quad .$$

We will see in a moment that w can be considered as an angle. The equation of motion in the new variables is

$$\dot{w} = \frac{\partial H}{\partial J} := \nu(J) \quad \text{so that} \quad w(t) = \nu t + \text{const} \quad .$$

To see the significance of the new constants ν, let one of the q's, say q_v, move through a complete cycle of period τ_v. Then w_i increases by

$$\Delta w_i = \tau_v \nu_i = \oint \frac{\partial w_i}{\partial q_v} dq_v = \oint \frac{\partial^2 W}{\partial q_v \partial J_i} dq_v$$

$$= \frac{\partial}{\partial J_i} \oint p_v dq_v = \frac{\partial}{\partial J_i} J_v = \delta_{iv} \quad \text{(not summed)} \quad .$$

Thus ν_v is the frequency of the periodic motion of the vth degree of freedom, and if it is possible to find $H(J)$ without solving the equations of motion one can also find the frequency.

Trying this out on the harmonic oscillator in one dimension ($m = 1$, angular frequency ω), J is found from the area of the ellipse

$$E = \tfrac{1}{2}(p^2 + \omega^2 x^2)$$

of Fig. 5.1 and then (6.10) gives

$$J = \frac{2\pi E}{\omega} \quad , \quad H(J) = \frac{\omega J}{2\pi} \quad , \quad \dot{w} = \frac{\partial H}{\partial J} = \frac{\omega}{2\pi}$$

$$W = \frac{1}{2}\left[\sqrt{\omega} x \sqrt{J/\pi - \omega x^2} + \frac{J}{\pi} \sin^{-1} \sqrt{\pi\omega/J} x \right]$$

$$p = \frac{\partial W}{\partial x} = \omega\sqrt{J/\pi\omega - x^2} \quad , \quad w = \frac{\partial W}{\partial J} = \frac{1}{2\pi} \sin^{-1} \sqrt{\pi\omega/J} x$$

whence

$$x = \sqrt{J/\pi\omega} \sin \omega t \quad , \quad p = \omega\sqrt{J/\pi\omega} \cos \omega t \quad .$$

These formulas are exact, but the method is also useful for approximations.

The Complex Phase Plane. Harmonic oscillators are important because they are involved, as a starting point, in so many calculations in molecular physics, solid state, and the quantum theory of fields, and that is why this book has approached them from so many angles. In Sect. 7.9, for example, it was useful to think of the phase plane of an oscillator as a complex plane in order to calculate the effects of a perturbation. (To see that it really was useful, try solving the same problem in some more obvious way.) Here I will briefly show what that treatment had to do with action-angle variables and show how they can be introduced into quantum mechanics.

First, the classical version. Write

$$\Pi := p + i\omega q = \sqrt{\omega J/\pi} e^{2\pi i w}$$

so that

$$H = \tfrac{1}{2}(p^2 + \omega^2 q^2) = \omega J/2\pi$$

as before. With this,

$$p = \sqrt{\omega J/\pi} \cos 2\pi w \quad , \quad q = \sqrt{J/\pi\omega} \sin 2\pi w \quad .$$

To verify that the J and w so defined are indeed canonically conjugate variables, use them (Problem 8.20) to show that $(p, q) = 1$. Next, still using J and w, it follows at once that

$$(J, e^{2\pi i w}) = -2\pi i e^{2\pi i w}$$

and this can be verified by calculating it by the use of p and q.

For the passage to quantum mechanics, use (5.61) to write

$$[J, e^{2\pi i w}] = 2\pi \hbar e^{2\pi i w}$$

where J and w are hermitian variables but it is most convenient to use $e^{2\pi i w}$, rather than w, because of the well-known difficulties in defining an angle in quantum mechanics. (Basically, the trouble is that the domain of a hermitian operator function of an angle w consists periodic functions of that angle such as $e^{2\pi i w}$, whereas w is not a periodic function of w.)

The oscillator is easily quantized. Remembering that

$$(p + i\omega q)(p - i\omega q) = p^2 + \omega^2 q^2 - \hbar\omega$$

write

$$H - \frac{1}{2}\hbar\omega = \frac{1}{2}(p + i\omega q)(p + i\omega q)^{\dagger} = \frac{1}{2}\frac{\omega}{\pi}\sqrt{J}e^{2\pi i w}(\sqrt{J}e^{2\pi i w})^{\dagger} = \frac{\omega J}{2\pi} \quad .$$

To find the eigenfunctions and eigenvalues of J, note that J resembles an angular momentum in the space whose coordinates are q and p/ω, and remember the theory of ordinary angular momentum. Assume there is a Φ for which $J\Phi = j\Phi$, and form a new $\Phi' := e^{2\pi i w}\Phi$. Then

$$J\Phi' = e^{2\pi i w}J\Phi + 2\pi \hbar e^{2\pi i w}\Phi = (j + 2\pi\hbar)\Phi' \quad .$$

The lowest possible value for j is zero, corresponding to an eigenfunction Φ_0, so that in general,

$$j = 2\pi n\hbar \quad , \quad n \in \mathbb{Z}_+$$

belonging to the eigenfunction $\Phi_n = e^{2\pi i n w}\Phi_0$, and the eigenvalues of H are $(n + \frac{1}{2})\hbar\omega$. It may be of interest to note the similarities and differences between this method of quantization and that developed in Sect. 2.6.

Quartic Oscillator. Another example is the quartic oscillator in Sect. 3.8, where an approximate $x(t)$ gives an approximate action integral. Using the results obtained there we easily find

$$J = (\Lambda\alpha_1^3)^{1/4} \quad , \quad \Lambda = \frac{2^{10} \cdot 35}{27\mu}$$

$$H = \alpha_1 = \Lambda^{-1/3}J^{4/3} \quad , \quad \nu = \frac{\partial H}{\partial J} = \frac{4}{3}(J/\Lambda)^{1/3}$$

and when ν is worked out it gives the period (3.58).

Coupled Oscillators. Suppose we have two harmonic oscillators with frequencies ω_1 and ω_2, coupled nonlinearly so that their Hamiltonian is

$$H(p_1, p_2, x_1, x_2) = H_{01}(p_1, x_1) + H_{02}(p_2, x_2) + \lambda x_1 x_2^2$$

where each H_0 is of the form

$$H_0 = \tfrac{1}{2}(p^2 + \omega^2 x^2) \quad .$$

First let $\lambda = 0$. The motion of the two oscillators will then, depending on the ratio of the frequencies, be a space-filling curve as shown in Fig. 8.1a or some Lissajous figure. Considering either one of the oscillators, assume a solution of the form $x = A \cos (\omega t + \beta)$. The energy and the action integral are then easily found,

$$J = \pi A^2 \omega \quad , \quad \alpha_1 = H = \tfrac{1}{2}A^2\omega^2$$

and in terms of the new canonical variables,

$$H = \frac{\omega J}{2\pi} \quad , \quad x(J, w) = \sqrt{\frac{J}{\pi\omega}} \cos 2\pi w \quad , \quad 2\pi w = \omega t + \beta \quad .$$

Each uncoupled oscillator can be described in this way, and the coupling is

$$H_{\text{int}} = \lambda\sqrt{\frac{J_1}{\pi\omega_1}\frac{J_2}{\pi\omega_2}} \cos 2\pi w_1 \cos^2 2\pi w_2$$

where $2\pi w_i = \omega_i t + \beta_i$. If there were no coupling the oscillators would continue with constant J's and therefore constant amplitudes. Suppose that λ is small. We expect the amplitudes to fluctuate slightly, but under what circumstances will there be large and permanent changes? Look at J_1:

$$\dot{J}_1 = -\frac{\partial H_{\text{int}}}{\partial w_1} = -\lambda\sqrt{\frac{J_1}{\pi\omega_1}\frac{J_2}{\pi\omega_2}}2\pi \sin (\omega_1 t + \beta_1) \cos^2 (\omega_2 t + \beta_2) \quad .$$

The time-dependent part has Fourier components at angular frequencies ω_1, $\omega_1 + 2\omega_2$, and $\omega_1 - 2\omega_2$. If ω_1 is not close to $2\omega_2$ these components will fluctuate and J_1 will change little. If it is close, integration over a time T gives

$$\Delta J_1 \overset{\propto}{\sim} \int_0^T \cos [(\omega_1 - 2\omega_2)t + \beta_1 - 2\beta_2]dt$$

so that if $(\omega_1 - 2\omega_2)T < 1$, J_1 will increase almost linearly with T due to resonance between the x_1 oscillator and the second harmonic of the x_2 oscillator. The amplitudes of the two motions will change greatly and the Lissajous figure will be replaced by some irregular motion.

Problem 8.20. To verify that J and w are conjugate variables, use them to evaluate the Poisson brackets (q, p) and $(J, e^{2\pi i w})$. Also evaluate the second of these using q and p.

Problem 8.21. Study the perturbation of regular motions in the Hénon-Heiles oscillator of Sect. 6.5 as follows: Decompose the Hamiltonian (6.53) into

$$H_{0x} = \tfrac{1}{2}(p_x^2 + x^2) + \tfrac{1}{3}x^3 \quad , \quad H_{0y} = \tfrac{1}{2}(p_y^2 + y^2) \quad , \quad H_{\text{int}} = -xy^2 \quad .$$

The x equation can be solved with a Jacobian elliptic function but approximate this with trigonometric functions as in (2.29). Then study the onset of secular perturbations as the energy is changed. (The difference between this and the example worked out above is that the frequencies of the uncoupled oscillators are the same only for very small amplitudes and as the energy increases resonances will be established. Your conclusions will be only qualitatively correct because the KAM theorem, which is proved with high-order perturbation theory, assures that the oscillations are more stable than at first appears.)

8.7 Canonical Perturbation Theory

The classical theory of perturbation can be given a handsome formal basis, but practical calculations tend to be long. First, the formal basis.

Assume as before that the Hamiltonian consists of a soluble part H_0 and a small part $V(p, q, t)$. Write a complete solution of the unperturbed problem as

$$p_n(\alpha, \beta, t) \quad , \quad q_n(\alpha, \beta, t) \quad , \quad \alpha_1 = E_0 \quad . \tag{8.48}$$

Assume[1] that these are soluble for the constants α_n and β_n, and call these constants

$$C_i = C_i(p, q, t) \quad i = 1, \dots 2f \tag{8.49}$$

reserving the name C_1 for β_1. Since φ_1 in (6.42) is $t + \beta_1$, the equation of motion for φ_1 is

$$\dot{\varphi}_1 = (\varphi_1, H_0) = (\beta_1, \alpha_1) = 1 \tag{8.50}$$

The rest of the C's are time-independent, so we can write

$$\dot{C}_i = (C_i, H_0) = \delta_{i1} \quad .$$

In the perturbed system the p's and q's are no longer the same functions of time, and the C's, still defined by (8.49), are no longer constant, for

$$\dot{C}_i = (C_i, H_0 + V) = (C_i, V) + \delta_{i1} \quad . \tag{8.51}$$

To evaluate this, suppose that $V(p, q, t)$ has been written via (8.48) as $V(C, t)$. Then

$$(C_i, V) = \frac{\partial C_i}{\partial q_n} \frac{\partial V}{\partial C_j} \frac{\partial C_j}{\partial p_n} - \frac{\partial C_i}{\partial p_n} \frac{\partial V}{\partial C_j} \frac{\partial C_j}{\partial q_n}$$

so that (8.51) is

$$\dot{C}_i = (C_i, C_j) \frac{\partial V}{\partial C_j} + \delta_{i1} \tag{8.52}$$

where i and j run from 1 to $2f$. The merit of this expression is that it is exact, and that by Poisson's theorem the Poisson brackets are independent of the choice of coordinates and momenta p, q used to evaluate them.

Equation (8.51) has a precise counterpart in quantum mechanics: if C_i is a time-independent dynamical variable that is constant in the unperturbed system, then in the perturbed system,

---

[1] See the remarks on solubility in Sect. 6.4.

$$\frac{d\langle C_i \rangle}{dt} = \frac{i}{\hbar} \langle [(\hat{H}_0 + \hat{V}), \hat{C}_i] \rangle = \frac{i}{\hbar} \langle [\hat{V}, \hat{C}_i] \rangle \quad . \tag{8.53}$$

The foregoing equations are all exact and must be untangled, since the C's are defined in terms of quantities pertaining to the perturbed system. If we write

$$C_i = C_i^{(0)} + C_i^{(1)} + \ldots \tag{8.54}$$

where the $C_i^{(0)}$ are calculated from the unperturbed system, (8.52) becomes

$$\dot{C}_i^{(1)} = (C_i^{(0)}, C_j^{(0)}) \frac{\partial V^{(0)}}{\partial C_j^{(0)}} + \ldots \tag{8.55}$$

where $V^{(0)}$ means $V(C^{(0)})$, and higher approximations can be obtained similarly.

In order to see how the formalism works out in practice, apply it first to a particularly simple and obvious situation.

Example 8.2. Perturbed Harmonic Oscillator. Let

$$H_0 = \frac{1}{2m}(p^2 + m^2\omega^2 x^2) \tag{8.56}$$

and suppose that as in Sect. 7.1, the perturbation consists of attaching an extra spring to it: $\omega \to \omega + \varepsilon$, $\omega^2 \to \omega^2 + 2\varepsilon\omega$, and therefore the Hamiltonian is changed by $\delta H = m\omega\varepsilon x^2$. Denoting the energy by α_1 write the solution of the unperturbed problem as

$$x(t) = \sqrt{\frac{2\alpha_1}{m\omega^2}} \cos \omega(t + \beta_1) \tag{8.57}$$

so that

$$\delta H = \frac{2\varepsilon\alpha_1}{\omega} \cos^2 \omega(t + \beta_1) \quad .$$

This fluctuates in time, so take the time average to see what change it will make in the long run:

$$\overline{\delta H} = \frac{\varepsilon\alpha_1}{\omega} \quad .$$

Since both the new β_1 and the old one satisfy (8.50), the change in β_1 produced by this perturbation satisfies

$$\frac{d\overline{\delta\beta_1}}{dt} = (\beta_1, \overline{\delta H}) = \varepsilon/\omega$$

so that on the average,

$$\overline{\delta\beta_1} = \varepsilon t/\omega + \text{const}$$

and $x(t)$ in (8.57) becomes

$$x(t) = \sqrt{\frac{2\alpha_1}{m\omega^2}} \cos [(\omega + \varepsilon)(t - t_0)] \quad . \tag{8.58}$$

There is no δx: we have done more than that and solved the problem of secular perturbations. (By a happy coincidence the answer is exact, but that is peculiar to this simple system.)

245

Problem 8.22. The motion of a free particle in two dimensions is perturbed by a constant force of arbitrary size. Find the exact solution by the foregoing perturbation procedure.

Perturbation theory is the basis of celestial mechanics [113]. Its main problem is to avoid secular perturbations; this is the more difficult since there actually are secular perturbations in the solar system arising from resonances. It happens, for example, that when Jupiter orbits the Sun exactly 5 times, Saturn orbits it 2.0134 times, and the near resonance produces a much larger exchange of energy between the two than would normally occur. Even so, it is small, since the force is weak and the periodicity of the resonance is 880 years. To obtain the accuracy required in astronomy or the space program requires very long calculations, but the methods are mostly simple in principle, as the example in the following section will show.

8.8 Newtonian Precession

The general-relativity precession is only a small part of the precession of the orbit of Mercury as viewed from the Earth. The largest part is accounted for by the precession of the coordinate system used, which is tied to the slowly-changing direction of the Earth's axis and will be evaluated in Sect. 9.5. Another part, to be discussed here, is contributed by the action of the other planets, and the part attributable to general relativity is only what is left after the others have been subtracted from the observed value.

To characterize the motion of a planet requires four numbers, essentially two coordinates and two momenta, but it is convenient to choose them as follows. First define the orbit by the magnitude and direction of its major axis a, and its eccentricity e. For this is enough to know a and the two components of the Laplacian integral A, since by (3.47), $A = \gamma e$. The direction of A is given by ϑ_a, Fig. 8.4, and ϑ_{a0} is the

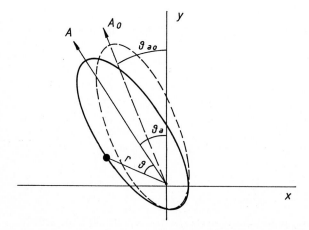

Fig. 8.4. Use of the vector A to calculate the precession of Mercury's orbit

value of ϑ_a at the time $t = 0$. Without perturbations ϑ_a would stay constant at ϑ_{a0}; we are going to calculate its rate of change. To locate the planet in its orbit assume that at some time $t = t_0$ the planet was at aphelion, and measure its subsequent displacement relative to that initial direction. The planet's position and motion at time t can be calculated from the four numbers

$$a, \ A_x, \ A_y, \ t_0 \ . \tag{8.59}$$

To calculate the Poisson brackets in (8.55) we need a set of coordinates and momenta. Which ones to choose is only a matter of convenience and does not affect the result, but the calculation is short if we remember the view of the Hamilton-Jacobi equation that was taken in Sect. 6.4, in which the characteristic function is taken to be the generator of a contact transformation to a set of coordinates β_n and momenta α_n with constant values. If the planetary mass is set equal to 1, they are (Problem 8.23)

$$\alpha_1 = E = -\frac{\gamma}{2a} \ , \qquad \alpha_2 = L = \sqrt{\gamma a(1 - e^2)}$$
$$\beta_1 = -t_0 \ , \qquad \beta_2 = \vartheta_{a0} \ . \tag{8.60}$$

In the unperturbed system the components of A are constant; evaluate them at time t_0:

$$A_x = -\gamma e \sin \beta_2 \ , \qquad A_y = \gamma e \cos \beta_2 \tag{8.61}$$

and the other variables are found from (8.60):

$$a = -\frac{\gamma}{2\alpha_1} \ , \qquad e^2 = 1 + \frac{2}{\gamma^2} \alpha_1 \alpha_2^2 = \frac{1}{\gamma^2}(A_x^2 + A_y^2) \ . \tag{8.62}$$

We can now calculate the required brackets almost at sight:

$$(A_x, A_y) = \frac{\partial A_x}{\partial \beta_2} \frac{\partial A_y}{\partial \alpha_2} - \frac{\partial A_y}{\partial \beta_2} \frac{\partial A_x}{\partial \alpha_2} = -2\alpha_1 \alpha_2 \quad \text{or} \tag{8.63}$$

$$(A_x, A_y) = \frac{\gamma}{a} \sqrt{\gamma a(1 - e^2)} \tag{8.64a}$$

and similarly

$$(a, t_0) = \frac{2a^2}{\gamma} \tag{8.64b}$$

$$(A_x, t_0) = \frac{a(1 - e^2)}{\gamma e^2} A_x \tag{8.64c}$$

$$(A_y, t_0) = \frac{a(1 - e^2)}{\gamma e^2} A_y \tag{8.64d}$$

$$(a, A_x) = (a, A_y) = 0 \ . \tag{8.64e}$$

The largest perturbation of the orbit of Mercury is created by Jupiter. Figure 8.5 shows the situation assuming for simplicity that both orbits lie in the same plane. We assume that Jupiter ($\mathbf{2}\!\!\!\downarrow$) moves in a circular orbit of radius a_2 with an angular

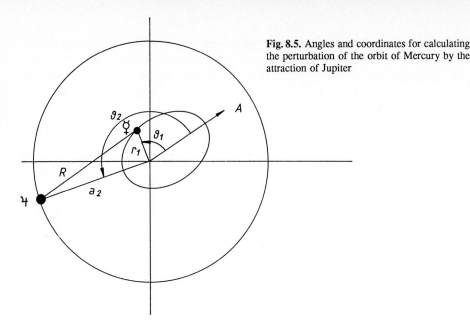

frequency Ω_2, while Mercury ($\mathrm{\small ☿}$) moves in an ellipse characterized by a_1 and e_1. The perturbing energy is

$$V = -\frac{\delta}{R} \qquad \delta = Gm_{\mathrm{\small 24}} \qquad . \tag{8.65}$$

In terms of the orbital variables this contains

$$\frac{1}{R} = \frac{1}{\sqrt{r_1^2 - 2r_1 a_2 \cos \vartheta + a_2^2}} \qquad , \qquad \vartheta = \vartheta_2 - \vartheta_1$$

$$= \frac{1}{a_2} \sum_{n=0}^{\infty} \left(\frac{r_1}{a_2}\right)^n P_n(\cos \vartheta) \tag{8.66}$$

where the P_n's are the Legendre polynomials, calculated by expanding the expression for $1/R$ in powers of r_1/a_2:

$$P_0(\cos \vartheta) = 1 \quad , \quad P_1(\cos \vartheta) = \cos \vartheta \quad , \quad P_2(\cos \vartheta) = \tfrac{1}{2}(3 \cos^2 \vartheta - 1)$$

and so forth. We will need only these three terms.

Since both $\cos \vartheta$ and r_1 are periodic functions of t, and since we are only interested in the cumulative effect after a long time, we can simplify everything by taking time averages as in Example 8.2. In the first term of the sum (8.66) we have

$$r_1(t) \cos \vartheta(t) = r_1(t)(\cos \vartheta_1 \cos \vartheta_2 + \sin \vartheta_1 \sin \vartheta_2)$$

in which $r_1 \cos \vartheta_1$ and $r_1 \sin \vartheta_1$ are periodic in t with period T_1 and $\cos \vartheta_2$ and $\sin \vartheta_2$ have period T_2. Since T_1 and T_2 are different and incommensurable, the entire term averages to zero, while by the same argument, $P_2(\cos \vartheta)$ can be replaced by

$$\overline{P_2} = \tfrac{1}{2}(\tfrac{3}{2} - 1) = \tfrac{1}{4} \qquad . \tag{8.67}$$

To complete the second term we need the time average of r_1^2. Averages of this kind are quite easily calculated; details are in the appendix to this chapter and the result is

$$\overline{r_1^2} = (1 + \tfrac{3}{2}e_1^2)a_1^2 \quad . \tag{8.68}$$

Thus effectively, the perturbing energy is

$$V = \overline{V} = -\frac{\delta}{a_2}\left[1 + \frac{1}{4}\left(1 + \frac{3}{2}e_1^2\right)\left(\frac{a_1}{a_2}\right)^2\right] \quad .$$

The precession is given by the time derivative of $\boldsymbol{A}$. The average rate of change of A_x is

$$\overline{\dot{A}_x} = (A_x, \overline{V}) = -\frac{3\delta a_1^2}{8a_2^3}(A_x, e_1^2)$$

by (8.64e). With (8.60), this is

$$\overline{\dot{A}_x} = -\frac{3}{8}\frac{\delta a_1^2}{\gamma^2 a_2^3}(A_x, A_x^2 + A_y^2)$$

$$= -\frac{3}{4}\frac{\delta a_1^2}{\gamma^2 a_2^3}(A_x, A_y)A_y = -\frac{3}{4}\frac{\delta a_1}{\gamma a_2^3}\sqrt{\gamma a_1(1 - e_1^2)}A_y \quad .$$

One revolution of Mercury takes a time

$$T_1 = \frac{2\pi a_1^{3/2}}{\gamma^{1/2}}$$

and in this time, on the average, A_x changes by

$$\Delta A_x = \overline{\dot{A}_x}T_1 = -\frac{3}{4}2\pi\frac{\delta}{\gamma}\left(\frac{a_1}{a_2}\right)^3\sqrt{1 - e_1^2}A_y \quad .$$

Choose axes so that at the moment of interest $\boldsymbol{A}$ lies along the y axis. Then the forward precession in one revolution is

$$-\frac{\Delta A_x}{A_y}\text{ radians} = -\frac{\Delta A_x}{A_y}\frac{360 \times 60^2}{2\pi}\text{ arc sec} \quad .$$

In one (Earth) century, Mercury makes 415.2 revolutions. The remaining parameters are

$$\frac{\delta}{\gamma} = \frac{m_{2\!\!1}}{m_\odot} = \frac{314.5m_\oplus}{329390m_\oplus} = 9.548 \times 10^{-4} \quad , \qquad \oplus = \text{Earth}$$

$$\frac{a_1}{a_2} = \frac{57.91 \times 10^6\,\text{km}}{778.3 \times 10^6\,\text{km}} = 7.441 \times 10^{-2}$$

$$e_1 = 0.2056$$

and we find for the precession in one century attributable to Jupiter

$$P = 155'' \quad .$$

The precession of Mercury's orbit caused by Jupiter and all the other planets is calculated as $531''$ per century. The precession of the equinoxes once in about 25 780 years adds $5026''$ to this. The observed value is $5600''$. The difference

$$5600'' - (531'' + 5026'') = 43''$$

agrees with the value calculated in Sect. 6.3.

The foregoing calculation is an example of a first-order perturbation. Higher orders are, as one might expect, more difficult and are described in works on celestial mechanics.

Problem 8.23. Calculate the brackets in (8.64).

Problem 8.24. The eccentricity of Jupiter's orbit is $e_2 = 0.056$. Refine the calculation done above to see whether this makes any difference.

Problem 8.25. Find the effect of the next nonvanishing term of the expansion (8.66).

Problem 8.26. Calculate (A_x, A_y) using Cartesian coordinates and momenta, and show that (8.64a) results.

Problem 8.27. The interaction with other planets not only causes a planet's orbit to precess; it also alters the position of the planet in its orbit. By how many seconds of arc in one revolution, on the average, does Jupiter change the orbital position of Mercury?

Problem 8.28. As seen from the earth, the moon travels in an approximately elliptical path in a plane (Fig. 8.6), slightly inclined to the plane of the earth's orbit around the sun (the plane of the ecliptic). Owing to the action of the sun, the plane of the moon's orbit wobbles – i.e., the angular momentum l precesses at a rate of one revolution in 18.61 y. Assuming that the earth orbits the sun in a circle of radius R, calculate this rate.

Take as variable the vector r running from the earth to the moon and show that the perturbing energy provided by interaction with the sun is

$$V = \frac{GM\mu}{2R}\left[3\left(\frac{R \cdot r}{R^2}\right)^2 - \frac{r^2}{R^2}\right] \tag{8.69}$$

where M is the sun's mass and μ is the reduced mass of earth and moon. Now, calculate the rate at which l precesses. The answer is

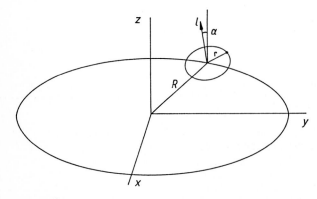

Fig. 8.6. Coordinates for Problem 8.28

250

$$\frac{T_{\text{p}}}{T_{\text{☽}}} = \frac{4}{3} \frac{m_{\oplus} + m_{\text{☽}}}{M} \left(\frac{R}{a_{\text{☽}}}\right)^3 \sec \alpha$$

where T_{p} is the period of the precession, $T_{\text{☽}}$ is one sidereal month[2], and the other variables are obvious. Numerical values are

$$m_{\oplus} + m_{\text{☽}} = 6.057 \times 10^{27}\,\text{g} \quad M = 1.971 \times 10^{33}\,\text{g}$$
$$R = 1.496 \times 10^{13}\,\text{cm} \quad a_{\text{☽}} = 3.8155 \times 10^{10}\,\text{cm}$$
$$\alpha = 5°8'43'' \quad T_{\text{☽}} = 27.32\,\text{d} \quad .$$

Appendix

Time Average of r^n in Kepler Orbits . We wish to evaluate

$$\overline{r^n} = \frac{1}{T} \oint r(t)^n dt \quad . \tag{8.70}$$

Start with (3.20),

$$\frac{dr}{dt} = \sqrt{\frac{2}{m}\left(E + \frac{\gamma}{r}\right) - \left(\frac{L}{mr}\right)^2}$$

and rewrite it using (2.75a, b) and (2.77),

$$-E = \frac{\gamma}{2a} \quad , \quad L^2 = (1 - e^2)\frac{m\gamma^2}{-2E} = (1 - e^2)\gamma ma \quad , \quad T = 2\pi\sqrt{\frac{m}{\gamma}}a^{3/2}$$

as

$$\frac{dr}{dt} = \frac{2\pi a}{rT}\sqrt{e^2 a^2 - (a - r)^2} \quad . \tag{8.71}$$

Next it will be convenient to introduce an angular variable ψ first used by Kepler in 1609.[3] Let

$$r = a(1 - e \cos \psi)$$

in terms of which (8.71) gives

$$dt = \frac{rT}{2\pi a}d\psi \quad .$$

Then substituting the last two relations into (8.70) gives

$$\overline{r^n} = \frac{a^n}{2\pi} \oint (1 - e \cos \psi)^{n+1} d\psi$$

where the integral is from 0 to 2π. The integral can be evaluated by simple trans-

---

[2] This is the time in which the moon circles the earth in a coordinate system determined by the fixed stars.

[3] Kepler's name for this variable was the eccentric anomaly, E. The letter is changed to avoid confusion with the energy.

formations of Laplace's integral for the Legendre polynomials [17, p.312]. A few values are

n	$\overline{(r/a)^n}$
3	$1 + 3e^2 + \frac{3}{8}e^4$
2	$1 + \frac{3}{2}e^2$
1	$1 + \frac{1}{2}e^2$
0	1
-1	1
-2	$(1 - e^2)^{-1/2}$
-3	$(1 - e^2)^{-3/2}$
-4	$\frac{1+e^2/2}{(1-e^2)^{5/2}}$

It is sometimes useful to know quantities like $\overline{r}$, the average value of the revolving radius vector. Clearly this lies along the axis, in the direction of aphelion. The component of r in this direction is $r \cos \vartheta$. Written in the units of this appendix, (3.27) is

$$r = \frac{a(1 - e^2)}{1 - e \cos \vartheta} \quad , \quad er \cos \vartheta = r - a(1 - e^2)$$

whence, from the table, the time average is

$$|\overline{r}| = \overline{r \cos \vartheta} = \frac{3}{2}ea$$

and similarly

$$\overline{(r \cos \vartheta)^2} = (\tfrac{1}{2} + 2e^2)a^2 \quad , \quad \overline{(r \sin \vartheta)^2} = \tfrac{1}{2}(1 - e^2)a^2 \quad .$$

9. The Motion of a Rigid Body

In the narrow world of the exact sciences there are several kinds of rigid body. There are nuclei, generally not spherical and precessing in the atomic field or some applied field. There are molecules, whose shapes may be very complex, and there are stars and planets interacting through gravitation. In the closely adjacent field of engineering, and especially space engineering, there is an immense variety of oscillating and rotating devices, some of which, such as the gyroscopic stabilizers used on space craft, are extremely sophisticated.

In every case the rigidity is an idealization. Nuclear and molecular spectra show series of lines associated with rigid-body motions among other series produced by motions in which the object changes its shape. Stars and planets are far from rigid, and the designers of real machinery are always conscious of the elastic properties of the materials they use. Nevertheless there are phenomena characteristic of rigid-body motion in general, and this chapter will investigate some of them. The discussion will be limited by a somewhat arbitrary restriction. Except for a few involved in inertial guidance (Sect. 9.4), man-made devices generally involve supports of some kind, fixed axes or restrictive gimbals by which forces are exerted from outside to constrain the system. Nuclei, molecules, and astronomical bodies float in the fields that influence their motion. We will consider only this latter class, skipping with regret (except for Problem 9.20) the theory of the spinning top, an object with one fixed point. For 200 years the top was analyzed mathematically in order to confirm the insights of childhood; more recently university students have been introduced to tops in order to confirm the insights of the blackboard. Instead, riding the crest of astrology, we will consider the precession of the equinoxes.

9.1 Angular Velocity and Momentum

The velocity of a point revolving around a fixed axis can be very simply described. Choose a point O on the axis to be the origin of coordinates and locate P by the vector r_0. Define the angular-velocity vector ω directed along the axis according to the right-hand rule, with the length equal to the numerical angular velocity ω. Then the velocity of P is given, in magnitude and direction, by

$$v = \omega \times r_0 \tag{9.1}$$

(Fig. 9.1) regardless of the location of O. Note that the components of the angular velocity ω are not introduced as time derivatives of anything. We shall return to this important point in Sect. 9.5.

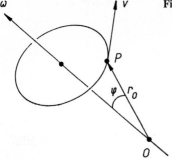

ω V

Fig. 9.1. Relation between linear velocity and angular velocity

If the axis itself is moving, so that the point O is in motion with velocity V, (9.1) becomes

$$v = V + \omega \times r_0 \quad . \tag{9.2}$$

In all the cases considered here the axis of rotation will pass through the center of mass, and we may choose coordinates so that the center of mass is at rest. If now O is chosen as the center of mass, V is zero and (9.1) remains valid even if ω changes its direction.

With this choice, evaluate the angular momentum of the particle P. It is

$$l = m r_0 \times v = m r_0 \times (\omega \times r_0) \quad \text{or}$$

$$l = m[r_0^2 \omega - (\omega \cdot r_0) r_0] \quad . \tag{9.3}$$

Note that l is not in general parallel to ω. It will in the sequel be convenient to introduce vector indices explicitly and write

$$l_i = m(r_0^2 \omega_i - r_{0i} r_{0j} \omega_j) \quad \text{(summed)} \quad \text{or}$$

$$l_i = m(r_0^2 \delta_{ij} - r_{0i} r_{0j}) \omega_j \quad . \tag{9.4}$$

The fundamental hypothesis about a rigid body (never really true) is that all parts of it maintain the same relative positions. This means that they all share the same ω. Thus the angular momentum of a rigid body is given by

$$L_i = I_{0ij} \omega_j \quad , \quad I_{0ij} = \sum m(r_0^2 \delta_{ij} - r_{0i} r_{0j}) \tag{9.5}$$

summed over all the masses, with the components r_{0k} appropriate to each mass. The formula looks nice but it is useless, for as the body moves all the r's change, and both I_{ij} and ω_j are functions of time. This can be remedied by introducing a new coordinate system.

The coordinates used so far, fixed in space and centered on O, have served only to specify the magnitudes and directions of the vectors r_0, v, and ω; we have said nothing about their rates of change. To be sure, v is the rate of change of r_0, but it was introduced by (9.1) and not as a time derivative. All the relations derived thus far hold equally well in a rotating coordinate system, for the lengths of vectors and the angles between them are unchanged in such a system. Of course, if we start thinking about the rates of change of vectors we must be careful, since a vector fixed in a rotating coordinate system appears to move when viewed from fixed axes. We shall see presently how to take this into account.

Choose axes fixed in the rigid body, with the origin at the center of mass as before. The interpretation of the vectors is unchanged: v represents a particle's velocity with respect to fixed axes, but the components of v are evaluated with respect to the moving axes. The position of a particle with respect to the moving axes is given by the constant vector r, and (9.5) becomes

$$L_i = I_{ij}\omega_j \quad , \qquad I_{ij} = \sum m(r^2\delta_{ij} - r_ir_j) \tag{9.6}$$

where the components I_{ij} are constant. This will greatly simplify the later analysis.

The same quantities occur in an evaluation of the body's kinetic energy, T:

$$T = \tfrac{1}{2}\sum m(\omega \times r)^2 = \tfrac{1}{2}\sum m[\omega^2 r^2 - (\omega \cdot r)^2]$$
$$= \tfrac{1}{2}\sum m(r^2\delta_{ij} - r_ir_j)\omega_i\omega_j \quad .$$

That is,

$$T = \tfrac{1}{2}I_{ij}\omega_i\omega_j \quad \text{or} \tag{9.7}$$

$$T = \tfrac{1}{2}L \cdot \omega \quad . \tag{9.8}$$

This recalls the fact that one can write $T = \tfrac{1}{2}p \cdot v$ for the motion of a particle, but here there is the difference that in general L and ω are not parallel.

9.2 The Inertia Tensor

An object like I_{ij}, bearing more than one index, is called a tensor if it has the right transformation properties. The requirement is that it should transform like a product of vectors, and since it is in our case formed from the product of vectors this criterion is satisfied automatically. A tensor can be regarded in two ways: as representing a linear operator or a thing. Considered as a linear operator, I transforms the vector ω into the vector L by rotating ω and changing its length. Considered as a thing, it represents the inertial properties of a solid object just as other physical properties are represented by various vectors and scalars. Written out in detail, the components are

$$I_{ij} = \begin{pmatrix} \sum m(y^2 + z^2) & -\sum mxy & -\sum mxz \\ -\sum mxy & \sum m(x^2 + z^2) & -\sum myz \\ -\sum mxz & -\sum myz & \sum m(x^2 + y^2) \end{pmatrix} \quad . \tag{9.9}$$

It is customary to label the rows of such an array by the first index and the columns by the second, but here it does not matter, since I is symmetrical,

$$I_{ij} = I_{ji} \quad . \tag{9.10}$$

Clearly also,

$$I_{ii} \text{ (summed)} = 2\sum mr^2 \quad . \tag{9.11}$$

This diagonal sum, called the trace of the tensor (The German *Spur* is also used) in invariant under rotations of the coordinate system. A single component of I_{ij} has already been encountered in (4.15) and the formula (4.16) is a restricted form of (9.6). The diagonal elements of I_{ij} are called coefficients of inertia and the off-diagonal ones are products of inertia.

It is the products of inertia that make calculations difficult. Suppose they were all zero. Then we would have

$$I_{ij} = \begin{pmatrix} I_x & & 0 \\ & I_y & \\ 0 & & I_z \end{pmatrix} \quad , \quad I_x = I_{xx} \quad , \quad \text{etc.} \tag{9.12a}$$

and in terms of unit vectors,

$$L = I_x \omega_x \hat{i} + I_y \omega_y \hat{j} + I_z \omega_z \hat{k} \tag{9.12b}$$

$$T = \tfrac{1}{2}(I_x \omega_x^2 + I_y \omega_y^2 + I_z \omega_z^2) \quad . \tag{9.12c}$$

This can always be achieved by the proper choice of axes with respect to the body in which they are fixed. The fact is proved in many books, e.g., [38], [116]. Rather than reproduce the details here, we remark only that things often have axes of symmetry and one merely needs to align the axes with them. The following example and problems will show how this occurs.

Example 9.1. A uniform slab in the shape of an isosceles triangle, Fig. 9.2. To evaluate $\sum mxy$, divide the slab in strips like s_1. For each value of y the strip gives equal contributions from positive and negative x, so $I_{xy} = 0$. For I_{xz} and I_{yz}, integrate first along the z axis. Since positive and negative values of z contribute equally, these also vanish. Thus even though the slab is not symmetrical with respect to y, its symmetry with respect to the other two axes suffices to diagonalize the inertia tensor.

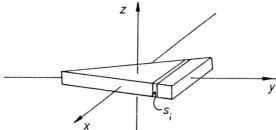

Fig. 9.2. Isosceles-triangular slab of uniform thickness

Problem 9.1. Prove that for any inertia tensor reduced to diagonal form,

$$I_x + I_y \geq I_z \quad , \quad I_y + I_z \geq I_x \quad , \quad I_z + I_x \geq I_y \quad . \tag{9.13}$$

Problem 9.2. A section of pipe with inside and outside radii r_1 and r_2 rolls down an inclined plane. Calculate the acceleration, and give the two limiting cases of a thin-walled pipe and a solid cylinder.

256

Problem 9.3. A shop sign painted on a board of width λ and height h hangs suspended from a horizontal bar. What is its period of oscillation?

Problem 9.4. Find the moment of inertia tensor of a triangular flat slab bounded by the points $(a,0,0)$, $(1,0,0)$ and $(0,1,0)$. Find a new coordinate system in which the tensor will be diagonal.

Problem 9.5. Provide the slab in Fig. 9.2 with literal dimensions and calculate its coefficients of inertia.

9.3 Dynamics in a Rotating Coordinate System

The basic differential equation governing rotating bodies is (4.11), but this, of course, refers to motion with respect to stationary axes. It is not hard to express it with respect to rotating axes, but before doing so I will digress on a few related topics.

Let a vector a be rigidly attached to a coordinate system S, coordinates xyz, which is rotating with angular velocity ω relative to a fixed system S_0, coordinates $x_0 y_0 z_0$. The tip of the vector, stationary in S, moves with respect to S_0 with a velocity

$$\dot{a}_0 = \omega \times a \quad (a \text{ fixed in } S) \quad . \tag{9.14a}$$

If a is moving with respect to S at a rate $\dot{a}$, then this velocity is added to that just given,

$$\dot{a}_0 = \omega \times a + \dot{a} \quad (a \text{ moving in } S) \quad . \tag{9.14b}$$

This relation will presently be used to transform the equations for rigid bodies. To see some of its further implications, suppose that the vector r is the position vector of a particle. Then with respect to the fixed axes the particle's velocity is

$$\dot{r}_0 = \omega \times r + \dot{r} \quad . \tag{9.15}$$

Write this as

$$v_0 = \omega \times r + v \quad . \tag{9.16}$$

Once more: v_0 is the velocity of the particle with respect to fixed axes; v with respect to moving axes. Now take the derivative again in the same way. The rate of change of v_0 with respect to fixed axes is the absolute acceleration,

$$a_0 := \dot{v}_{00} := \dot{\omega}_0 \times r + \omega \times \dot{r}_0 + \dot{v}_0$$

where $\dot{v}_0$ is a hybrid, the rate of change of v (not v_0) with respect to fixed axes. So that everything on the right will refer to moving axes use (9.14b) three times.

$$a_0 = (\omega \times \omega) \times r + \dot{\omega} \times r + \omega \times (\omega \times r) + \omega \times v + \omega \times v + \dot{v}$$

$$a_0 = \dot{\omega} \times r + \omega \times (\omega \times r) + 2\omega \times v + \dot{v} \quad . \tag{9.17a}$$

This can also be written as

$$a_0 = \dot{\omega} \times r + (\omega \cdot r)\omega - \omega^2 r + 2\omega \times v + \dot{v} \quad . \tag{9.17b}$$

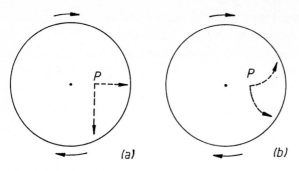

Fig. 9.3. (a) Paths of a marble rolling on a rotating turntable as viewed from the room. (b) As viewed by an observer rotating with the turntable

The first, third, and last terms should be obvious. The second corrects the third when r is not perpendicular to ω as it usually is in elementary physics, and the fourth term is the Coriolis acceleration. Like the centrifugal acceleration $\omega \times \omega \times r$, it is part of the price one pays for looking at the world from a merry-go-round. Put a piece of paper onto a phonograph turntable and roll an inked marble on it, Fig. 9.3. With respect to the room, the marble's path will be (approximately) straight, but the tracks it leaves on the turntable will curve towards the left, however it is started. If we write laws of dynamics to describe the motions of particles as seen in the rotating frame, they must contain forces corresponding to these accelerations. If the real force on the particle is $F = ma_0$ the apparent force is

$$m\dot{v} = F - m\dot{\omega} \times r - m\omega \times \omega \times r - 2m\omega \times v \quad . \tag{9.18}$$

The third term is the centrifugal force and the last is the Coriolis force.

The Coriolis force is very noticeable in the atmosphere; in the southern hemisphere the situation is qualitatively as shown in Fig. 9.3. If there is a high-pressure region at P, air flowing away from it is deflected towards the left and ends up circulating counterclockwise around P; a low-pressure region produces a clockwise circulation. In the northern hemisphere the relations are opposite. Knowing them, one can often deduce the direction of highs and lows from shifts in the wind as they go past.

Foucault's Pendulum. Another illustration of the Coriolis force is the behavior of a long pendulum hung carefully so that it can swing in any direction. Figure 9.4 shows a suitable set of coordinates, with respect to which the earth's angular velocity has the components

$$\omega_x = 0 \quad , \quad \omega_y = \Omega \cos \lambda \quad , \quad \omega_z = \Omega \sin \lambda \tag{9.19}$$

where Ω is the earth's angular velocity and λ is the latitude. For small displacements, the force on the pendulum is

$$F_x = -\frac{mg}{l} x \quad , \quad F_y = -\frac{mg}{l} y$$

and omitting the negligible centrifugal force gives

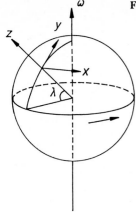

Fig. 9.4. Coordinates for Foucault's pendulum

$$m\dot{\boldsymbol{v}} = \boldsymbol{F} - 2m\boldsymbol{\omega} \times \boldsymbol{v} \quad \text{or} \tag{9.20}$$

$$m\ddot{x} = F_x - 2m(\omega_y \dot{z} - \omega_z \dot{y})$$

$$m\ddot{y} = F_y - 2m(\omega_z \dot{x} - \omega_x \dot{z}) \quad.$$

Since the pendulum's vertical motion is also negligible, these are

$$\ddot{x} = -\frac{g}{l}x + 2\Omega \sin \lambda \, \dot{y}$$

$$\ddot{y} = -\frac{g}{l}y - 2\Omega \sin \lambda \, \dot{x} \quad.$$

To solve these neatly, consider the xy plane as a complex plane with a complex coordinate

$$x + \mathrm{j}y = \zeta \quad, \quad \mathrm{j}^2 = -1 \quad.$$

The equation for ζ is

$$\ddot{\zeta} = -\omega_0^2 \zeta - 2\mathrm{j}\Omega \sin \lambda \, \dot{\zeta} \quad, \quad \omega_0^2 = g/l \quad.$$

This is solved by setting $\zeta = A\mathrm{e}^{\mathrm{i}\nu t}$ where the i that gives the time phase of motion is a $\sqrt{-1}$ that has nothing to do with the j that locates points on the plane tangent to the earth. There are, of course, two square roots of -1; i and j may designate the same one or different ones. The point is that the designations are made independently. This gives

$$\nu^2 = \omega_0^2 + 2\mathrm{i}\mathrm{j}\Omega_\lambda \nu \quad, \quad \Omega_\lambda = \Omega \sin \lambda$$

which has two solutions

$$\nu_{\substack{1\\2}} = \mathrm{i}\mathrm{j}\Omega_\lambda \pm \sqrt{\omega_0^2 + \Omega_\lambda^2}$$

$$\simeq \mathrm{i}\mathrm{j}\Omega_\lambda \pm \omega_0$$

since we are neglecting terms in Ω^2. The general solution is

259

$$\zeta = Ae^{i\nu_1 t} + Be^{i\nu_2 t}$$
$$= e^{-j\Omega_\lambda t}(Ae^{i\omega_0 t} + Be^{-i\omega_0 t}) \quad .$$

The pendulum must be carefully released from rest:

$$\zeta(0) = A + B \quad , \quad \dot\zeta(0) = 0$$

from which

$$A = \frac{1}{2}\left(1 - ij\frac{\Omega_\lambda}{\omega_0}\right)\zeta(0) \quad , \quad B = \frac{1}{2}\left(1 + ij\frac{\Omega_\lambda}{\omega_0}\right)\zeta(0) \quad \text{and}$$

$$\zeta(t) = e^{-j\Omega_\lambda t}\zeta(0)\left(\cos\omega_0 t + j\frac{\Omega_\lambda}{\omega_0}\sin\omega_0 t\right) \quad . \tag{9.21}$$

The second term in the parentheses is very small. Essentially, the pendulum swings at a frequency ω_0 in a plane that rotates clockwise with the angular velocity Ω_λ. There is no use mounting such a pendulum on the equator.

Léon Foucault tried his pendulum first on a modest scale in 1851 and then installed a heavy iron sphere suspended by a 67-m wire from the dome of the Panthéon in Paris. It created something of a sensation as people realized that this was the first direct and public dynamical proof of the Copernican hypothesis that it is the earth that turns and not the sky. (Today one says more carefully that a description that takes the sky as a reference system requires simpler equations of motion than one centered on the earth. To say that one turns and the other does not is not a physical statement.)

Hamiltonian Formulation. The Hamiltonian in a rotating system contains a kinematical term. By (5.59) the change in a dynamical variable a corresponding to an infinitesimal rotation of coordinates around the ith axis is

$$\Delta a_0 = \Delta\vartheta_i(a, L_i)$$

or, dividing by Δt,

$$\dot a_0 = \omega_i(a, L_i)$$

and this gives

$$\dot a_0 = (a, \boldsymbol{\omega} \cdot \boldsymbol{L}) + \dot a$$

when we include the change $\dot a$ with respect to the rotating axes. This is the change in a with respect to fixed axes, given by the Hamiltonian as usual,

$$\dot a_0 = (a, H) \quad .$$

Thus the change in a with respect to rotating axes can be written as

$$\dot a = (a, H_{\text{eff}})$$

in which the effective Hamiltonian [see (5.31)] is

$$H_{\text{eff}} = H - \boldsymbol{\omega} \cdot \boldsymbol{L} \quad . \tag{9.22}$$

This is useful where, as in the example of the Foucault pendulum, the angular velocity ω is a parameter of the rotating system. Presently we will consider situations in which it is a dynamical variable, and in Sect. 9.6 we will see how to write down the Hamiltonian for that case.

Problem 9.6. Two planets of mass M revolve around their common center of gravity. Show that it is possible to introduce a particle of negligible mass, positioned so that the three form an equilateral triangle of side l, that rotates with them (Fig. 9.5).

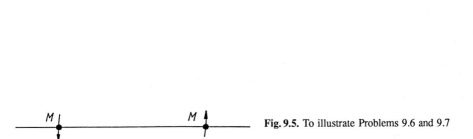

Fig. 9.5. To illustrate Problems 9.6 and 9.7

Problem 9.7*. Transforming the coordinates to the rotating coordinate system, show that the particle's orbit is stable for small displacements of its position.

Problem 9.8. If the Mississippi River is about 1 km wide at latitude 45° and the current flows at about 7 m/s, how much higher is the water level on the west bank than on the east?

Problem 9.9. In 1833 [54], Ferdinand Reich studied free fall in a mine shaft in Saxony. The shaft was 188 m deep and in 106 observations he found that a test object fell an average 28 mm east of the point directly below where it was dropped. Is this in accordance with theory? If so, it was the first experimental verification of Copernicus.

Problem 9.10. A ball is thrown vertically upward with initial velocity v. Where will it strike the ground?

Problem 9.11. What is the shape of the curves described by (9.21)? Show the function of the small second term.

Problem 9.12. Use the Hamiltonian (9.22) to derive the equations for the Foucault pendulum.

Problem 9.13. A hockey puck is started with velocity v_0 in an arbitrary direction across level and frictionless ice on a lake. What value of v_0 will give a path with a radius of curvature equal to 50 m?

Larmor's Theorem. If an atom is immersed in a magnetic field it begins to turn without changing its shape; this is the basis of semi-classical theories of the Zeeman effect. The general statement is Larmor's theorem:

If a system of identical charged particles, interacting among each other and with a central mass, is placed in a magnetic field B, its motion becomes (except for centrifugal effects) the superposition of a constant rotation around the central mass and the accelerations produced by the other forces. The precessional angular velocity is

$$\omega = -\frac{e}{2m}B \quad .$$ (9.23)

The proof is a direct application of (9.18). Let the total force on a particle be given by

$$F = f + ev \times B \quad .$$ (9.24)

Then in the rotating system, (9.20) gives for each particle

$$m\dot{v} = f + v \times (2m\omega + eB) + \ldots$$

and if (9.23) is satisfied only nonmagnetic forces remain. The theorem is applicable in practice only to atoms because except for the stationary central mass, all the particles must have the same ratio e/m.

9.4 Euler's Equations

After these digressions we can quickly formulate the equations for the angular velocity of a rigid body subject to an arbitrary moment of force, M. The equation of motion is (4.11) with the origin at the center of mass so that $\bar{r} = 0$. In the notation of Sect. 9.3,

$$\dot{L}_0 = M \quad .$$

With respect to axes fixed in the body, this is

$$\dot{L} + \omega \times L = M$$ (9.25)

and if the axes are oriented so that I is diagonal, it is

$$I_x\dot{\omega}_x + (I_z - I_y)\omega_y\omega_z = M_x$$
$$I_y\dot{\omega}_y + (I_x - I_z)\omega_x\omega_z = M_y$$
$$I_z\dot{\omega}_z + (I_y - I_x)\omega_x\omega_y = M_z \quad .$$ (9.26)

These equations, which sum up the theory of rigid bodies in a remarkably symmetrical and appealing form, were given by Euler in 1760 with the comment "Summa totius theoriae motus corporum rigidorum his tribus formulis satis simplicibus continetur."[1] A word of caution. As I have mentioned before, the angular velocities ω_i were introduced in (9.1) as kinematical vectors, not as time derivatives of anything. If Euler's equations are used to analyze the behavior of a spinning object with torques applied, the components of M depend on the object's angular orientation. To

---

[1] The whole theory of the motion of rigid bodies is contained in these three rather simple formulas.

specify what angular velocities have to do with angles requires further development, postponed to Sect. 9.5, but if the torques are zero the equations can be integrated as they stand.

The first step of the integration in this case is to note that Euler's equations contain the conservation of energy and angular momentum. The angular momentum L is constant with respect to space-fixed axes, of course, and to specify the individual components requires angular coordinates we are not ready to discuss, but even in the body-fixed coordinates, L^2 is constant (Problem 9.14). We thus have

$$I_x\omega_x^2 + I_y\omega_y^2 + I_z\omega_z^2 = 2T(\text{const}) \tag{9.27a}$$

$$I_x^2\omega_x^2 + I_y^2\omega_y^2 + I_z^2\omega_z^2 = L^2(\text{const}) \tag{9.27b}$$

and it is often convenient to write the first of these as

$$\frac{L_x^2}{I_x} + \frac{L_y^2}{I_y} + \frac{L_z^2}{I_z} = 2T \quad . \tag{9.27c}$$

To find the general solution of the Euler equations we solve for ω_y and ω_z, say, in terms of ω_x and substitute them into the first Euler equation, which yields an elliptic integral of the form (2.19). The details are left for Problem 9.15.

Asymmetric Rotator. An ordinary blackboard eraser is an example of an object with three different coefficients of inertia. If thrown into the air it becomes what is known as an asymmetric top, and with a little practice the following results may be demonstrated in the classroom.

Toss the eraser so that it spins nearly around one of the axes of symmetry, which we shall name the third axis. The axis will precess about a fixed direction in space, and sharp eyes will detect that if the long axis is the one chosen, the precession is in the opposite direction from the spin, whereas if the shortest axis is chosen, the precession is the other way. The following analysis will show why.

By the conventions of the experiment, the initial angular momentum around one of the axes, say z, is much greater than that around the other two, so that $\omega_z \gg \omega_x$ and ω_y, and Euler's third equation becomes

$$I_z\dot{\omega}_z \simeq 0 \quad . \tag{9.28}$$

Thus ω_z is approximately constant; call it Ω, and the other two equations are

$$I_x\dot{\omega}_x + (I_z - I_y)\Omega\omega_y = 0$$

$$I_y\dot{\omega}_y + (I_x - I_z)\Omega\omega_x = 0 \quad . \tag{9.29}$$

These equations can be solved by the substitutions

$$\omega_x = ae^{\lambda t} \quad , \quad \omega_y = be^{\lambda t} \tag{9.30}$$

and the condition for solubility turns out to be

$$\lambda^2 = \frac{(I_z - I_y)(I_x - I_z)}{I_x I_y}\Omega^2 \quad . \tag{9.31}$$

263

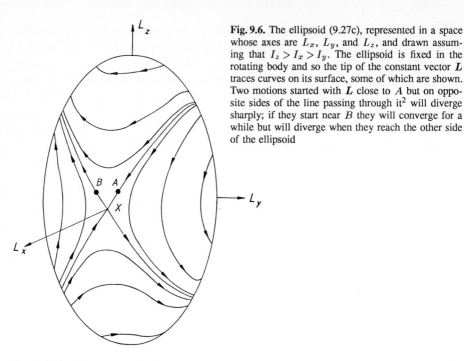

Fig. 9.6. The ellipsoid (9.27c), represented in a space whose axes are L_x, L_y, and L_z, and drawn assuming that $I_z > I_x > I_y$. The ellipsoid is fixed in the rotating body and so the tip of the constant vector L traces curves on its surface, some of which are shown. Two motions started with L close to A but on opposite sides of the line passing through it[2] will diverge sharply; if they start near B they will converge for a while but will diverge when they reach the other side of the ellipsoid

The four possible situations are

$$I_z \leq I_x, I_y \quad \text{or} \quad I_x, I_y \leq I_z \quad , \quad \lambda^2 \leq 0$$
$$I_x < I_z < I_y \quad \text{or} \quad I_y < I_z < I_x \quad , \quad \lambda^2 > 0 \quad .$$

If $\lambda^2 < 0$ the motion is periodic, while if $\lambda^2 > 0$ the initial motion is unstable and the axis wanders through large angles. The rest of the solution is left for Problem 9.15. Figure 9.6 explains the instability.

Symmetric Rotator. If two of the moments of inertia are equal (or nearly so), the situation simplifies. Setting $I_y = I_x$ in (9.26) gives $\dot{\omega}_z = 0$, or $\omega_z = \Omega$, a constant. In the remaining two equations introduce the abbreviation

$$\omega_p := \frac{I_z - I_x}{I_x} \Omega$$

and solve them by setting $\omega_x + i\omega_y = \zeta$. (There is no occasion for both an i and a j because the motion is pure precession.) This gives

$$\dot{\zeta} - i\omega_p \zeta = 0$$

so that

$$\zeta(t) = \zeta(0)e^{i\omega_p t} \tag{9.32}$$

which shows the precession of the angular-velocity vector with respect to the body-fixed coordinates.

---

[2] A line passing through a hyperbolic fixed point is called a *separatrix*.

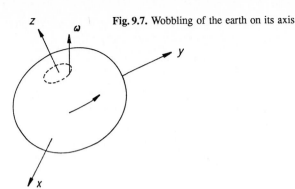

Fig. 9.7. Wobbling of the earth on its axis

Something of the sort is seen in the motion of the earth, Fig. 9.7. Careful measurements have shown that the earth's axis of rotation wanders counterclockwise around the North Pole in an irregular path never more than about 5 m from the Pole. Since

$$\frac{I_z - I_x}{I_x} = \frac{1}{304} \qquad (9.33)$$

(we will see in Sect. 9.5 how this value is determined), and since the earth's rotation is counterclockwise as seen from the Pole, the sign of the precession is correct. The mean period is about 50 percent longer than the 300 days predicted from (9.33). The irregularities are explained by the irregular loading of atmospheric tides, and the discrepancy in the period by the earth's lack of rigidity.

As seen from outside, the precession is in the same direction as from inside. Let k be a unit vector along the axis of symmetry, which we call the z axis. Then by (9.14a), k moves with respect to space-fixed axes according to

$$\dot{k}_0 = \omega \times k \quad , \quad k \parallel z \quad \text{or}$$

$$\dot{k}_{0_x} = \omega_y \quad , \quad \dot{k}_{0_y} = -\omega_x \quad .$$

Thus

$$\dot{k}_0 := \dot{k}_{0x} + i\dot{k}_{0y} = \omega_y - i\omega_x = -i\zeta(t)$$

and integration using (9.32) gives

$$k_0 = -\frac{\zeta(0)}{\omega_p} e^{i\omega_p t} \quad . \qquad (9.34)$$

The sense of the precession is the same as that of Ω or opposite according as $I_z > I_x$ or $I_z < I_x$.

Problem 9.14. Show that Euler's equations imply the constancy of T and L^2 as defined in (9.27).

Problem 9.15. Complete the solution of Euler's equations in terms of the Jacobian elliptic functions sn, cn, and dn. It will be convenient to assign the labels 1, 2, 3 to the axes by assuming that $L^2 > 2I_3 T$.

Problem 9.16. Suppose that in the reduction leading to (9.32) I_y had been very nearly but not quite equal to I_x. How would the results have been different?

It is possible to absorb the preceding mathematics and still not know much about how the symmetric rotator actually moves, since the entire description is given with respect to axes fixed in the rotator. We need to translate it back into laboratory coordinates. This can be done using the Euler angles to be discussed in Sect. 9.5 but one perhaps learns more from a clever construction given by Louis Poinsot in 1851.

First we need some geometrical facts. In the laboratory system, the angular momentum vector L is fixed in space. Around it, at a rate ω_p, precesses the vector k that represents the body's axis of symmetry. We can write L as

$$L = I_x \omega_x \hat{i} + I_x \omega_y \hat{j} + I_z \omega_z \hat{k} = I_x \omega + (I_z - I_x)\omega_z \hat{k} = I_x(\omega + \omega_p \hat{k})$$

and this shows that the three vectors L, ω, and k lie in the same plane. Next, the components of ω and k in the direction of L are constant. The first is obvious from

$$\omega \cdot L = 2T$$

where T is the constant kinetic energy. Further,

$$L^2 = I_x(L \cdot \omega + \omega_p L \cdot k) = I_x(2T + \omega_p L \cdot k) \quad . \tag{9.35}$$

Since L^2 is constant, $L \cdot k$ must be constant also, and therefore the vectors k and ω must both precess around L at the constant rate ω_p.

To see how the rotator actually behaves we must consider the meaning of ω. Figure 9.8a shows three successive positions of the vector ω, at times separated by Δt. P_1 is a point on the rotator that lies on the axis ω_1, so that at the initial moment the rotator is turning around P_1. P_2 is a point on the rotator chosen so that after a time Δt, when ω has moved to ω_2, P_2 will lie on the new axis, and so on. Now let $\Delta t \to 0$. The broken lines in the figure become arcs of two circles, the lower one fixed in space and the upper one, fixed in the body, rolling along the lower one. This tells how the body moves. Figure 9.8b shows the situation.

To see how to read the diagram it will help to focus on a simple example. In the first volume of his memoirs [117] Richard Feynman tells how someone at Cornell throws a plate into the air, spinning with a slight wobble, and he notices that "the medallion rotates twice as fast as the wobble rate" and is finally able to show why this is so. Let us check this statement. Replace the plate by a thin disc of mass M and radius a, spinning around its z axis. It is then easy to show that $I_x = Ma^2/4$ and $I_z = Ma^2/2$, so that the wobble rate $\omega_p = \Omega$. Thus L splits the angle between ω and k, the rolling cone is twice the size of the fixed one, and a little thought shows that it rotates at a rate $\omega_p/2 = \Omega/2$. So Feynman, remembering the event after many years, put the $\frac{1}{2}$ in the wrong place. (Note that there is an approximation here: Ω is $\omega_z = \omega \cos \vartheta$, not ω, and I am assuming that the angle ϑ in the figure is small so that they are nearly the same.)

Problem 9.17. Work out what will be seen if a cylinder of length l, radius a, and mass M is thrown into the air, wobbling slightly. (This contains the plate as a special case.) Draw the diagram analogous to Fig. 9.8b that applies when the cylinder is long, work out what happens when $l = 3a$, and try it.

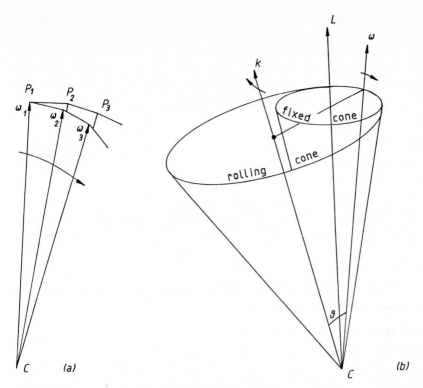

Fig. 9.8. (a) Poinsot's construction to show that a cone fixed in a symmetrical rotator rolls without slipping on the fixed cone defined by the precessing vector ω. **(b)** The two cones when $\omega_p > 0$. C is the center of mass

The Gyrocompass. In 1852, the year after he had demonstrated the rotation of the earth with a pendulum, it occurred to Foucault that it could be demonstrated in a smaller space using a gyroscope. In its simplest form, Fig. 9.9, the device consists of a gyroscope suspended so that its axes can turn around the vertical.

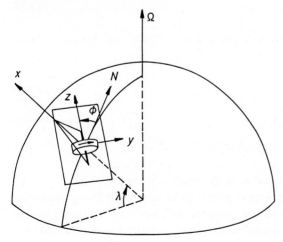

Fig. 9.9. Foucault gyroscope suspended above the earth's surface

Since $I_x = I_y$ in the gyroscope, we need not fix the x and y axes in the wheel. If we let it spin in the xy plane the coefficients of inertia I_x and I_y remain constant as long as the axis of spin is aligned with the z axis, and to make them constant was the only reason for introducing the rotating axes in the first place. Remembering that the orientation of the axes xyz changes as the earth rotates, we can write down the components of the vector ω that describes the change. If Ω is the earth's angular velocity,

$$\omega_x = \Omega \sin \lambda + \dot{\phi}$$
$$\omega_y = \Omega \cos \lambda \sin \phi \qquad\qquad (9.36)$$
$$\omega_z = \Omega \cos \lambda \cos \phi$$

where ϕ describes the turning of the gyroscope's axis.

To find the equations of motion in the non-spinning coordinates we note that if $I_x = I_y$, (9.6) becomes

$$L_x = I_x \omega_x \quad , \quad L_y = I_x \omega_y \quad , \quad L_z = I_3(\omega_z + s)$$

where s is the spin angular velocity. The transition from moving to fixed coordinates is given as before by (9.14b), and we find

$$I_x \dot{\omega}_x + (I_z - I_x)\omega_y \omega_z + I_z s \omega_y = M_x$$
$$I_x \dot{\omega}_y + (I_x - I_z)\omega_x \omega_z - I_z s \omega_x = M_y$$
$$I_z(\dot{\omega}_z + \dot{s}) = M_z \quad . \qquad\qquad (9.37)$$

With the gyroscope suspended as in Fig. 9.9, $M_x = M_z = 0$, and since s corresponds to thousands of revolutions per minute while ω_z corresponds to less than one revolution per day, we can take s to be constant. With (9.36), the vanishing of M_x then gives

$$I_x \ddot{\phi} + (I_z - I_x)\Omega^2 \cos^2 \lambda \sin \phi \cos \phi + I_z s \Omega \cos \lambda \sin \phi = 0 \quad .$$

The middle term is negligible, and if ϕ is small the rest reduces to

$$\ddot{\phi} + \left(\frac{I_z}{I_x} s \Omega \cos \lambda \right) \phi = 0 \quad . \qquad\qquad (9.38)$$

At equilibrium $\phi = 0$ and the gyroscope points north; around this direction it oscillates with a period

$$P = 2\pi \sqrt{\frac{I_x}{I_z s \Omega \cos \lambda}} \quad . \qquad\qquad (9.39a)$$

As the ready reckoner will at once agree,

$$\Omega = \frac{2\pi \times 366.25}{365.25 \times 24 \times 60^2} = 7.292 \times 10^{-5} \, \text{s}^{-1} \qquad\qquad (9.39b)$$

and with s of the order of $10^2 \, \text{s}^{-1}$, P is of the order of 1 minute.

Foucault's demonstration apparatus never worked very well. The gyrocompass in a modern inertial navigation system is equipped with devices to damp the oscillations, to keep it spinning, and to compensate disturbances that result from changes of speed and course; it has largely replaced the magnetic compass where accurate bearings are needed.

Problem 9.18. A gyroscope spinning around the y axis has $I_x = I_y < I_z$. Using the non-spinning coordinate system evaluate the moment M when the direction of the gyroscope's axis is changed and show that the usual phenomena of the gyroscope are explained.

Problem 9.19. In what direction does the gyrocompass point at equilibrium if it is being carried north at a velocity v? Give a numerical estimate.

9.5 The Precession of the Equinoxes

It has been known since Hipparchus ($\sim 125\,\mathrm{BC}$), and possibly long before that, that the North Pole traces out a circle in the sky, or to put it another way, that the equinoxes move backwards around the ecliptic at a rate of about 50 arc seconds per year. Figure 9.10 shows the plane of the ecliptic, containing the earth's orbit, intersecting the equatorial plane along the line $\Omega\mho$. Astronomers call this the line of nodes. The sun moves clockwise around the earth in the plane of the ecliptic. At Ω, the ascending node, it passes above the equatorial plane; at $\mho$, the descending node, it passes below it. The ascending node points at present towards a point in the Zodiacal constellation of Pisces. The Age of Aquarius is to begin when it enters Aquarius (Problem 9.21).

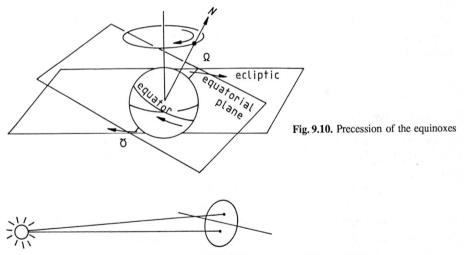

Fig. 9.10. Precession of the equinoxes

Fig. 9.11. The sun exerts a torque on a spheroidal earth because the near half is attracted more strongly than the far half

The precession is a gyroscopic effect that arises because the earth is not a sphere, Fig. 9.11. The calculation of the precessional velocity will be done in four parts: deriving the equations of motion, deriving the potential function for the interaction of another body with the spheroidal earth, solving the equations, and evaluating the effects produced by the sun and the moon.

269

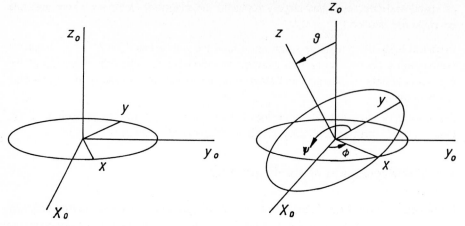

Fig. 9.12. The Euler angles are generated by rotating the xy plane and then tipping the z axis

Euler's Angles. We have first to establish the relations between angular velocity and angles. The best way to do this is due to Euler: define the orientation of the rotating body by means of angles and then express the components of ω in terms of them and their derivatives. Figure 9.12 shows an appropriate construction in two stages. To get from the space-fixed axes to axes oriented with respect to the spinning body, but not spinning with it, first rotate the xy plane so that x coincides with the line of nodes; then rotate the yz plane so as to achieve the desired inclination of the z axis. If the angles ϑ and ϕ are changing, and if the body's angular orientation relative to the line of nodes is ψ, the components of ω in the moving frame can be written down by inspection,

$$\omega_x = \dot{\vartheta}$$
$$\omega_y = \dot{\phi} \sin \vartheta$$
$$\omega_z = \dot{\phi} \cos \vartheta + \dot{\psi} \quad . \tag{9.40}$$

(If two of the coefficients of inertia are not equal, so that it is necessary to fix the axes in the body, the third angle ψ has to designate a rotation of the entire coordinate system around the z axis; see for example [38].)

We can now proceed in either of two ways: find differential equations for ϑ, ϕ, and ψ in terms of the moment M by Euler's equations in the form (9.26), or write down the Lagrangian for the system and follow the route of Lagrange, Hamilton, or Hamilton and Jacobi to a solution. We will use Lagrange's equations.

The kinetic energy of the earth's rotation is

$$T = \tfrac{1}{2}[I_x(\omega_x^2 + \omega_y^2) + I_z\omega_z^2]$$

which is

$$T = \tfrac{1}{2}[I_x(\dot{\vartheta}^2 + \sin^2 \vartheta \dot{\phi}^2) + I_z(\dot{\phi} \cos \vartheta + \dot{\psi})^2] \quad . \tag{9.41}$$

The potential energy depends on the situation; next we need a formula to describe the interaction between the sun and the earth.

The Interaction Potential. First, evaluate the contribution of the sun. If the earth is considered at rest at the origin, the sun moves around it in a path we can take to be a circle of radius R traversed at a constant angular velocity $-n$. (The negative sign is because it travels clockwise.) Its position is given by

$$x_0 = R \cos nt \quad , \quad y_0 = -R \sin nt \quad . \tag{9.42}$$

In the moving coordinate system the sun's coordinates are related to these by

$$
\begin{aligned}
x &= x_0 \cos \phi + y_0 \sin \phi \\
y &= (-x_0 \sin \phi + y_0 \cos \phi) \cos \vartheta \\
z &= (x_0 \sin \phi - y_0 \cos \phi) \sin \vartheta \quad .
\end{aligned}
\tag{9.43}
$$

(These are best derived by carrying out in succession the two rotations shown in Fig. 9.11.) With (9.42), these take an obvious form,

$$
\begin{aligned}
x &= R \cos (\phi + nt) \\
y &= -R \sin (\phi + nt) \cos \vartheta \\
z &= R \sin (\phi + nt) \sin \vartheta \quad .
\end{aligned}
\tag{9.44}
$$

Now calculate the interaction potential when the sun is at this point in the sky, Fig. 9.13. The distance from the sun to the element of mass dm is

$$l = \sqrt{(r - r')^2} \quad , \quad r \cdot r = R^2$$

and the potential is

$$V = -GM \int \frac{dm}{l}$$

integrated over the spheroid of the earth. Since $r' \ll R$, perform Legendre's expansion in l:

$$\frac{1}{l} = \frac{1}{\sqrt{r'^2 - 2r' \cdot r + R^2}} = \frac{1}{R} \left[1 + \frac{r' \cdot r}{R^2} + \frac{3(r' \cdot r)^2 - r'^2 R^2}{2R^4} + \dots \right] \quad .$$

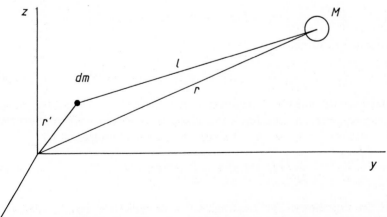

Fig. 9.13. Coordinates for calculating the attraction of a mass M by a spheriodal earth

271

The interaction potential is

$$V(r) = -\frac{GM}{R} \int \varrho(x', y', z') \left[1 + \frac{1}{R^2}(x'x + y'y + z'z) \right.$$

$$+ \frac{3}{2R^4}(x'^2 x^2 + y'^2 y^2 + z'^2 z^2 + 2x'y'xy + 2y'z'yz + 2x'z'xz)$$

$$\left. - \frac{r'^2}{2R^2} + \ldots \right] dx' dy' dz' \quad .$$

Assume that the earth's density is symmetrically distributed in the sense that $\varrho(x') = \varrho(-x')$, etc. Then all terms containing a single x', y', or z' go out and there is left

$$V(r) = -\frac{GM}{R} \int \varrho(x', y', z') \left\{ 1 + \frac{1}{2R^4}[(3x'^2 - r'^2)x^2 + (3y'^2 - r'^2)y^2 \right.$$

$$\left. + (3z'^2 - r'^2)z^2] \right\} dx' dy' dz' \quad .$$

The first term of the integral is the point-mass potential

$$V_0 = -\frac{GMm}{R} \quad , \qquad m = \int \varrho d_3 r'$$

while the rest are closely related to the coefficients of inertia. From

$$I_x = \int (r'^2 - x'^2)\varrho d_3 r' \quad \text{and}$$

$$I_x + I_y + I_z = 2 \int r'^2 \varrho d_3 r' \tag{9.45}$$

we readily deduce that

$$\int \varrho(3x'^2 - r'^2)d_3 r' = I_y + I_z - 2I_x$$

which, since $I_x = I_y$, is $I_z - I_x$. Similar results hold for analogous expressions. Thus finally,

$$V(r) = -\frac{GMm}{R} - \frac{GM}{2R^5}(I_z - I_x)(x^2 + y^2 - 2z^2) + \ldots$$

or, in terms of the Euler angles,

$$V(r) = -\frac{GMm}{R} - \frac{GM}{2R^3}(I_z - I_x)[1 - 3 \sin^2 (\phi + nt) \sin^2 \vartheta] \quad . \tag{9.46}$$

The sun's annual motion is very fast with respect to the precession, so the expression can be averaged over time. (This amounts to replacing the sun by a ring of the same mass encircling the earth.) Since $\sin^2 (\phi + nt)$ averages to 1/2,

$$\overline{V}(r) = -\frac{GMm}{R} - \frac{GM}{2R^3}(I_z - I_x)\left(1 - \frac{3}{2} \sin^2 \vartheta\right) \quad . \tag{9.47}$$

Solution of the Equations. The Lagrangian that describes the earth's rotation is formed from (9.41) and (9.47):

$$L = \tfrac{1}{2}[I_x(\dot{\vartheta}^2 + \sin^2 \vartheta \dot{\phi}^2) + I_z(\dot{\phi} \cos \vartheta + \dot{\psi})^2]$$
$$+ \frac{GM}{2R^3}(I_z - I_x)(1 - \frac{3}{2}\sin^2 \vartheta) \tag{9.48}$$

omitting the irrelevant first term of (9.47). The coordinates ϕ and ψ are cyclic, so that we have at once two constant momenta

$$\frac{\partial L}{\partial \dot{\phi}} = (I_x \sin^2 \vartheta + I_z \cos^2 \vartheta)\dot{\phi} + I_z \Omega \cos \vartheta := p_\phi (\text{const})$$

$$\frac{\partial L}{\partial \dot{\psi}} = I_z(\dot{\phi} \cos \vartheta + \Omega) := p_\psi (\text{const}) \tag{9.49}$$

where Ω is written for the daily rotational rate $\dot{\psi}$. Since Ω is more rapid than $\dot{\phi}$ by a factor of $365 \times 26\,000$, these are very nearly

$$p_\phi \simeq I_z \Omega \cos \vartheta \quad , \quad p_\psi \simeq I_z \Omega$$

so that ϑ is very nearly constant. This is consistent with the observed motion, in which whatever wobbles there may once have been are now very small, so the approximation is good enough.

Neglecting $\dot{\phi}$ compared to Ω, we find that the equation for ϑ is

$$I_x \ddot{\vartheta} + I_z \Omega \dot{\phi} \sin \vartheta + \frac{3GM}{2R^3}(I_z - I_x) \sin \vartheta \cos \vartheta \simeq 0$$

and if the precession is uniform with ϑ constant, this gives for the precessional angular velocity

$$\dot{\phi} = -\frac{3GM}{2R^3} \frac{I_z - I_x}{I_z} \frac{\cos \vartheta}{\Omega} \quad .$$

According to Kepler's third law,

$$\frac{GM}{R^3} = \frac{4\pi^2}{T^2}$$

where T is one year. Thus

$$\dot{\phi} = -6\pi^2 \frac{\Omega}{(\Omega T)^2} \frac{I_z - I_x}{I_z} \cos \vartheta$$

where the negative sign means that the precession is backward along the ecliptic. The relevant data, remembering (9.39b), are

$$\frac{I_z - I_x}{I_z} = \frac{1}{304} \quad , \quad \vartheta = 23°27' \quad , \quad \Omega = 7.29 \times 10^{-5}\,\text{s}^{-1}$$

and, for the sun's motion, $\Omega T \simeq 2\pi \times 366$. This gives a precessional period of about 80,000 years, much too long. The calculation is saved when it is realized that the moon, though much less massive than the sun, is also much closer to the earth and actually contributes more to the precession. When its contribution is included (Problem 9.20) we find a value near the observed value

$$P_{\text{obs}} = \frac{2\pi}{\dot{\phi}_{\text{obs}}} = 25\,780\,\text{y} \quad . \tag{9.50}$$

Problem 9.20. Calculate the contribution of the moon to the precession of the equinoxes.

Problem 9.21. According to the horoscopes published in newspapers and magazines, the sun is in Aries, the first sign of the zodiac, from March 21 to April 19. It is actually in Aries from about April 17 to May 17 (depending on a somewhat arbitrary assignment of boundaries). Assuming that the tables used by these astrologers take no account whatever of precession, when were they written down?

Problem 9.22. Find a chart of the sky that shows the vernal equinox and estimate when the Age of Aquarius will begin.

Problem 9.23. Set up the equations of motion for the top shown in Fig. 9.14 and calculate the rate of uniform precession when it is set spinning with angular velocity Ω. Now study the effect of small deviations from uniform precession. (Large deviations involve one in elliptic integrals.) Analyze and sketch the behavior of a top that is set spinning rapidly (this makes the deviations small) with the axis held stationary and then released.

Fig. 9.14. Spinning top

Problem 9.24. If one did not believe relativity theory, a possible explanation for the precession of planetary perihelia is that the sun is slightly oblate. Assuming a solar oblateness that gives the correct value for Mercury (43.1″ per century, $e = 0.2056$, $T = 88$ days), use (9.46) with $\vartheta = 0$ to find the precession to be expected for the earth in one century (5.0 ± 1.2″ obs., $e = 0.0167$). Equation (6.35) will be useful.

9.6 Quantum Mechanics of a Rigid Body

Although the quantum theory of rigid-body motion is a hybrid of quantum dynamical laws and purely classical ideas of rigidity and structure, it nevertheless provides insight into both molecular and nuclear spectra, and thus has some contact with reality. In analyzing it we will also be able to develop more clearly some ideas about rotating coordinates in quantum mechanics.

In the work thus far we have started from Euler's or Lagrange's equations of motion, but quantum mechanics starts most naturally from a Hamiltonian. That given

274

in (9.22) will not do here because it applies to situations where the angular velocity ω is given; here it is a dynamical variable and we must start again. Before doing so, however, we pause to study the commutation relations for angular momentum as expressed in a rotating system of coordinates.

The Commutation Relations for Angular Momentum. In order to explain the origin of the angular-momentum operator we derived in Sect. 5.4 a special case of it: the operator that gives infinitesimal angular displacements applied to functions of x, y and z. Quantum mechanically, we were dealing with orbital quantities in the coordinate picture, and the commutation relations for the components of L follow directly from those for the linear momentum.

For the derivation to come a more direct and general approach is provided by the theory of this chapter. Let a be an arbitrary vector fixed in a rotating coordinate system. Then in a time δt its tip will move a distance

$$\delta a_0 = \omega \times a \delta t \tag{9.51}$$

with respect to the fixed frame. It will be convenient for the moment to consider the variation of the scalar quantity

$$A = \lambda \cdot a \tag{9.52}$$

where the λ_i are a set of arbitrary non-transforming constants. Further, introduce the infinitesimal angular displacements during the time δt:

$$\delta \vartheta_i = \omega_i \delta t \quad , \quad i = x, y, z \quad . \tag{9.53}$$

Then by the rule for permuting dots and crosses, Problem 9.26,

$$\delta A_0 = \lambda \cdot \delta \vartheta \times a = -\delta \vartheta \cdot \lambda \times a \quad . \tag{9.54}$$

In quantum mechanics, suppose that $\hat{a}$ is some vectorial dynamical variable, in general a function of coordinates, momenta, spins, etc., and $\hat{A} = \lambda \cdot \hat{a}$. In the form of (5.51) express the change in $\hat{A}$ in terms of commutation with an operator $\hat{J}$ defined by

$$\delta \hat{A}_0 = i \delta \vartheta \cdot [\hat{J}, \hat{A}] \quad . \tag{9.55}$$

The geometrical meaning of a rotation requires that (9.54) hold in quantum mechanics also, so that since $\delta \vartheta$ is arbitrary,

$$[\hat{J}, \lambda \cdot \hat{a}] = i \lambda \times \hat{a} \quad . \tag{9.56}$$

This can be written in component form if we introduce the totally antisymmetric tensor

$$\varepsilon_{ijk} = \pm 1 : \quad \varepsilon_{ijk} = -\varepsilon_{jik} = -\varepsilon_{ikj} \quad , \quad \varepsilon_{123} = 1 \quad . \tag{9.57}$$

That is, $\varepsilon_{ijk} = +1$ when the sequence ijk can be obtained from 123 by an even number of permutations and -1 otherwise. Using this any vector product can be written in component form as

$$(A \times B)_i = \varepsilon_{ijk} A_j B_k \tag{9.58}$$

and (9.56) becomes

$$\lambda_j[\hat{J}_i, \hat{a}_j] = \mathrm{i}\varepsilon_{ijk}\lambda_j \hat{a}_k \quad .$$

With λ_j arbitrary, it follows that

$$[\hat{J}_i, \hat{a}_j] = \mathrm{i}\varepsilon_{ijk}\hat{a}_k \quad . \tag{9.59}$$

This is the familiar cyclic relation

$$[\hat{J}_x, \hat{a}_y] = \mathrm{i}\hat{a}_z \quad \text{etc.} \tag{9.60}$$

and if, as a special case, $\hat{a} = \hat{J}$, it gives

$$[\hat{J}_x, \hat{J}_y] = \mathrm{i}\hat{J}_z \quad \text{etc} \quad . \tag{9.61}$$

A familiar example of operators that satisfy this relation is the components of orbital angular momentum. Real angular momentum contains a factor of $\hbar$ which the reader can insert where necessary.

The theory of rigid-body motion is expressed in terms of rotating coordinate frames. In such a frame, the rotation can be described in a way perfectly symmetrical to what we have just done with respect to fixed frames: consider the apparent change $\dot{a}$ in the rotating frame of a vector that is fixed in space. By (9.14b),

$$\dot{a}_0 = 0 = \omega \times a + \dot{a} \quad \text{or}$$

$$\dot{a} = -\omega \times a \quad . \tag{9.62}$$

Let $\hat{K}$ be the operator in the rotating frame that generates angular displacements in the same way that $\hat{J}$ does in the fixed frame. The derivation goes through exactly as before except for the negative sign; we find that

$$[\hat{K}_i, \hat{a}_j] = -\mathrm{i}\varepsilon_{ijk}\hat{a}_k \quad \text{and} \tag{9.63}$$

$$[\hat{K}_i, \hat{K}_j] = -\mathrm{i}\varepsilon_{ijk}\hat{K}_k \tag{9.64a}$$

$\hat{K}$ represents angular momentum in the rotating frame just as $\hat{J}$ represents it in the fixed frame.

Problem 9.25. The preceding derivations were carried out in the notations of quantum mechanics, but they have exact counterparts in classical mechanics. Work out the corresponding relations, compare them with (5.61), and in particular, show that to (9.64a) corresponds

$$(K_i, K_j) = -\varepsilon_{ijk}K_k \quad . \tag{9.64b}$$

Problem 9.26. Verify (9.58) from the definition (9.57) and, for practice, use these relations to derive the vector formulas

$$a \times [b \times c] = (a \cdot c)b - (a \cdot b)c \quad \text{and}$$

$$a \cdot [b \times c] = [a \times b] \cdot c \quad .$$

Problem 9.27. To see the dynamical content of (9.61), consider a rotating system with a magnetic moment given by $M = \gamma L$, where γ is the gyromagnetic ratio. In classical physics a magnetic field B will exert a torque on it equal to $-B \times M$. In quantum mechanics the interaction is described by a term $-B \cdot M$ in the Hamiltonian. Find $d\langle L \rangle / dt$ and compare it with the classical formula.

Problem 9.28. Solve the equations set up in the preceding problem; that is, solve the equation of motion of L in classical physics and find the eigenfunctions and eigenvalues in quantum physics. Explain the relation between the two solutions which may seem, at first sight, to be so different as to be almost unrelated.

The Equations of Motion. Although rigidity is even more of an idealization on the microscopic scale that it is with laboratory apparatus, we can define a rigid body in terms of the classical Hamiltonian

$$H = \frac{1}{2}\left(\frac{K_x^2}{I_x} + \frac{K_y^2}{I_y} + \frac{K_z^2}{I_z} \right) + V \tag{9.65}$$

where the potential is expressed in terms of the Euler angles. The justification for this form is that it leads to the right equations of motion (Problem 9.29). We shall consider here only a symmetric top ($I_y = I_x$) in force-free motion. For this the quantum-mechanical H, transcribed from (9.65), can be written as

$$\hat{H} = \frac{1}{2}\left[\frac{1}{I_x}\hat{K}^2 + \left(\frac{1}{I_z} - \frac{1}{I_x} \right)\hat{K}_z^2 \right]\hbar^2 \quad . \tag{9.66}$$

The eigenfunctions and eigenvalues of $\hat{K}_z$ and $\hat{K}^2$ correspond closely to those of $\hat{J}_z$ and $\hat{J}^2$; the negative sign in the commutation relation makes no difference to the eigenvalue spectrum and we can at once write down the eigenvalues of $\hat{H}$ as

$$E_{k,m_k} = \frac{1}{2}\hbar^2\left[\frac{1}{I_x}k(k+1) + \left(\frac{1}{I_z} - \frac{1}{I_x} \right)m_k^2 \right] \quad \text{where} \tag{9.67a}$$

$$k = 0, 1, 2, \dots \quad \text{and} \quad -k \le m_k \le k \quad . \tag{9.67b}$$

Because the classical theory from which these expressions originate is essentially a theory of the orbital motion of particles, one does not expect to find half-odd values of k, and in experiments none are found.

Rotational States of Nuclei. In trying to apply the abstract formalism developed above to situations in the real world, we at once run into complications. Suppose for example that the rotator is a spherically symmetric nucleus with $I_x = I_y = I_z$. The nucleus consists of particles, each one of which moves in a spherically symmetric potential provided by the others. But what does it mean to a particle to say that the nucleus as a whole rotates? Nothing: a rotating spherically symmetrical potential well is indistinguishable from one that does not rotate. Thus neither the individual nucleons nor the nucleus made up of them behaves any differently in a spherical nucleus imagined to be rotating than in the same nucleus at rest. Spherical nuclei (those with closed shells) have no rotational spectra. The same applies when $I_x = I_y$ in the restricted sense that the symmetrical object does not rotate about its axis of symmetry and the possible states[3] are therefore only those with $m_k = 0$:

$$E_{k,0} = \frac{\hbar^2}{2I_x} k(k+1) \quad . \tag{9.67c}$$

Rotational effects are most evident (i.e., single-nucleon effects are least evident) in the rotation spectra of spheroidal nuclei with even numbers of neutrons and protons. For such nuclei, the allowed values of k are even. This can be understood if we note that the plane of symmetry through the nucleus divides it into halves each of which contains an even number of particles. Thus if the halves are exchanged particle by particle, an even number of changes of sign takes place and the resulting wave function is invariant under the interchange; that is, even in k. Experimentally the odd levels do not occur and the rotating nucleus spins down, losing two units of spin at a time through electric quadrupole radiation. The gamma energy is thus

$$E_\gamma = E_{k,0} - E_{k-2,0} = \frac{\hbar^2}{I_x}(2k-1) \quad .$$

There is a class of nuclei called *yrast* (from the Swedish word for "dizziest") that show remarkable rigidity and spin very fast. Table 9.1 shows gamma-ray energies from ^{152}Dy observed in six successive transitions starting at $k = 60$ [119], together with energies calculated from the first entry assuming that I_x is independent of k, i.e. perfect rigidity. At this rate of spin the nuclei are very elongated, with major axis about twice the minor axis. When the spin drops to about 22 the rigidity largely disappears.

Table 9.1. Gamma-ray Energies From Rotational Transitions of ^{152}Dy

k	$E_{\gamma \text{ obs}}$ (keV)[*]	$E_{\gamma \text{ calc}}$ (keV)
60	1449.4 ± 0.6	(1449.4 ± 0.6)
58	1401.7 ± 0.4	1400.7 ± 0.6
56	1353.0 ± 0.3	1352.0 ± 0.6
54	1304.7 ± 0.3	1303.2 ± 0.6
52	1256.6 ± 0.3	1254.5 ± 0.6
50	1208.7 ± 0.3	1205.8 ± 0.6

[*] From [119]

Problem 9.29. Use (9.64b) to derive Euler's equations (9.26) from the classical Hamiltonian (9.65).

Problem 9.30. Show that the quantum-mechanical Hamiltonian formed by transcribing (9.65) does not commute with $\hat{K}_z$ if all I's are different, and that therefore energy eigenfunctions of this asymmetrical rotator cannot be taken to be eigenfunctions of $\hat{K}^2$ and $\hat{K}_z$.[4]

---

[3] The reader may already have encountered an example of this situation in considering the specific heat of a diatomic gas, to which molecular rotation contributes two degrees of freedom and not the three that one would expect if the molecule could rotate about its axis. See [118, Chap. 11].

[4] How to determine the energy levels in this case is shown in [120, Sect. 101]. A brilliant and very complete discussion is [121].

Problem 9.31. The usual expression for the radius of a nucleus is $r = r_0 A^{1/3}$, where $r_0 = 1.2\,\mathrm{fm}$. Compare the moment of inertia calculated from the data in Table 9.1 with that estimated on the assumption that the nucleus is a roughly uniform sphere of radius r.

9.7 Spinors

We have seen in Chap. 5 one way to introduce a change of coordinate axes. If it is an infinitesimal rotation around the third axis, for example, we write

$$f(x_1, x_2) \to (1 + \frac{i\varepsilon}{\hbar}\hat{J}_z)f(x_1, x_2) = f(x_1 - \varepsilon x_2, x_2 + \varepsilon x_1) \tag{9.68}$$

corresponding to a clockwise rotation of the coordinate axes through an angle ε. The corresponding operation on a vector (not the axes) is (9.51), in which if $\omega_z \delta t = \varepsilon$, $\omega_x = \omega_y = 0$, we have

$$\delta a_{x0} = -\varepsilon a_y \quad , \quad \delta a_{y0} = \varepsilon a_x \quad , \quad \delta a_{z0} = 0$$

corresponding to a clockwise rotation. One can think of such a rotation as a counter-clockwise rotation of the basis or a clockwise rotation of the vector. In the following discussion we will describe rotations by telling what they do to vectors and, later, to a geometric object called a spinor that can be used to represent a vector.

Orthogonal Transformations. A linear transformation of a vector can be written in the form

$$a'_i = c_{ij}a_j \quad \text{(summed)} \tag{9.69a}$$

or equivalently in matrix notation

$$a' = ca \quad . \tag{9.69b}$$

A rotation in any number of dimensions is defined as a transformation that leaves the length of a vector invariant:

$$a'_i a'_i = a_i a_i = c_{ij}a_j c_{ik}a_k$$

and if this is to hold for any vector a_i we must have

$$c_{ij}c_{ik} = \delta_{jk} \quad . \tag{9.70a}$$

Introducing the notation T to denote the transpose of a matrix, we can write this as

$$c^{\mathrm{T}}_{ji}c_{ik} = \delta_{jk}$$

or, in matrix notation,

$$c^{\mathrm{T}}c = 1 \quad \text{and} \tag{9.70b}$$

$$c^{\mathrm{T}} = c^{-1} \quad . \tag{9.70c}$$

Transformations so restricted are called orthogonal transformations.

279

Problem 9.32. Show that another property of an orthogonal transformation is

$$c_{ij}c_{kj} = \delta_{ik} \tag{9.70d}$$

and show that the matrix that rotates axes in the xy plane satisfies (9.70a, d).

Problem 9.33. In order to see why the transformation is called orthogonal, show that two orthogonal vectors a and b are transformed by it into two orthogonal vectors.

Problem 9.34. An infinitesimal rotation in 3 dimensions is defined by (9.51) and (9.53),

$$a \rightarrow a' = a + \delta\vartheta \times a \quad . \tag{9.71a}$$

Writing $c_{ij} = \delta_{ij} + \gamma_{ij}$, or $c = 1 + \gamma$, show that (9.71a) is represented by the matrix

$$\gamma = \begin{pmatrix} 0 & -\delta\vartheta_z & \delta\vartheta_y \\ \delta\vartheta_z & 0 & -\delta\vartheta_x \\ -\delta\vartheta_y & \delta\vartheta_x & 0 \end{pmatrix} \quad . \tag{9.71b}$$

In order to describe the infinitesimal rotations of a space vector we shall introduce three matrices J_i by writing (9.71) as

$$\delta a_0 = -i(\delta\vartheta \cdot J)a \quad \text{with} \tag{9.72a}$$

$$J_x = \begin{pmatrix} 0 & 0 & 0 \\ 0 & 0 & -i \\ 0 & i & 0 \end{pmatrix} \quad , \quad J_y = \begin{pmatrix} 0 & 0 & i \\ 0 & 0 & 0 \\ -i & 0 & 0 \end{pmatrix} \quad , \quad J_z = \begin{pmatrix} 0 & -i & 0 \\ i & 0 & 0 \\ 0 & 0 & 0 \end{pmatrix}$$

$$\tag{9.72b}$$

Problem 9.35. Show that these matrices can be extracted from (9.71b).

Problem 9.36. Show that the matrices J_i satisfy the commutation relations (9.61). It is for this reason that we have called them J. The negative sign in (9.72a) reflects the fact that whereas $\hat{J}$ in (9.68) rotates the basis clockwise, J in (9.72) rotates a vector clockwise.

Problem 9.37. Since the matrices J_i satisfy the commutation relations (9.61), they represent an angular momentum. What is the value of this angular momentum? That is, what is J^2? Find the eigenvalues of J_z. (You may need to consult a book such as [122] to see how to find the eigenvalues of a matrix.)

Spinors. It has happened several times in physics that mathematical devices that were first introduced in order to simplify calculations have turned out to contain profound physics. The Hamilton-Jacobi theory is an example. Another example is the Cayley-Klein parameters, by which Klein was able to simplify some calculations in the theory of the top [38], [123–125]. In the hands of Pauli and Dirac these parameters turned out to be the key to a theory of particles of spin $\frac{1}{2}$. It is not necessary to repeat the theory of rigid-body motion in the new variables, but I will show how they arose and what they have to do with spin. The approach will be similar to that followed earlier: to describe the rotation of the rigid body by telling what happens to an arbitrary vector a embedded in it.

A spinor is a vector with 2 complex components that can be used to represent an ordinary 3-dimensional vector with real components in the following way. The spinor u that represents a vector a is defined by the relations

$$a_x := 2\,\mathrm{Re}\,(u_1^* u_2) \quad , \quad a_y := 2\,\mathrm{Im}\,(u_1^* u_2) \quad , \quad a_z := |u_1|^2 - |u_2|^2 \qquad (9.73)$$

from which it follows at once that

$$a^2 = (|u_1|^2 + |u_2|^2)^2 \quad . \qquad (9.74)$$

Thus any transformation that represents a rotation preserves the complex length $|u|^2$ of the spinor u, defined as

$$|u_1|^2 + |u_2|^2 = u^\dagger u \quad . \qquad (9.75)$$

Here the dagger represents the hermitian conjugate or adjoint, defined as the matrix formed by taking the complex conjugate and the transpose: for any matrix c,

$$c^\dagger = c^{*\mathrm{T}} \quad . \qquad (9.76)$$

The vector a is overspecified by u because whereas a has 3 independent real components, u has 2 complex ones, defined by 4 real numbers. The nature of the redundancy is clear from (9.73): if a spinor u_α ($\alpha = 1, 2$) represents a certain vector a, then the spinor $\mathrm{e}^{\mathrm{i}\psi} u_\alpha$ represents exactly the same vector.

Transformations of u. An arbitrary linear transformation of u has the form

$$u \rightarrow u' = Au \qquad (9.77)$$

where the matrix A depends on four complex numbers,

$$a = \begin{pmatrix} \alpha & \beta \\ \gamma & \delta \end{pmatrix} \quad . \qquad (9.78)$$

The adjoint transforms by the adjoint matrix,

$$u^{\dagger\prime} = u^\dagger A^\dagger$$

and so if (9.75) is to be invariant we must have

$$u^{\dagger\prime} u' = u^\dagger A^\dagger A u \quad .$$

This is satisfied for all u provided that

$$A^\dagger A = 1 \quad . \qquad (9.79)$$

Such transformations are called unitary, and if A is written in the form (9.78) the elements are thereby restricted (Problem 9.38) by the relations

$$|\alpha|^2 + |\gamma|^2 = |\beta|^2 + |\delta|^2 = 1 \quad , \quad \alpha\beta^* + \gamma\delta^* = 0 \quad . \qquad (9.80)$$

But the A so defined is still too general, since if A is unitary, so is $A' = \mathrm{e}^{\mathrm{i}\phi} A$ where ϕ is any real phase, and we have already seen (but compare Problem 9.39) that A' and A give rise to the same transformation of the components of a. It follows from (9.79) that the determinant of A is of the form $\mathrm{e}^{\mathrm{i}\chi}$ with χ real (Problem 9.40) and so we can fix the phase of A simply by requiring that χ take some particular value; zero is the obvious one. Thus we make A not only unitary but also unimodular; i.e., its determinant is 1:

$$A^{\dagger}A = 1 \quad , \quad \det A = 1 \tag{9.81}$$

and we will make these transformations correspond to rotations of a. Any 2×2 matrix that satisfies (9.81) has the form

$$A = \begin{pmatrix} \alpha & \beta \\ -\beta^* & \alpha^* \end{pmatrix} \quad , \quad |\alpha|^2 + |\beta|^2 = 1 \tag{9.82}$$

but rather than investigate the general properties of such an A, we look first at its infinitesimal transformations.

Infinitesimal Transformations. The most general matrix in the infinitesimal neighborhood of unity that satisfies (9.82) has the form

$$A = \begin{pmatrix} 1 + i\varepsilon_3 & \varepsilon_2 + i\varepsilon_1 \\ -\varepsilon_2 + i\varepsilon_1 & 1 - i\varepsilon_3 \end{pmatrix} \tag{9.83}$$

where the ε_i are real infinitesimals and quantities of order ε^2 are neglected. We wish to see what transformations of a are induced by such transformations of u. It is found (Problem 9.41) by means of (9.73) that they are

$$a \rightarrow a' = a + 2\varepsilon \times a \tag{9.84}$$

where ε is the vector whose components are ε_i. Comparison with (9.71a) shows that these are the rotations we have already seen, with

$$\varepsilon_i = \tfrac{1}{2}\delta\vartheta_i \quad . \tag{9.85}$$

The general form (9.83) can be written as

$$A = 1 - i\left[\varepsilon_1 \begin{pmatrix} 0 & 1 \\ 1 & 0 \end{pmatrix} + \varepsilon_2 \begin{pmatrix} 0 & -i \\ i & 0 \end{pmatrix} + \varepsilon_3 \begin{pmatrix} 1 & 0 \\ 0 & -1 \end{pmatrix}\right]$$

and this, with (9.85) allows us to write the infinitesimal unitary unimodular transformations of u in a form exactly analogous to that of (9.72a),

$$\delta u = -i(\delta\vartheta \cdot \tfrac{1}{2}\sigma)u \tag{9.86}$$

where the matrices σ_i are the Pauli matrices

$$\sigma_1 = \begin{pmatrix} 0 & 1 \\ 1 & 0 \end{pmatrix} \quad , \quad \sigma_2 = \begin{pmatrix} 0 & -i \\ i & 0 \end{pmatrix} \quad , \quad \sigma_3 = \begin{pmatrix} 1 & 0 \\ 0 & -1 \end{pmatrix} \quad . \tag{9.87}$$

These matrices have the property that $\sigma_i = \sigma_i^{\dagger}$; such matrices are called hermitian. They generate rotations of a via the spinor u just as the matrices J generate them directly, and they satisfy the relations

$$\sigma_i^2 = 1 \quad , \quad \sigma_1\sigma_2 = -\sigma_2\sigma_1 = i\sigma_3 \quad \text{etc.} \tag{9.88}$$

continuing in cyclic order, so that

$$[\sigma_1, \sigma_2] = 2i\sigma_3 \quad \text{etc} \quad . \tag{9.89}$$

The spin operator in quantum mechanics is

$$\hat{S}_i = \tfrac{1}{2}\hbar\sigma_i$$

and it satisfies the commutation relations that define an angular momentum,

$$[\hat{S}_x, \hat{S}_y] = i\hbar\hat{S}_z \quad \text{etc} \quad .$$

In terms of the Pauli matrices (9.73) takes a very simple form,

$$a_i = u^{\dagger}\sigma_i u \quad . \tag{9.90}$$

Thus given u we can find the corresponding vector and, perhaps unexpectedly, the operator that does this is the spin operator. (This is a good excuse for calling u a spinor). Shortly we will see how to find u given $\boldsymbol{a}$.

If we look at the finite transformations corresponding to the infinitesimal (9.83) we encounter a remarkable feature of spinors. Let the rotation be around the third axis so that $\varepsilon_1 = \varepsilon_2 = 0$ and

$$u \rightarrow \begin{pmatrix} 1 + i\varepsilon_3 & 0 \\ 0 & 1 - i\varepsilon_3 \end{pmatrix} u \quad , \qquad \varepsilon_3 = \frac{1}{2}\delta\vartheta_3 \quad .$$

As in Sect. 5.5, the finite transformation is formed by iteration,

$$A(\vartheta_3) = \lim_{N \to \infty} \begin{pmatrix} 1 + \frac{i}{2}\frac{\vartheta_3}{N} & 0 \\ 0 & 1 - \frac{i}{2}\frac{\vartheta_3}{N} \end{pmatrix}^N \quad \text{or}$$

$$A(\vartheta_3) = \begin{pmatrix} e^{i\vartheta_3/2} & 0 \\ 0 & e^{-i\vartheta_3/2} \end{pmatrix} \tag{9.91}$$

and there are similar formulas for the other two elementary rotations, Problem 9.42. If the angle of rotation is 2π, so that $\boldsymbol{a}$ returns to its original orientation, the components of u are multiplied by -1, and it is not until α has rotated through 4π that u returns to its original value. It is natural to ask whether this mathematical oddity has any counterpart in the physical world. The answer, perhaps surprisingly, is yes. A simple demonstration popularized by Dirac (Fig. 9.15) shows that it is indeed possible to perform an experiment which tells whether a rotation through 2π has been made but is unable to detect a rotation through 4π.

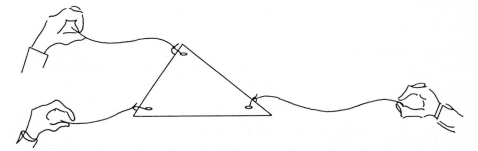

Fig. 9.15. Dirac's toy for illustrating the difference between rotations of 2π and 4π

Attach three pieces of string to the corners of a triangular piece of cardboard and give them to three hands to hold. A fourth hand now rotates the cardboard through 4π around a horizontal axis. If the triangle is held still it is now possible to disentangle the strings without letting go the ends so that no evidence of the rotations remains, but if the rotation is through 2π the strings cannot be disentangled. That this fact is indeed a physical manifestation of spinor transformations can be shown by a subtle but straightforward topological argument [126, 127].

The subject of this book is dynamics, not topology, and we may ask whether there is any dynamical phenomenon which could detect a rotation of 2π but not one of 4π. Such an experiment has been carried out [128]. An iron foil is magnetized in opposite directions along parallel strips. Slow neutrons passing through the foil have their wave functions effectively rotated, and the change in phase can be determined by the interference pattern. The results show clearly the sign change under rotation through 2π.

Problem 9.38. Show that the most general form of a unitary A is (9.78) restricted by (9.80).

Problem 9.39. Show that if A yields a certain transformation of a, then $A' = e^{i\phi}A$ yields the same transformation.

Problem 9.40. Show from the properties of determinants that if D is the determinant of A, then (9.79) requires

$$|D|^2 = 1 \quad \text{whence} \quad D = e^{i\chi}$$

for some real χ.

Problem 9.41. Show that if A given by (9.83) acts on u, the corresponding transformation of a is (9.84).

Problem 9.42. With the derivation of (9.91) as a model or in some other way, find the matrices A corresponding to finite rotations of a around the first and second coordinate axes.

Spinors and Vectors. Our next task is to find u given a. From (9.73) and (9.74) we have

$$\frac{a_x - ia_y}{a - a_z} = \frac{2u_1 u_2^*}{2|u_2|^2} = \frac{u_1}{u_2}$$

which is satisfied by

$$u_1 = \mu(a_x - ia_y) \quad , \quad u_2 = \mu(a - a_z)$$

with μ chosen so that (9.74) is also satisfied:

$$|u_1|^2 + |u_2|^2 = |\mu|^2(a_x^2 + a_y^2 + a^2 - 2aa_z + a_z^2) = 2|\mu|^2 a(a - a_z) = a$$

so that

$$u_1 = \frac{(a_x - ia_y)e^{i\chi}}{\sqrt{2(a - a_z)}} \quad , \quad u_2 = \sqrt{\frac{1}{2}(a - a_z)}e^{i\chi} \tag{9.92}$$

where χ is an arbitrary phase common to both. These formulas take a more transparent form if we express a in terms of the usual polar angles ϑ and ϕ:

$$a_x = a \sin \vartheta \cos \phi \quad , \quad a_y = a \sin \vartheta \sin \phi \quad , \quad a_z = a \cos \vartheta$$

whence

$$a_x - ia_y = ae^{-i\phi} \sin \vartheta \quad .$$

We find (Problem 9.46)

$$u_1 = \sqrt{a}e^{-i\phi/2} \cos \vartheta/2 \quad , \quad u_2 = \sqrt{a}e^{i\phi/2} \sin \vartheta/2 \tag{9.93}$$

when χ has been set equal to $\phi/2$ for symmetry. Note again that u changes sign when $\phi = 2\pi$ and returns to its original value when $\phi = 4\pi$.

Basis Vectors. In defining a 3-vector one specifies a triplet of basis vectors and writes, in a common notation,

$$a = a_x \hat{i} + a_y \hat{j} + a_z \hat{k} \tag{9.94a}$$

where $\hat{i}$, $\hat{j}$, and $\hat{k}$ are unit vectors whose directions are defined in terms of three parameters: two for the direction of $\hat{k}$ and one for the orientation of $\hat{i}$ relative to $\hat{k}$; $\hat{j}$ is then determined by (usually) the right-hand convention.[5] Spinors also require a basis, and (9.93) shows how it can be attached to the 3-basis just mentioned. The basis vectors are often called α and β, so that a typical spinor is

$$u = u_1 \alpha + u_2 \beta \quad . \tag{9.94b}$$

By (9.93), α and β correspond to vectors pointing parallel and antiparallel, respectively, to $\hat{k}$. This requires two parameters, and the third is a phase, adjusted so that real spinor coordinates with zero phase will lie in the xz plane.

Note that ϕ is only the relative phase of α and β. One can still multiply a pair of spinor coordinates by an arbitrary $e^{i\chi/2}$, but clearly from (9.90) that this does not affect the vector that the spinor determines. Does an absolute phase signify anything? Suppose the basic three-dimensional vectorial object is not a thing like a pointed stick but a thing like an arrow. The difference is that one can see an arrow rotate because the feathers destroy its circular symmetry. To specify the orientation of the arrow requires not only the direction of its shaft but also the angle of a feather with respect to a reference direction perpendicular to the arrow. Suppose first that the arrow points along the z direction. The corresponding unit spinor is $u = \binom{1}{0}$. Then the rotation (9.91) multiplies u by $e^{i\psi/2}$ and rotates the coordinate axes through ψ around the direction of the arrow or, alternatively, rotates the arrow the other way. If the arrow points in some other direction the new components u_i are multiples of the old u_1 and carry the same factor $e^{i\psi/2}$. It follows that the three-dimensional object corresponding to a unimodular spinor is an arrow, not a vector, and that ψ gives the arrow's axial rotation away from the reference direction. Partly for historical reasons, this object is not a familiar item in geometry and I have had to invent a name for it.

---

[5] Of course this prescription assumes the existence of an absolute standard of directions, perhaps the galactic coordinate system.

Arrows are convenient in representing the motion of a rigid body, but the resulting theory [124] is not very simple and there is no need to go into it here. A further generalization is possible: if the restriction to unimodular transformations is dropped the resulting larger group of transformations can be associated with Lorentz transformations [129], and so two-dimensional spinors of this more general kind have applications in relativistic theory as well. Just as unimodular spinors are useful in representing the spin states of particles of spin 1/2, so relativistic spinors are used to construct the four-component object that represents the spin and charge states of an electron in Dirac's relativistic quantum mechanics.

Problem 9.43. Show that if an arrow pointing along the z axis is rotated by $e^{i\psi/2}$ the same exponential rotates an arrow pointing in any other direction.

Problem 9.44. Derive the formulas (9.92).

Problem 9.45. Derive the formulas (9.93).

9.8 Particles with Spin

The angular-momentum properties of particles are associated with the angular dependence of their wave functions; for example, the vectorial wave function $\psi = \mathbf{r} f(r)$ describes a particle of spin 1. A spinor is another geometric object with directional properties; let us see how else it can be put into a wave function.

The wave function

$$\psi = u f(r) \tag{9.95}$$

is 2-component object. The operator that represents the angular momentum when the wave function is a 2-component spinor is

$$\hat{S} = \tfrac{1}{2}\hbar\boldsymbol{\sigma} \tag{9.96}$$

and the expectation values of the components of S are formed using the spinor u and its hermitian conjugate $u^\dagger$. It follows from the principles of quantum mechanics that the object represented in this way has an intrinsic angular momentum equal to $\tfrac{1}{2}$. (Since $f(r)$ is spherically symmetric, there is in this case no orbital angular momentum.) The components u_1 and u_2 are the two independent eigenfunctions of S_z:

$$S_z \alpha = \tfrac{1}{2}\hbar\alpha \quad , \quad S_z \beta = -\tfrac{1}{2}\hbar\beta \tag{9.97}$$

and thus they represent the states in which the spin is certainly up or down. In the state described by (9.93), the probabilities of the values $S_z = \hbar/2$ and $-\hbar/2$ are $\cos^2 \vartheta/2$ and $\sin^2 \vartheta/2$ respectively. All this is obvious mathematically; what is not obvious mathematically is that nature has given us things to play with that correspond to the formalism. It has been found that a number of the familiar particles such as e, μ, n, p, Λ, Σ, and Ξ have spin 1/2, while several others such as Δ, Σ^*, Ξ^*, and Ω^- have spin 3/2. It is a relatively simple matter to find the wave equations, relativistic and nonrelativistic, for particles of spin, 0, 1/2, and 1. Higher spins present a stubborn problem.

Nothing in the foregoing argument provides any information as to the dynamics of the systems described; that requires further hypothesis. We got the classical Hamiltonian for rigid-body motion by putting the body together out of point particles each of which obeys Newton's laws. Clearly no such simple construction is possible here, but if a baryon consists of three quarks and a meson consists of a quark and an antiquark, together with other fields, and if quarks have spin $\frac{1}{2}$, there is some hope of understanding the states of spinning particles. Experimentally, families of particles of mass M and spin J can be arranged according to the linear relation

$$M^2 \approx J \times 1\,\text{GeV}^2$$

and this is qualitatively explained by the "bag model" of particle structure [130].

The interaction between spin and an external electromagnetic field is introduced through the hypothesis known as minimal electromagnetic coupling, as follows: Write the field-free nonrelativistic Hamiltonian in the form

$$H_0 = \frac{\hat{p}^2}{2m} = \frac{(\boldsymbol{\sigma} \cdot \hat{\boldsymbol{p}})^2}{2m} \tag{9.98}$$

where the equivalence of the two forms follows from the algebraic properties (9.88). Then use the rule (5.19) that includes a magnetic field by the substitution

$$\boldsymbol{p} \to \boldsymbol{p} - e\boldsymbol{A} \tag{9.99}$$

and finally, add the scalar potential eV. On simplifying $[\boldsymbol{\sigma} \cdot (\hat{\boldsymbol{p}} - e\boldsymbol{A})]^2$ we find

$$H = \frac{(\hat{\boldsymbol{p}} - e\boldsymbol{A})^2}{2m} - \frac{e}{m}\hat{\boldsymbol{S}} \cdot \boldsymbol{B} + eV \tag{9.100}$$

in which the middle term represents the coupling of a particle with gyromagnetic ratio e/m to the magnetic field. We see that the electron's magnetic interaction emerges out of the hypothesis (9.98). It should be noted that we have merely expressed one hypothesis in terms of another, though the step (9.99) can be justified by fundamental arguments [Ref. 131, p. 188].

Historically, (9.100) was written down by Pauli in 1927. The "derivation" given here made its way into the literature long afterwards, and its only merit is that it gives a way of understanding the electron's gyromagnetic ratio without becoming involved in arguments in which the electron is pictured as a spinning distribution of charge. The real value of these considerations lies in their generalization from rotational transformations to Lorentz transformations, which may be viewed as rotations in four dimensions. This was done by Dirac in 1928 and the resulting relativistic wave equation is now one of the most solidly based relations in all of physics. The matrices now have 4 rows and columns, but the same gyromagnetic ratio emerges, as well as the existence and properties of positrons.

Problem 9.46. Derive (9.100) from (9.98) and (9.99).

Problem 9.47. Show from (9.100) that $d\langle \boldsymbol{S}\rangle/dt$ belongs to the equation of motion that classical physics gives for the same situation.

Problem 9.48. Can hydrogen nuclei in a hydrocarbon serve as a gyrocompass?

Spin in a Magnetic Field. Equation (9.90) looks remarkably like the quantum-mechanical formula for the expectation value of the spin of something described by a two-component wave function, and yet nothing was said about quantum mechanics in writing it down. It represents the vector a in terms of a spinor quantity, rotations being performed by a unitary transformation of u that can be written in the form (9.82) or, by Stone's theorem, as

$$u(t) = e^{iK(t)}u_0$$

where K is a hermitian matrix operating on u. Or alternatively, as in the Heisenberg representation of quantum mechanics, one can put the time dependence into σ by writing

$$\tilde{\sigma}(t) = e^{-iK(t)}\sigma e^{iK(t)} \quad . \tag{9.101}$$

But if one tries to use this procedure in classical mechanics, even in simple cases, the result is complicated and unnatural; classical mechanics does not want to be expressed that way.

To see how it works in quantum mechanics look at the simple case of a spin of $\frac{1}{2}$ with gyromagnetic ratio γ precessing in a magnetic field B directed along the z axis. The equation of motion is

$$i\hbar\dot{u} = Hu \quad , \quad H = -\gamma\hbar BS_z = -\tfrac{1}{2}\gamma\hbar B\sigma_z \quad . \tag{9.102}$$

This is easily solved (Problem 9.49) to give

$$u_x = e^{i\omega t}u_{x0} \quad , \quad u_y = e^{-i\omega t}u_{y0} \quad , \quad \omega = \tfrac{1}{2}\gamma B \tag{9.103}$$

but it is perhaps more interesting to solve the equation as

$$u(t) = e^{-iHt/\hbar}u_0$$

and calculate directly

$$e^{-i\gamma Bt\sigma_z/2}\begin{pmatrix}\sigma_x\\\sigma_y\\\sigma_z\end{pmatrix}e^{i\gamma Bt\sigma_z/2} = \begin{pmatrix}\sigma_x\cos\gamma Bt - \sigma_y\sin\gamma Bt\\\sigma_x\sin\gamma Bt + \sigma_y\sin\gamma Bt\\\sigma_z\end{pmatrix} \quad .$$

(Problem 9.50. Note the disappearance of the $\frac{1}{2}$). In (9.90) this gives the components of the vector S,

$$S_x = \cos\gamma Bt\, S_{x0} - \sin\gamma Bt\, S_{y0} \quad , \quad S_y = \sin\gamma Bt\, S_{x0} + \cos\gamma Bt\, S_{y0} \quad ,$$
$$S_z = S_{z0} \quad . \tag{9.104}$$

The expectation value of the spin gyrates with angular frequency γB around the z axis. Note that in this derivation σ plays two parts. It belongs to the mathematical machinery by which vectors in three-dimensional space are represented by complex vectors in a space of two dimensions,[6] and it also appears in the Hamiltonian operator

---

[6] Read these words with scepticism. Just because it is natural and customary to describe space in terms of real numbers in three dimensions does not mean that that is the only, or even the best way to do it. The description in terms of complex numbers in two dimensions contains more physics and the same sensory experience. It can readily be extended to describe spacetime and it is the proper framework in which to understand the *Klein-Opat* experiment [128] as well as fundamental physical principles such as Dirac's equation.

as a physical variable. Once we are in the two-dimensional "spin space" there is nothing specifically quantum-mechanical about σ.

Trying to interpret (9.104) in classical terms we ask: What are the characteristics of the thing that is spinning, and if it is spinning, where in the Hamiltonian is the expression for its kinetic energy? A classical version of the theory cannot ignore these questions. For maximum simplicity, consider a magnetized ball that is set spinning around the axis of the magnet. In laboratory coordinates the Lagrangian is then

$$L = \tfrac{1}{2} I \omega^2 - \boldsymbol{M} \cdot \boldsymbol{B} \tag{9.105}$$

where $\boldsymbol{M}$ is the ball's magnetic moment, which for purposes of this comparison we assume is oriented parallel to the spin.

It is easy to complete the calculation in Euler angles. With $I_z = I_x = I$ the kinetic energy can be read from (9.41), and so (9.105) is

$$L = \tfrac{1}{2} I(\dot{\vartheta}^2 + \dot{\varphi}^2 + 2 \cos \vartheta \, \dot{\varphi} \dot{\psi} + \dot{\psi}^2) - MB \cos \vartheta$$

and with $\dot{\psi}$ written as Ω the uniform precession (Problem 9.51) is at the rate

$$\dot{\varphi} = \frac{MB}{I\Omega} \ . \tag{9.106}$$

Since $M/I\Omega$ is the ball's gyromagnetic ratio γ this is the same as the earlier result (9.104), but there is a fundamental difference. Whereas (9.104) is the general solution of (9.102), (9.106) is a particular solution of the classical equation of motion. There are many other motions possible other than simple precession; they correspond to the nutations of a top. The difference illustrates one of the ways in which the classical description is more specific than the quantal one.

Problem 9.49. Show that (9.103) solves (9.102).

Problem 9.50. Derive equation (9.104).

Problem 9.51. Show that (9.106) is a solution of the classical equations of motion.

Problem 9.52. The solution (9.106) is a very special one. Study the effect of small oscillations in the angle ϑ. In particular, calculate and sketch graphically the behavior of the ball if it is set spinning around the magnetic axis with Ω large and the axis held stationary and then released. (The solution of Problem 9.23 will be useful here.) Finally, still in the small-angle approximation, explore what happens when the ball spins around an axis other than that of the magnet.

10. Continuous Systems

If one considers the motions of a continuous system – a stretched string or the air in a room – it might at first seem that one would have to deal with the infinite limit of an N-particle system and that vast complexities would arise. This would of course be true if one wanted to understand everything the individual molecules are doing, but we will not be so ambitious. The right approach is to forget all about discreteness and deal from the beginning with continuous functions of the coordinates. This does not always work. If for example we want to study the elastic vibrations of a crystal we must start from the intermolecular forces. Arguments of this kind go back to Cauchy in 1822 and are of a much higher order of difficulty than any to be attempted here. We shall consider a few examples of mechanical vibrations – stretched strings, vibrating membranes, sound waves – and then an example of a field that can be treated in much the same way – the matter field ψ. Finally, a simple matter field will be quantized.

10.1 Stretched Strings

The simple one-dimensional system of a string fastened at the ends and vibrating transversely is a good introduction to the subject, since the extension to more than one dimension is only a matter of detail. In the continuum limit the string has a number f of degrees of freedom that becomes infinite, and its configuration can be specified by giving the displacement $u(t)$ corresponding to every point x along the string. This suggests that f is not only infinite but that it has the power of the continuum – like the points in a line, the degrees of freedom cannot even be counted. But we do not need that kind of generality, for the string is continuous and every point moves very much like the other points in its immediate neighborhood; this restriction will greatly simplify the analysis. Label the points on the string by the variable x and specify the displacement by the continuous function $u(x,t)$ (Fig. 10.1a) which, in the case considered here, must vanish at the two ends:

$$u(0,t) = u(l,t) = 0 \quad . \tag{10.1}$$

Another way to characterize the string is to express $u(x,t)$ as a Fourier decomposition,

$$u(x,t) = \sum_{-\infty}^{\infty} a_m(t)e^{imKx} \quad , \qquad K = \frac{\pi}{l} \tag{10.2}$$

where the value of K is explained in Fig. 10.1b. Because the displacement $u(x,t)$

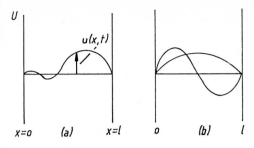

is a real number, we have

$$u(x,t) = u^*(x,t) = \sum_{-\infty}^{\infty} a_m^* e^{-imKx} = \sum_{-\infty}^{\infty} a_{-m}^* e^{imKt}$$

so that the coefficients will satisfy

$$a_{-m}^*(t) = a_m(t) \quad . \tag{10.3}$$

In the Fourier representation the $a_m(t)$ are the generalized coordinates of the system. They are infinite in number, but it is a countable infinity, the infinity of the integers and not that of the points on a line. This results from the assumed continuity of $u(x,t)$ as a function of x, and it makes things easier. The condition (10.3) is something of an inconvenience, since one prefers to have the various coordinates move independently of each other, but we can overcome it by treating a more general case in which (10.3) is not used and $u(x,t)$ is free to take on complex values, and then selecting out the desired special cases by imposing boundary conditions: if $u(x,0)$ and $\dot{u}(x,0)$ are real, for example, then $u(x,t)$ will remain real at all subsequent times.

It will be instructive to study the problem in four ways:

	x descr.	Fourier descr.
Lagrangian	1	3
Hamiltonian	2	4

which I will take up in order.

10.2 Four Modes of Description

1) Lagrange's Equations, x Description. Let τ be the tension of the cord, and pull the cord aside into a shape given by $u(x,t)$. The old length was l. The new length is

$$l' = \int ds = \int_0^l \sqrt{1 + u'^2(x,t)} dx \simeq \int_0^l \left[1 + \frac{1}{2} u'^2(x,t)\right] dx \tag{10.4}$$

where $u' = \partial u/\partial x$. We have made the approximation of assuming that the displacement is small. This is called the linear approximation, since it leads to a linear equation of motion. Problem 10.2 shows that things become much more complicated when nonlinear terms are included.

The slightly displaced string is stretched by an amount

$$l' - l = \frac{1}{2} \int_0^l u'^2 \, dx$$

and the work required to do this is the potential energy,

$$V = \frac{\tau}{2} \int_0^l u'^2 \, dx \quad . \tag{10.5a}$$

If ϱ is the linear mass density, which we assume constant (but see Problem 10.26), then the kinetic energy is obviously

$$T = \frac{\varrho}{2} \int_0^l \dot{u}^2 \, dx \tag{10.5b}$$

where $\dot{u} := \partial u / \partial t$.

The Lagrangian of the string is then

$$L = \int_0^l \left(\frac{\varrho}{2} \dot{u}^2 - \frac{\tau}{2} u'^2 \right) dx \quad . \tag{10.6}$$

This is the sum of contributions from each element of length along the string. The action is given by

$$S = \int_{t_0}^t L \, dt = \int_0^l dx \int_{t_0}^t dt \, \mathcal{L}(u', \dot{u}) \tag{10.7a}$$

where $\mathcal{L}$, the Lagrangian density, is

$$\mathcal{L} = \tfrac{1}{2} (\varrho \dot{u}^2 - \tau u'^2) \quad . \tag{10.7b}$$

In general (see Prob. 10.3), $\mathcal{L}$ is a function $\mathcal{L}(u, u', \dot{u})$. S is also the sum of the actions belonging to successive elements dx of the string, each of which moves in response to time-dependent forces exerted by the adjacent elements. That for small amplitudes $(\tau u'^2/2) dx$ is the potential energy produced by these forces is true but not immediately obvious. To show generally that the Lagrangian of a continuous mechanical system can be written in terms of the system's total kinetic and potential energies is a little long [132, Chap. 8] and we shall not pursue the matter here. The reason is that it is of no general use to be able to do so, for the main utility of variational methods today is in the theory of fields, where one considers the whole field at once; one does not put it together, as with the string, out of individual elements. (In Maxwell's theory we do not begin by focusing on what happens in a volume element $dx \, dy \, dz$.)

The equation of motion of the string are derived from varying S. The variation δu vanishes at t_0 and t by the terms of Hamilton's principle and at $x = 0$ and $x = l$ because the string is not free to move there. Hamilton's principle $\delta S = 0$ is thus a problem in one dependent and two independent variables, which has been shown in Problem 1.21 to lead to the Euler-Lagrange equation

$$\frac{\delta \mathcal{L}}{\delta u} := \frac{\partial \mathcal{L}}{\partial u} - \frac{\partial}{\partial x}\frac{\partial \mathcal{L}}{\partial u'} - \frac{\partial}{\partial t}\frac{\partial \mathcal{L}}{\partial \dot{u}} = 0 \quad . \tag{10.8}$$

In our particular case $\mathcal{L}$ is independent of u, and the equation of motion is

$$\tau u'' - \varrho \ddot{u} = 0 \quad \text{or}$$

$$u'' - \frac{1}{v^2}\ddot{u} = 0 \quad , \qquad v^2 = \tau/\varrho \quad . \tag{10.9}$$

This is d'Alembert's equation and its solutions are extremely general, for any doubly differentiable function of the form

$$u(x,t) = f(x - vt) + g(x + vt) \tag{10.10}$$

satisfies it. That the sum of two solutions is again a solution is the result of linearity, and f and g represent waves of arbitrary waveform moving to right and left respectively at the speed v. In this approximation the waveform does not change as the wave progresses and the velocity is independent of amplitude and frequency.

The normal modes of a system are the motions in which the motion of each part is periodic with the same frequency. If the frequency is ω, the normal modes resulting from (10.9) are the solutions of

$$u'' + \frac{\omega^2}{v^2}u = 0 \tag{10.11}$$

specified by the appropriate boundary conditions.

Problem 10.1. For many field calculations boundary conditions like (10.1) are inappropriate because they introduce reflected waves. An alternative is the periodic boundary condition $u(l) = u(0)$. This changes the value of K in (10.2) and also the conditions on δu in Hamilton's principle. Show how the theory developed above is changed.

Problem 10.2. Find the wave equation that results if the wave amplitude is great enough so that the first three terms in the expansion (10.4) must be used.

Problem 10.3. A stretched string is attached at intervals to a straight bar by springs of unstressed length d, Fig. 10.2. Find the wave equation and give a solution. Assume the springs are close enough together so that their effect is essentially continuous.

Problem 10.4. The normal-mode vibrations of a string are those in which each part of it moves sinusoidally with the same frequency. Find the normal modes and their frequencies for the stretched string and also for the modification shown in Fig. 10.2.

Problem 10.5. Show that if the equation of motion is satisfied the integral S vanishes, i.e. potential and kinetic energies are, on the average, equal. Assume that the

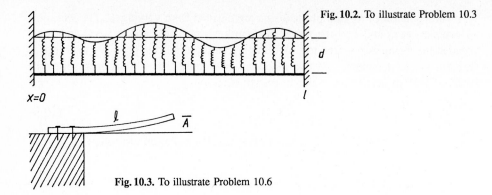

Fig. 10.2. To illustrate Problem 10.3

$x=0$

Fig. 10.3. To illustrate Problem 10.6

time integral extends over a whole number of complete cycles and that either the string is fastened at the ends or else its ends are joined to form a loop.

Problem 10.6*. A bent board (Fig 10.3) stores potential energy because the fibres on one side of the median plane are stretched and those on the other side are compressed. Assuming that the fibres obey Hooke's law, show that the stored energy is given by an expression of the form

$$V = c \int_0^l u''^2 dx \quad .$$

Find Lagrange's equation for the free oscillations of a uniform diving board. The boundary conditions are $u = u' = 0$ at the left end and $u'' = 0$ at the right because no torque is applied there. Find the frequency of the lowest mode of oscillation.

Problem 10.7. Find the equation for longitudinal vibrations of a string fixed at the ends if a static tension f produces a stretch efl in a string of length l and linear density λ.

2) Hamilton's Equations, x Description. Define the momentum π conjugate to the variable u by

$$\pi(x, t) = \frac{\partial \mathcal{L}}{\partial \dot{u}}$$

and the Hamiltonian density $\mathcal{H}$ by

$$\mathcal{H} = \pi \dot{u} - \mathcal{L} \quad . \tag{10.12}$$

From (10.8) the equation of motion for π is

$$\dot{\pi} = \frac{\partial \mathcal{L}}{\partial u} - \frac{\partial}{\partial x} \frac{\partial \mathcal{L}}{\partial u'} \tag{10.13}$$

while for arbitrary variations in u, $\dot{u}$, and u',

$$\delta \mathcal{H} = \pi \delta \dot{u} + \dot{u} \delta \pi - \frac{\partial \mathcal{L}}{\partial u} \delta u - \frac{\partial \mathcal{L}}{\partial \dot{u}} \delta \dot{u} - \frac{\partial \mathcal{L}}{\partial u'} \delta u' = \dot{u} \delta \pi - \frac{\partial \mathcal{L}}{\partial u} \delta u - \frac{\partial \mathcal{L}}{\partial u'} \delta u' \quad .$$
$$\tag{10.14}$$

Thus we can regard $\mathcal{H}$ as a function of π, u, and u', and (10.13) becomes

$$\dot{\pi} = -\frac{\partial \mathcal{H}}{\partial u} + \frac{\partial}{\partial x}\frac{\partial \mathcal{H}}{\partial u'} := -\frac{\delta \mathcal{H}}{\delta u} \tag{10.15a}$$

where the functional derivative is of the same form as that in (1.40b) but with x as independent variable, while (10.14) gives

$$\dot{u} = \frac{\partial \mathcal{H}}{\partial \pi} = \frac{\delta \mathcal{H}}{\delta \pi} \quad . \tag{10.15b}$$

These are Hamilton's equations for a one-dimensional continuous system.

We can easily verify that $\mathcal{H}$ is the energy density by seeing that its linear integral H is conserved. For any density $a(u, u', \pi, \pi')$ with

$$A = \int_0^l a \, dx$$

we have

$$\frac{dA}{dt} = \int_0^l \left(\frac{\partial a}{\partial u}\dot{u} + \frac{\partial a}{\partial u'}\dot{u}' + \frac{\partial a}{\partial \pi}\dot{\pi} + \frac{\partial a}{\partial \pi'}\dot{\pi}' \right) dx = \int_0^l \left(\frac{\delta a}{\delta u}\dot{u} + \frac{\delta a}{\delta \pi}\dot{\pi} \right) dx$$

plus terms evaluated at the boundaries which vanish if A is conserved. By (10.15) this is

$$\frac{dA}{dt} = \int_0^l \left(\frac{\delta a}{\delta u}\frac{\delta \mathcal{H}}{\delta \pi} - \frac{\delta a}{\delta \pi}\frac{\delta \mathcal{H}}{\delta u} \right) dx \quad . \tag{10.16}$$

The resemblance to (5.64) containing Poisson brackets is obvious and the conservation of energy follows at once.

In the present case, from (10.7b),

$$\pi = \varrho \dot{u} \quad \text{and} \tag{10.17}$$

$$\mathcal{H} = \frac{\pi^2}{2\varrho} + \frac{\tau}{2}u'^2 \tag{10.18a}$$

while the total energy is

$$H = \int_0^l \mathcal{H} \, dx = \int_0^l \left(\frac{\varrho}{2}\dot{u}^2 + \frac{\tau}{2}u'^2 \right) dx \tag{10.18b}$$

as one could have written with much less trouble from (10.5).

Problem 10.8. Find Hamilton's equations and solve them for the lowest normal mode of the system of Fig. 10.2.

Problem 10.9. Find Hamilton's equations for a nonuniform cord in which ϱ is a function of x.

3) Lagrange's Equations, Fourier Description. In this description the coordinates are the Fourier amplitudes $a_m(t)$ and the velocities are their time derivatives. As already mentioned, we will not impose the reality conditions (10.3) on the coordinates but allow them to follow from the boundary conditions. But if u is complex it is not obvious how to minimize or maximize a complex S: is $-2i$ smaller or larger than 1? The solution is to make S automatically real by defining

$$\mathcal{L} = \frac{\varrho}{2}|\dot{u}|^2 - \frac{\tau}{2}|u'|^2 \tag{10.19}$$

noting that if, as we anticipate, u is real, the new $\mathcal{L}$ is the same as the old one. Putting (10.2) into (10.19) gives

$$L = \int_0^l \sum_{m=-\infty}^{\infty} \sum_{m'=-\infty}^{\infty} \left(\frac{\varrho}{2} \dot{a}_m \dot{a}_{m'}^* - \frac{\tau}{2} m m' K^2 a_m a_{m'}^* \right) e^{i(m-m')Kx} dx \quad .$$

The integral of $e^{i(m-m')Kx}$ vanishes unless $m' = m$, and it is then equal to l:

$$L = l \sum_{m=-\infty}^{\infty} \left(\frac{\varrho}{2}|\dot{a}_m|^2 - \frac{\tau}{2} m^2 K^2 |a_m|^2 \right) \quad . \tag{10.20}$$

We could take Re a_m and Im a_m as the coordinates, but it is more natural to take a_m and a_m^* (Problem 10.13). The equation

$$\frac{\delta L}{\delta a_m^*} := \frac{\partial L}{\partial a_m^*} - \frac{d}{dt}\frac{\partial L}{\partial \dot{a}_m^*} = 0$$

gives

$$l\left(-\frac{\tau}{2} m^2 K^2 a_m - \frac{d}{dt}\frac{\varrho}{2}\dot{a}_m \right) = 0$$

or, with (10.9),

$$\ddot{a}_m + (mKv)^2 a_m = 0 \tag{10.21}$$

for each m, and $\delta L/\delta a_m = 0$ gives the complex conjugate. The problem of the continuous cord is thus reduced to that of a set of independent simple harmonic oscillators. That is, the $a_m(t)$ are the normal-mode coordinates. The solution of (10.21) is

$$a_m(T) = A_m e^{im\omega_0 t} + B_m e^{-im\omega_0 t} \quad , \qquad \omega_0 := Kv \tag{10.22}$$

where A_m and B_m are complex constants to be determined by the initial conditions and the conditions at the boundaries. We have solved the problem in Fourier variables with about as much labor as it took to state it in terms of x. To complete the solution, we must put in the initial conditions. Suppose we are given the initial configuration $u(x,0)$ and the initial velocity $u(x,0)$ of the cord at each point. We have

$$u(x,0) = \sum_{-\infty}^{\infty}(A_m + B_m)e^{imKx} \tag{10.23a}$$

$$\dot{u}(x,0) = i\omega_0 \sum_{-\infty}^{\infty} m(A_m - B_m)e^{imKx} \quad . \tag{10.23b}$$

By Fourier analysis we can find the coefficients $A_m + B_m$ and $m(A_m - B_m)$ and

296

then solve for A_m and B_m. If $u(x,0)$ and $\dot{u}(x,0)$ are real, then a_m obeys (10.3) (Problem 10.15). A detailed example is worked out in Sect. 10.3.

4) Hamilton's Equations, Fourier Description. The momentum conjugate to a_m is

$$p_m := \frac{\partial \mathcal{L}}{\partial a_m} = \frac{l\varrho}{2}\dot{a}_m^*$$

and, of course, the momentum conjugate to a_m^* is the complex conjugate of this, p_m^*. The Hamiltonian is easily seen to be

$$H = \sum_{-\infty}^{\infty} \left(\frac{2}{l\varrho}|p_m|^2 + \frac{l\tau}{2}m^2K^2|a_m|^2 \right) \quad . \tag{10.24}$$

Hamilton's equations lead at once to (10.20), and the energy comes out as

$$E = K^2 l\tau \sum_{\infty}^{\infty} m^2(|A_m|^2 + |B_m|^2) \tag{10.25}$$

when a_m is given by (10.22). The time-dependent terms have cancelled and E is constant.

Problem 10.10. Derive the Hamiltonian (10.24).

Problem 10.11. Show that Hamilton's equations are equivalent to Lagrange's equations for the system just discussed.

Problem 10.12. Evaluate E with particular attention to the cancellation of time-dependent terms.

Problem 10.13. Show that the equations obtained from (10.20) by varying a_m and a_m^* *independently* are equivalent to those obtained by varying Re a_m and Im a_m.

Problem 10.14. A vibrating string is gradually tightened until the tension is twice what it was before. How are the new frequency and amplitude of oscillation related to the old ones?

Problem 10.15. Show that if $u(x,0)$ and $\dot{u}(x,0)$ are real, then the Fourier amplitudes derived from (10.23) satisfy (10.3) automatically.

10.3 Example: A Plucked String

In this section we shall work out an example illustrating the general argument given above and propose a few corollary problems. Figure 10.4 shows the initial position of a string pulled to one side and released from rest at $t = 0$.

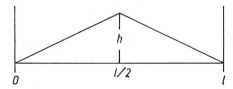

Fig. 10.4. String displaced at the middle

297

$$u(x,0) = \begin{cases} 2hl^{-1}x & (0 < x < l/2) \\ 2hl^{-1}(l-x) & (l/2 < x < l) \end{cases} \quad , \quad \dot{u}(x,0) = 0 \quad .$$

Since $\dot{u}(x,0) = 0$, the time dependence of the normal coordinates is given by cosines, and

$$u(x,t) = \sum_{-\infty}^{\infty} A_m e^{imKx} \cos mKvt \quad . \tag{10.26}$$

The constants can now be chosen so as to satisfy the boundary conditions. In order that $u(x,t) = 0$ at $x = 0$ and l, the exponentials must go together to form sines,

$$u(x,t) = \sum_{1}^{\infty} B_m \sin (mKx) \cos (mKvt) \quad , \quad K := \pi/l \quad . \tag{10.27}$$

With this,

$$u(x,0) = \sum_{1}^{\infty} B_m \sin (mKx)$$

and B_m is readily found from

$$\begin{aligned} B_m &= \frac{2}{l} \int_{0}^{l} u(x,0) \sin (mKx) dx \\ &= \frac{8h}{\pi^2 m^2} \sin \frac{m\pi}{2} \quad . \end{aligned}$$

Thus $B_m = 0$ for m even, and

$$B_m = (-1)^{(m-1)/2} \frac{8h}{\pi^2 m^2} \quad , \quad m \text{ odd} \quad . \tag{10.28}$$

The motion of the string is given by

$$\begin{aligned} u(x,t) = \frac{8h}{\pi^2} \Big(& \sin Kx \cos Kvt - \frac{1}{3^2} \sin 3Kx \cos 3Kvt \\ & + \frac{1}{5^2} \sin 5Kx \cos 5Kvt - \ldots \Big) \quad . \end{aligned} \tag{10.29}$$

The wavelengths are $2\pi/K$, $2\pi/3K$, ..., or $2l$, $2l/3$, ..., as is clear from an elementary construction. The intensities of the various harmonics depend on the squares of the Fourier components of u and therefore vary as 1, $1/3^4$, ... Why the even terms vanish can be understood by sketching some of the even modes and comparing them with Fig. 10.4.

Euler's Sums. The Fourier expansion just used has an interesting mathematical application. With $0 \le x \le l/2$ and $t = 0$, it is

$$\sum_{\substack{n \text{ odd} \\ 1}}^{\infty} \frac{8h}{\pi^2 n^2} \sin \frac{n\pi}{2} \sin \frac{n\pi x}{l} = \frac{2hx}{l} \quad . \tag{10.30}$$

298

Let $x = l/2$:

$$\sum_{\substack{n \text{ odd} \\ 1}}^{\infty} \frac{8h}{\pi^2 n^2} \sin^2 \frac{n\pi}{2} = \sum_{\substack{n \text{ odd} \\ 1}}^{\infty} \frac{8h}{\pi^2 n^2} = h$$

so that

$$S_0 := \sum_{n=1,3,5,\dots} \frac{1}{n^2} = \frac{\pi^2}{8} \quad.$$

Let

$$S := \sum_{n=1}^{\infty} \frac{1}{n^2} \quad, \quad S_e := \sum_{n=2,4,6,\dots} \frac{1}{n^2} \quad. \tag{10.31}$$

Then

$$S = S_0 + S_e \quad.$$

But also $S_e = \frac{1}{4} S$ (Problem 10.17). Thus $S = \frac{4}{3} S_0$ and

$$S = \frac{\pi^2}{6} \quad, \quad S_0 = \frac{\pi^2}{8} \quad, \quad S_e = \frac{\pi^2}{24} \quad. \tag{10.32}$$

These sums are not especially easy to derive in other ways. They were obtained by Euler in 1737.

Problem 10.16. Set $x = l/4$ in (10.30) and derive a sum not reducible to those just found.

Problem 10.17. Show that $S_e = \frac{1}{4} S$.

Problem 10.18. Evaluate $\sum_{1}^{\infty} n^{-4}$.

Problem 10.19. What is the harmonic content of the sound produced when a string is held at the point $x = x_0$ and released from rest? What if $x_0 = l/n$ where n is an integer?

Problem 10.20. A piano string is set in motion by being struck, not plucked. Assuming that the hammer hits a piano string at $x = x_0$, find the harmonic content of the sound produced.

Problem 10.21. Show that the ninth harmonic of a given note is dissonant with the note. Where should the hammer strike the string in order not to excite it? Are the manufacturers of grand pianos aware of this fact?

Problem 10.22. An electron is miraculously introduced into a potential well with infinite sides with a wave function of the form of Fig. 9.4, properly normalized. What is the probability that if the electron's energy is subsequently measured it will turn out to be $(n\pi/l)^2 \hbar^2/2m$?

Problem 10.23. Use a computer to plot $u(x,t)$ for several values of t. You may be surprised at the result. (It was known to Euler in 1772.) Now use the identity $2 \sin A \cos B = \sin(A + B) + \sin(A - B)$ to sum the series (10.29). (You may want to find the sum for $\partial u/\partial x$ and then integrate to find u.) Show how this result could have been obtained directly from (10.10) without the detour through Fourier analysis. (This is how Euler did it.) Sketch your solution for $t = 0$ and show how the form of the string changes as time advances. Can you explain at least some of its features by physical arguments? Of what use is the Fourier analysis?

Problem 10.24*. The vibrations of a stiff wire satisfy

$$\varepsilon^2 y^{iv} - y'' = (\omega/c)^2 y$$

where the coefficient of the fourth derivative measures the stiffness. Assuming that a wire of length l is mounted rigidly at the ends, find an estimate for the frequency of its nth mode.

Answer: $\omega_n = (1 + 2\varepsilon/l)n\pi c/l + O(\varepsilon^2)$.

10.4 Practical Use of Variation Principles

I showed in Sect. 3.8 that Maupertuis's principle makes some computations easier in point mechanics, and in continuum mechanics one can do something similar. Consider for example the problem of finding the lowest vibrational mode of a string. (We know of course what it is, and that will be useful in assessing the approximation.) In the double-integral variational principle

$$\delta \iint \left(\frac{\varrho}{2}|\dot{u}|^2 - \frac{\tau}{2}|u'|^2 \right) dx\, dt = 0$$

let $u \propto e^{-i\omega t}$ with ω to be determined. The integration over t is now trivial, and we have

$$\delta \int \left(\frac{\varrho}{2}\omega^2|u|^2 - \frac{\tau}{2}|u'|^2 \right) dx = 0 \quad . \tag{10.33}$$

Suppose the string stretches from $x = -1$ to $x = +1$, and let $\lambda = \varrho\omega^2/\tau$ be the unknown quantity that gives ω. Then

$$\delta \int_{-1}^{1} (\lambda|u|^2 - |u'|^2)dx = 0 \tag{10.34}$$

with $\delta u = 0$ at the ends. The calculus of variations (Problem 10.27) leads to the equations we have already solved, but it will be instructive to pretend we don't know how.

To solve the problem approximately, invent a curve with a single antinode, satisfying the boundary conditions and containing free parameters. Calculate the action and then find what values of the parameters make it a minimum. Try for example

$$u = (1 - x^2)(c_0 + c_1 x^2) \quad . \tag{10.35}$$

This is zero at $x = \pm 1$ and is symmetrical right and left. The value of the integral in (10.34) is

$$S = (\tfrac{16}{15}\lambda - \tfrac{8}{3})c_0^2 = (\tfrac{32}{105}\lambda - \tfrac{16}{15})c_0 c_1 + (\tfrac{16}{315}\lambda - \tfrac{88}{105})c_1^2$$

and the equations $\partial S/\partial c_0 = 0$ and $\partial S/\partial c_1 = 0$ give

$$(2\lambda - 5)c_0 + (\tfrac{2}{7}\lambda - 1)c_1 = 0$$
$$(2\lambda - 7)c_0 + (\tfrac{2}{3}\lambda - 11)c_1 = 0 \quad . \tag{10.36}$$

In order for these to have any solution at all, their determinant must vanish, and this gives the eigenvalue condition

$$\lambda^2 - 28\lambda + 63 = 0 \tag{10.37}$$

whose smallest root (lowest frequency) is

$$\lambda = 14 - \sqrt{133} = 2.46744 \quad .$$

The exact value is

$$\lambda = \frac{\pi^2}{4} = 2.46740 \quad .$$

The fractional error is about 16×10^{-6}.

Notes: a) They don't always come out this well. Clearly the choice of a trial function was a happy one, so much so that the result is of interest in connection with algebraic approximations to π.

b) We have really only found one parameter of u, namely c_1/c_0 since normalization is undetermined, but the homogeneous approach with the extra unknown is generally simpler.

c) The great accuracy with which the eigenvalue has been determined is not matched by a corresponding accuracy in the eigenfunction. Substituting λ into (10.36) gives the approximation

$$u_{\text{var}} = c_0(1 - 1.2207x^2 + 0.2207x^4) \quad .$$

The exact solution is

$$u_{\text{exact}} = c_0 \cos \frac{\pi}{2} x = c_0(1 - 1.2337x^2 + 0.2537x^4 + \ldots) \quad .$$

At $x = \tfrac{1}{2}$, for example, with $c_0 = 1$,

$$u_{\text{var}}(\tfrac{1}{2}) = 0.7086 \quad , \quad u_{\text{exact}}(\tfrac{1}{2}) = 0.7071 \tag{10.38}$$

with a fractional error of about 2×10^{-3}. We shall see in Sect. 10.7 that the fractional error in a variational eigenfunction is of the order of the square root of that in the eigenvalue. In this case $\sqrt{16 \times 10^{-6}} = 4 \times 10^{-3}$.

Problem 10.25. The justification given for writing (10.34) was not a proof. Fill the gap by deriving the wave equation, writing down a solution that depends periodically on t, and showing that it satisfies (10.34).

Problem 10.26. The density of a string secured at $x = \pm 1$ is given by

$$\lambda(x) = \lambda_0 \cos^2 \frac{\pi x}{4} \quad (-1 \le x \le +1) \quad .$$

Estimate the frequency of the lowest normal mode.

Problem 10.27. Show that the solution of (10.34) by the calculus of variations leads to the results obtained earlier.

Problem 10.28. Suppose that instead of (10.35) the trial function is chosen to be simply $1 - x^2$, with no adjustable constant. Use the result of Problem 10.5 to make an estimate of λ. Is the estimate any good?

Problem 10.29. Try to improve the example by choosing a better form for $u(x)$.

Problem 10.30. Assuming a reasonable form for the lowest eigenfunction, carry out a variational calculation to find the fundamental frequency of the stiff wire in Problem 10.24.

10.5 More Than One Dimension

The equation for the vibration of a drumhead follows from the same argument as that used to derive the equation for the string. Let the drumhead's vertical displacement be $u(x, y, t)$. Then if we consider an element of area $dx\,dy$ to be drawn on it when $u = 0$, the x direction will be stretched by a factor $1 + (\partial u/\partial x)^2/2$ as in (10.4), and similarly for the y direction, so that the area will change to

$$\left[1 + \frac{1}{2} \left(\frac{\partial u}{\partial x} \right)^2 \right] \left[1 + \frac{1}{2} \left(\frac{\partial u}{\partial y} \right)^2 \right] dx\,dy$$

and the increase in area is thus

$$\Delta A = \frac{1}{2} \left[\left(\frac{\partial u}{\partial x} \right)^2 + \left(\frac{\partial u}{\partial y} \right)^2 \right] dx\,dy \quad .$$

If the drumhead is in isotropic tension the potential energy stored will be proportional to this quantity,

$$V = \frac{\mu}{2} \iint \left[\left(\frac{\partial u}{\partial x} \right)^2 + \left(\frac{\partial u}{\partial y} \right)^2 \right] dx\,dy = \frac{\mu}{2} \iint (\nabla u)^2 dx\,dy \quad . \tag{10.39}$$

Problem 10.31. Let the drumhead's area density be σ, and finish deriving its equation of motion

$$\frac{\partial^2 u}{\partial x^2} + \frac{\partial^2 u}{\partial y^2} - \frac{1}{v^2} \frac{\partial^2 u}{\partial t^2} = 0 \quad . \tag{10.40}$$

How does v^2 depend on μ and σ?

Changes of Variables. Most drumheads are round, and the boundary condition is therefore given most conveniently if u is considered as a function of r and ϑ, with

$$u(a, \vartheta) = 0 \tag{10.41}$$

where a is the radius of the drum. The variables in the Laplacian operator in (10.40) may be changed by direct calculation, or one may look up the necessary formula in tables, but it is also useful to know how to make the transformation in the variational integral. The components of ∇u are the spatial rates of change of u in two orthogonal directions, $\partial u/\partial x$ and $\partial u/\partial y$. Choose the directions of increasing r and ϑ as two new orthogonal directions:

$$(\nabla u)_r = \frac{\partial u}{\partial r} \quad , \quad (\nabla u)_\vartheta = \frac{\partial u}{r\partial\vartheta} = \frac{1}{r}\frac{\partial u}{\partial\vartheta} \quad .$$

Thus (10.39) becomes

$$V = \frac{\mu}{2} \iint \left[\left(\frac{\partial u}{\partial r}\right)^2 + \frac{1}{r^2}\left(\frac{\partial u}{\partial\vartheta}\right)^2 \right] r\, dr\, d\vartheta$$

and the action integral is

$$S = \iint \left\{ \frac{\sigma}{2}\left(\frac{\partial u}{\partial t}\right)^2 - \frac{\mu}{2}\left[\left(\frac{\partial u}{\partial r}\right)^2 + \frac{1}{r^2}\left(\frac{\partial u}{\partial\vartheta}\right)^2 \right] \right\} r\, dr\, d\vartheta\, dt \quad .$$

The general problem is of the form

$$\delta \iiint \mathcal{L}(u, u_r, u_\vartheta, u_t)\, r\, dr\, d\vartheta\, dt = 0 \quad .$$

where $u_r := \partial u/\partial r$, etc. Carrying out the variation gives

$$\iiint \left(r\frac{\partial\mathcal{L}}{\partial u}\delta u + r\frac{\partial\mathcal{L}}{\partial u_r}\delta u_r + r\frac{\partial\mathcal{L}}{\partial u_\vartheta}\delta u_\vartheta + r\frac{\partial\mathcal{L}}{\partial u_t}\delta u_t \right) dr\, d\vartheta\, dt \quad .$$

Integrate the last three terms by parts:

$$\iiint \left[r\frac{\partial\mathcal{L}}{\partial u} - \frac{\partial}{\partial r}\left(r\frac{\partial\mathcal{L}}{\partial u_r}\right) - \frac{\partial}{\partial\vartheta}\left(r\frac{\partial\mathcal{L}}{\partial u_\vartheta}\right) - \frac{\partial}{\partial t}\left(r\frac{\partial\mathcal{L}}{\partial u_t}\right) \right] \delta u\, dr\, d\vartheta\, dt$$

$$+ \iint \left[r\frac{\partial\mathcal{L}}{\partial u_r}\delta u \right]_{r=0}^{r=R} d\vartheta\, dt + \iint \left[r\frac{\partial\mathcal{L}}{\partial u_\vartheta}\delta u \right]_{\vartheta=0}^{\vartheta=2\pi} dr\, dt$$

$$+ \iint \left[r\frac{\partial\mathcal{L}}{\partial u_t}\delta u \right]_{t=0}^{t} dr\, d\vartheta = 0 \quad .$$

If u and u' are continuous at $r = 0$, and $\delta u = 0$ at $r = R$, where R may be infinite, constant, or a function of ϑ, the first term in the second line vanishes. (It may be necessary to think for a moment to see why this is so.) The second term vanishes because $\vartheta = 2\pi$ is the same line as $\vartheta = 0$, and the third term vanishes as usual. We have left

$$\frac{\delta(r\mathcal{L})}{\delta u} = 0 \tag{10.42}$$

with r, ϑ, and t as independent variables, as the Euler-Lagrange equation. In the present case,

$$\mathcal{L} = \frac{\sigma}{2}u_t^2 - \frac{\mu}{2}\left(u_r^2 + \frac{1}{r^2}u_\vartheta^2 \right) \tag{10.43}$$

and (10.42) gives

$$\sigma u_{tt} - \mu\left(u_{rr} + \frac{1}{r}u_r + \frac{1}{r^2}u_{\vartheta\vartheta} \right) = 0 \quad . \tag{10.44}$$

The expression in parentheses is the Laplacian operator in two dimensions, and the method is capable of immediate generalization to any kind of coordinate system in any number of dimensions. If the new coordinates are $q_1, q_2, \ldots q_r$, given as functions of the Cartesian $x_1, x_2, \ldots x_r$, find the Jacobian

$$J = \frac{\partial(x)}{\partial(q)}$$

so that the volume element becomes

$$dx_1 \ldots dx_r = J \, dq_1 \ldots dq_r \quad .$$

The action integral is

$$S = \int \mathcal{L}[u(q), \ldots] J \, dq_1 \ldots dq_r \, dt \tag{10.45}$$

and this gives at once the Euler-Lagrange equation for u when $J\mathcal{L}$ is taken as the Lagrangian. Of course, there are many other systems than the drumhead considered here that are described by a Laplacian operator; they include continuous mechanical systems as well as fields described by Maxwell's and Schrödinger's equations. We shall discuss the latter separately below.

Wave Equation in Polar Coordinates. The wave equation in the surface is

$$u_{rr} + \frac{1}{r} u_r + \frac{1}{r^2} u_{\vartheta\vartheta} - \frac{1}{v^2} u_{tt} = 0 \quad , \quad v^2 := \mu/\sigma \quad . \tag{10.46}$$

Since an arbitrary waveform can be formed of normal-mode vibrations by Fourier superposition, it is sufficient to consider vibrations at angular frequency ω, for which

$$u_{rr} + \frac{1}{r} u_r + \frac{1}{r^2} u_{\vartheta\vartheta} + \lambda^2 u = 0 \quad , \quad \lambda^2 := \omega^2/v^2 \quad . \tag{10.47}$$

(this is sometimes called Helmholtz's equation.) Polar coordinates are one of the 11 varieties for which the wave equation is separable (Sect. 6.7). If we set

$$u(r, \vartheta) = R(r)\Theta(\vartheta)$$

then (10.47) becomes

$$\left(R'' + \frac{1}{r} R' + \lambda^2 R \right) \Theta + \frac{1}{r^2} R\Theta'' = 0 \quad \text{or}$$

$$\frac{r^2}{R} \left(R'' + \frac{1}{r} R' + \lambda^2 R \right) = -\frac{\Theta''}{\Theta} \quad .$$

Since the left side is independent of ϑ and the right side is independent of r, both must be constant.

Set the constant equal to m^2. The ϑ equation is

$$\Theta'' = -m^2 \Theta \quad \text{whence}$$

$$\Theta = A_m \sin m\vartheta + B_m \cos m\vartheta \tag{10.48}$$

and continuity on traversing a complete circle requires m to be real and an integer.

The r equation is now

$$R'' + \frac{1}{r}R' + \left(\lambda^2 - \frac{m^2}{r^2}\right)R = 0 \quad . \tag{10.49}$$

This is Bessel's equation [122]. It has two classes of solutions, roughly analogous to the sine and cosine solutions of (10.11),

$$R = C_m J_m(\lambda r) + D_m Y_m(\lambda r) \quad . \tag{10.50}$$

The functions Y_m, called Bessel functions of the second kind, are all infinite at $r = 0$ and so have no place here. The Bessel functions of the first kind, J_m, were (predictably) first studied by Euler, in 1764, which was 60 years before Bessel's memoir. They are represented by the series

$$J_m(x) = \frac{(x/2)^m}{m}\left[1 - \frac{(x/2)^2}{1(m+1)} + \frac{(x/2)^4}{1 \cdot 2(m+1)(m+2)} - \cdots\right] \tag{10.51}$$

and for $x \gg m$ are given asymptotically by

$$J_m(x) \sim \sqrt{\frac{2}{\pi x}}\cos\left(x - \frac{\pi}{4} - \frac{m\pi}{2}\right) \quad . \tag{10.52}$$

Figure 10.5 shows the first few J_m's as a function of x. They are extensively tabulated [133] and are available in computer subroutines.

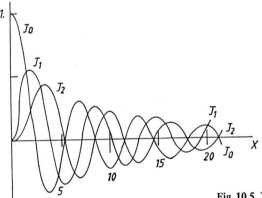

Fig. 10.5. Three Bessel functions $J_m(x)$

The solution we have found is of the form

$$u(r, \vartheta) = J_m(\lambda r)(A_m \sin m\vartheta + B_m \cos m\vartheta)$$

and the complete solution multiplies this by a sinusoidal function of t. Since drums are usually struck into motion, we shall look at solutions $u(r, \vartheta, t)$ that satisfy

$$u(r, \vartheta, 0) = 0 \quad , \quad \dot{u}(r, \vartheta, 0) = v_0(r, \vartheta) \tag{10.53}$$

so that

$$u(r, \vartheta, t) = \sum_{m=0}^{\infty} J_m(\lambda r)(A_m \sin m\vartheta + B_m \cos m\vartheta) \sin \omega_m t \quad . \tag{10.54}$$

First, we must find the frequencies ω_m. They are determined by the requirement that $J_m(\lambda a) = 0$ (a is the radius of the drumhead). For each m, $J_m(x)$ has an infinite number of zeros; the first few are listed in Table 10.1. Let x_{mn} be the nth zero of $J_m(x)$. Then $J_m(\lambda a) = 0$ implies that $\lambda a = x_{mn}$ or, by (10.47)

$$\omega_{mn} = \frac{v}{a} x_{mn} \quad . \tag{10.55}$$

These are the frequencies present in the sound of a drum. They are not in the ratios of small integers (or any integers) and therefore are not in any harmonic series.

Table 10.1. Zeros of $J_m(x)$

m	first	second	third	fourth
0	2.40	5.52	8.65	11.79 (11.78)
1	3.83	7.02	10.17	13.32 (13.35)
2	5.14	8.42	11.62	14.80 (14.92)
3	6.38	8.76	13.02	16.22 (16.49)

Numbers in parentheses are calculated from (10.52).

To satisfy the initial condition write

$$\dot{u}(r, \vartheta, 0) = \sum_{m,n} \omega_{mn} J_m(\lambda_{mn} a)(A_{mn} \sin m\vartheta + B_{mn} \cos m\vartheta) = v_0(r, \vartheta) \tag{10.56}$$

where the extra indices n number the zeros of each J_m. For the stretched string we solved the corresponding equations (10.22) by Fourier inversion. The same is possible here because the Helmholtz equation (10.47) has the form of an eigenvalue equation with eigenvalues λ_{mn}. The eigenvalues are all real and different, and the differential operator is hermitian; therefore the eigenfunctions are all orthogonal. A more advanced investigation shows that they form a complete set. It follows that the A_{mn} and B_{mn} can be determined from a knowledge of $v(r, \vartheta)$ and the problem, from the formal point of view, is solved.

Problem 10.32. A round drum is hit in the exact center. What frequencies are heard? Give numerical values for the first few and an approximate formula for the rest. Roughly how many frequencies will be heard? You may wish to experiment beating a drum with implements of different sizes and degrees of hardness.

Problem 10.33. What frequencies are present when a square drum is beaten? Are they harmonically related? How would a square drum sound compared with a round one?

Problem 10.34. Estimate the two lowest zeros of $J_0(x)$ by a variational solution of Bessel's equation. The correct values are 2.40483 and 5.52008.

10.6 Waves in Space

This section discusses sound waves as an example of waves in general. Not all waves obey the same equation of motion – electromagnetic waves, for example, have polarization characteristics that are described by making the wave function a vector, but the mathematics to be sketched here is encountered in aerodynamics, electromagnetic theory, and quantum mechanics.

Start with the equation derived in Sect. 4.6 that describes the propagation of sound waves,

$$\nabla^2 p - \frac{1}{v^2}\frac{\partial^2 p}{\partial t^2} = 0 \quad , \qquad v^2 = \frac{\partial p}{\partial \varrho} \quad . \tag{10.57a}$$

This is d'Alembert's equation for the pressure, and the density ϱ satisfies the same equation.

To find $\partial p/\partial \varrho$ requires some hypothesis about the gas. Newton assumed it obeys Boyle's law and derived $\partial p/\partial \varrho = p_0/\varrho_0$ where the subscript denotes the background value. This assumes that the temperature remains constant as the sound wave goes through. But for waves in the ordinary frequency range the variations are far too rapid for this and we do better to assume that the compressions and rarefactions take place adiabatically. The adiabatic law is $pV^\gamma = \text{const}$ or

$$p = K\varrho^\gamma$$

where γ is the ratio of the specific heat at constant pressure to that at constant volume, equal to 1.403 for air at $0°\,\text{C}$ and one atmosphere. Thus the computed speed of sound is

$$v = \sqrt{\gamma p_0/\varrho_0} \quad . \tag{10.57b}$$

At a pressure of 76 cm of mercury at $0°\,\text{C}$

$$p_0 = 1.0132 \times 10^5\,\text{nm}^{-2} \quad , \qquad \varrho_0 = 1.2929\,\text{kg\,m}^{-3}$$

and

$$v = 331.6\,\text{ms}^{-1}$$

in agreement with observation.

The Lagrangian density for sound waves is

$$\mathcal{L} = \frac{1}{2}\left[(\nabla p)^2 - \frac{1}{v^2}(\partial p/\partial t)^2\right] \tag{10.58}$$

since one easily verifies (Problem 10.37) that it leads to the d'Alembert equation (10.57a).

Solution of the Wave Equation. This is a long and complex topic, for there are many kinds of solutions according to the boundary conditions imposed, and for a general discussion we refer to other books [122, 134]. Here I will only sketch a few lines to show the relation to the earlier treatment of the drumhead on the one hand and to a basic knowledge of quantum mechanics on the other.

The variables can be separated in each of the 11 Stäckel systems (Sect. 6.10), and each one leads to appropriate sets of functions. Cartesian coordinates are easy, but there are few rectangular boxes in nature. Here I will discuss only spherical polar coordinates. The wave equation in these coordinates can be derived by changing variables in the Lagrangian integral (Sect. 10.5), and this yields

$$\frac{\partial^2 p}{\partial r^2} + \frac{2}{r}\frac{\partial p}{\partial r} + \frac{1}{r^2 \sin \vartheta}\frac{\partial}{\partial \vartheta}\left(\sin \vartheta \frac{\partial p}{\partial \vartheta}\right) + \frac{1}{r^2 \sin^2 \vartheta}\frac{\partial^2 p}{\partial \phi^2} - \frac{1}{v^2}\frac{\partial^2 p}{\partial t^2} = 0 \quad (10.59)$$

We consider only a single frequency ω, so that the last term becomes $(\omega/v)^2 p$.

Begin by separating the radial variable from the angular ones:

$$p(r, \vartheta, \phi) = R(r)Y(\vartheta, \phi) \quad \text{where}$$

$$r^2\left(R'' + \frac{2}{r}R' + \lambda^2 R\right)Y$$

$$+ \left[\frac{1}{\sin \vartheta}\frac{\partial}{\partial \vartheta}\left(\sin \vartheta \frac{\partial Y}{\partial \vartheta}\right) + \frac{1}{\sin^2 \vartheta}\frac{\partial^2 Y}{\partial \phi^2}\right]R = 0 \quad , \qquad \lambda = \frac{\omega}{v} \quad .$$

If we divide by RY the variable r is separated out, and introducing a constant of separation K gives

$$R'' + \frac{2}{r}R' + \left(\lambda^2 - \frac{K}{r^2}\right)R = 0 \quad \text{and} \tag{10.60}$$

$$\frac{1}{\sin \vartheta}\frac{\partial}{\partial \vartheta}\left(\sin \vartheta \frac{\partial Y}{\partial \vartheta}\right) + \frac{1}{\sin^2 \vartheta}\frac{\partial^2 Y}{\partial \phi^2} + KY = 0 \quad . \tag{10.61}$$

If the quantum-mechanical operator for the square of the orbital angular momentum (omitting the $\hbar$) is transformed into polar coordinates (see any book on quantum mechanics), the result is

$$\hat{L}^2 = -\frac{1}{\sin \vartheta}\frac{\partial}{\partial \vartheta}\sin \vartheta \frac{\partial}{\partial \vartheta} - \frac{1}{\sin^2 \vartheta}\frac{\partial^2}{\partial \phi^2}$$

so that (10.61) is the eigenvalue equation

$$\hat{L}^2 Y(\vartheta, \phi) = KY(\vartheta, \phi) \quad . \tag{10.62}$$

We know that the eigenvalues K are

$$K = l(l+1) \quad l = 0, 1, 2, \ldots \tag{10.63}$$

and it is part of the standard lore of quantum mechanics that the functions $Y(\vartheta, \Phi)$ are the spherical harmonics

$$Y_l^m(\vartheta, \phi) = P_l^m(\cos \vartheta)e^{im\phi} \tag{10.64}$$

where $m = -l, \ldots +l$ and the P_l^m are the associated Legendre functions. There is a second set of solutions of (10.61) in which P_l^m is replaced by the functions of the second kind, Q_l^m. These are almost never of use in quantum mechanics because they become infinite at $\vartheta = 0$. They are however essential to many calculations in acoustics and electromagnetic theory, notably in the design of loudspeakers and antennas.

With (10.63) we return to the radial wave equation (10.60). It is almost but not quite of the form of Bessel's equation, but if we write

$$R(r) = \frac{z(r)}{\sqrt{\lambda r}}$$

it changes to

$$z'' + \frac{1}{r}z' + \left[\lambda^2 - \frac{(l+1/2)^2}{r^2}\right]z = 0 \quad . \tag{10.65}$$

This is Bessel's equation of order $l + \frac{1}{2}$, and it has two solutions, $J_{l+1/2}(\lambda r)$, which is finite at the origin, and $Y_{l+1/2}(\lambda r)$, which is not. The radial function R is therefore of the form

$$R(r) = j_l(\lambda r) \quad \text{or} \quad y_l(\lambda r) \quad \text{where} \tag{10.66}$$

$$j_l(\lambda r) := \frac{J_{l+1/2}(\lambda r)}{\sqrt{\lambda r}} \quad , \quad y_l(\lambda r) := \frac{Y_{l+1/2}(\lambda r)}{\sqrt{\lambda r}} \tag{10.67}$$

are called spherical Bessel and Neumann functions. They are all expressible in terms of sines and cosines:

$$j_0(x) = \frac{\sin x}{x} \qquad\qquad y_0(x) = -\frac{\cos x}{x}$$

$$j_1(x) = \frac{\sin x}{x^2} - \frac{\cos x}{x} \qquad y_1(x) = -\frac{\cos x}{x^2} - \frac{\sin x}{x} \quad . \tag{10.68}$$

These functions are appropriate to the solution of standing-wave problems; incoming and outgoing waves are formed from linear combinations such as

$$j_0(x) \pm iy_0(x) = \mp\frac{i}{x}e^{\pm ix} \quad \text{etc.} \tag{10.69}$$

which are called Hankel functions.

These few remarks are enough to show why most of the mathematical apparatus of the quantum theory of angular momentum lay ready to hand when the theory was first invented. The angular part of the solutions of the wave equation in polar coordinates is standard, and only the radial equations, when quantum mechanics came along, were somewhat unfamiliar.

Problem 10.35. Derive the Laplacian operator for spherical coordinates by changing variables in the variational integral.

Problem 10.36. It is possible to compute the average molecular velocity in a gas from its bulk properties by kinetic theory (Sect. 4.5), and the sound velocity from (10.57b). How do they compare?

Problem 10.37. Show that (10.58) is the Lagrangian density for sound waves and find the corresponding Hamiltonian density. Find the numerical factor that makes this an energy density.

Problem 10.38. Starting from the general formula for energy density derived in the previous problem, find a general expression for the acoustic energy transmitted per second through a unit area. Specialize it to a plane wave and show that the result is reasonable. (If you do not know how to do this, look at the analogous derivation of the current density in elementary quantum mechanics.)

Problem 10.39. Assume that when sound waves reach a solid surface the air molecules next to the surface do not move (this is not quite true), and formulate the appropriate wave equation and boundary conditions. Using these, discuss the motion of sound waves down a pipe of rectangular cross section $a \times b$ and find the lowest frequency that will propagate. Using the result of Problem 10.38, evaluate the sound energy passing down the pipe. What happens if the sound frequency presented to the pipe is lower than this limit? (The theory of electromagnetic waveguides is similar to this except that the boundary conditions require a little more thought.)

Problem 10.40. Solve the preceding problem for a pipe of circular cross section, radius a.

10.7 The Matter Field

In his first paper on wave mechanics [135] *Schrödinger* made a strange proposal. Recognizing that the Hamilton-Jacobi action is something like a phase, yet obviously trying not to introduce complex numbers into physical quantities, he proposed to write W in the form

$$W = K \ln \psi \tag{10.70}$$

where K is a constant to be determined. In terms of ψ, the Hamilton-Jacobi equation is

$$H\left(x_i, \frac{K}{\psi} \frac{\partial \psi}{\partial x_i}\right) = E \quad . \tag{10.71}$$

In Cartesian coordinates,

$$(\nabla \psi)^2 - \frac{2m}{K^2}[E - V(r)]\psi^2 = 0 \quad . \tag{10.72}$$

But this is only a transcription of classical physics; if one goes ahead and solves it one gets nothing new. Schrödinger proposed instead a physical principle: ψ is to be chosen so as to make the volume integral of the quantity on the left a minimum and not necessarily zero. We have therefore a variational problem of the form

$$\delta \int \left\{ (\nabla \psi)^2 - \frac{2m}{K^2}[E - V(r)]\psi^2 \right\} dv = 0 \tag{10.73}$$

which yields the differential equation

$$\nabla^2 \psi + \frac{2m}{K^2}[E - V(r)]\psi = 0 \quad . \tag{10.74}$$

The discrete eigenvalues of this equation, as everybody knows, coincide with the Balmer levels when $V(r) = -e^2/r$ and $K = \hbar$.

In a note added in proof, Schrödinger suggests a better formulation: Write the integral to be varied as

$$\int \left[K^2 T\left(\boldsymbol{r}, \frac{\partial \psi}{\partial \boldsymbol{r}} \right) + V(\boldsymbol{r})\psi^2 \right] dv \tag{10.75}$$

where T is the classical expression for kinetic energy, subject to the condition that ψ remain normalized to unity,

$$\int \psi^2 dv = 1 \quad . \tag{10.76}$$

This is an equivalent, but neater formulation in terms of an isoperimetric problem.

Problem 10.41. The strange form of (10.70) reflects Schrödinger's reluctance to introduce complex numbers into mechanics. Starting instead with $\psi = \exp(iW/K)$, show how he could have phrased his argument so as to introduce a complex ψ.

Schrödinger's Variational Principle. In Schrödinger's revised version, the integral to be varied is

$$W' = \int \left[\frac{\hbar^2}{2m}(\nabla\psi)^2 + V(\boldsymbol{r})\psi^2 + \lambda\psi^2 \right] dv \tag{10.77}$$

where λ is the Lagrange multiplier. This is of the same form as (10.73) and so of course the results are the same, but it now becomes a little clearer that the desired functions ψ are those for which the integral (10.76) exists at all – since if it exists it can be set equal to 1. That is, a ψ corresponding to a stationary state must be normalizable.

We have seen in Sect. 7.7 how Schrödinger's second paper explored the relation between (10.74) and the Hamilton-Jacobi equation, and introduced what is now called the time-dependent Schrödinger equation with which we began the study of mechanics in Chap. 1. His variational principle for (10.77) was a new discovery, similar in form to that given in (10.58) for a typical spatial wave phenomenon, but having no logical connection with it, since Schrödinger's ψ does not refer to any mechanical property of the system represented. The variational principle does, however, reflect a very general property of eigenvalue problems as is easily seen.

Suppose $\hat{K}$ represents any hermitian operator and f is an eigenfunction belonging to the eigenvalue κ:

$$\hat{K}f = \kappa f \quad . \tag{10.78}$$

Now write κ as

$$\kappa = \frac{\int f^* \hat{K} f \, dq}{\int f^* f \, dq} \tag{10.79}$$

and imagine that f is varied slightly from the exact eigenfunction. The quantity κ given by (10.79) is no longer an eigenvalue, but it will have a value nearby. Varying f and κ in the equation

$$\kappa \int f^* f \, dq = \int f^* \hat{K} f \, dq$$

gives

$$\delta\kappa \int f^* f \, dq + \kappa \int (\delta f^* f + f^* \delta f) dq = \int \delta f^* \hat{K} f \, dq + \int f^* \hat{K} \delta f \, dq$$

and by the use of (10.78) and the hermiticity of $\hat{K}$ this reduces to

$$\delta\kappa \int f^* f \, dq = 0 \quad \text{or} \quad \delta\kappa = 0 \quad . \tag{10.80}$$

Thus if δf is of the order of some small parameter ε, $\delta\kappa$ is of the order of ε^2 or smaller; this fact has already been noted in connection with (10.34).

We can also prove very easily that calculating a ground-state energy by variational means always gives too high a value. Let ψ_n be the exact (but unknown) eigenfunctions of the Hamiltonian $\hat{H}$, and let some varied function ψ be expanded in terms of them:

$$\psi = \sum_{n=0}^{\infty} c_n \psi_n \tag{10.81}$$

where if ψ and the ψ_n are normalized,

$$\sum_{n=0}^{\infty} |c_n|^2 = 1 \quad . \tag{10.82}$$

The expectation value of $\hat{H}$ in the state ψ is

$$\langle H \rangle = \sum_{m,n} \int c_m^* \psi_m^* H c_n \psi_n dq = \sum_{n=0}^{\infty} |c_n|^2 E_n \quad . \tag{10.83}$$

If E_0 is the ground-state energy then

$$\langle H \rangle = \sum |c_n|^2 E_n \geq \sum |c_n|^2 E_0 \quad \text{or}$$

$$\langle H \rangle \geq E_0 \tag{10.84}$$

so that E_0 is a lower bound on $\langle H \rangle$.

Problem 10.42. Knowing the form of the wave function for the ground state of a simple harmonic oscillator, make at least two variational approximations to that of an anharmonic oscillator with $V(x) = \kappa x^4/4$. Which approximation is the best?

Example 1. The Helium Atom. Examples of the use of Schrödinger's variational principle for finding stationary states are given in every text on quantum mechanics, but it may be useful to outline one such calculation, omitting computational details which at first sight obscure the simplicity of what is being done. The system is a quantum-mechanical analogue of that discussed in Sect. 7.6.

The problem is to estimate the ground-state energy and wave function of a helium atom. It is a three-body problem and therefore completely intractable in classical mechanics because the details of motion are so complicated. But in quantum mechanics we do not follow the details; the indeterminacy principle deprives them of meaning. Instead, we treat it as a system with a very simple behavior.

The idea is to assume that the wave function for each electron in the ground state is not very different from that of an electron in a hydrogen-like atom of suitable but unknown nuclear charge. Let the charge be ζe. We shall solve the problem by varying the parameter ζ. The normalized wave function for two electrons in the ground state is

$$\psi(r_1, r_2) = \frac{\alpha^3}{\pi} e^{-\alpha(r_1+r_2)} \quad , \qquad \alpha := \zeta \frac{me^2}{\hbar^2} \quad . \tag{10.85}$$

The complete Hamiltonian for the two-electron system is

$$\hat{H} = -\frac{\hbar^2}{2m}(\nabla_1^2 + \nabla_2^2) - \frac{Ze^2}{r_1} - \frac{Ze^2}{r_2} + \frac{e^2}{|r_1 - r_2|} \tag{10.86}$$

where Z is the nuclear charge, and the expectation value of this operator in the assumed state is found to be

$$U = 2\left[\zeta^2 - 2\left(Z - \frac{5}{16}\right)\zeta\right]\frac{me^4}{2\hbar^2} \tag{10.87}$$

where $me^4/2\hbar^2 = 1\,\text{Rydberg} = 13.606$ eV. U goes through a minimum when ζ is chosen equal to $Z - 5/16$; we can explain this picturesquely by saying that each electron is shielded from the nucleus by the other electron 5/16 of the time. The energy is then

$$E = U_{\min} = -2(Z - \tfrac{5}{16})^2\,\text{Ry} \quad . \tag{10.88}$$

For helium this is $-5.695\,\text{Ry} = -77.49$ eV. The experimental value is -78.98 eV; the error is 1.9 %. This is perhaps not impressive by comparison with what we could achieve in simple vibration problems, but comparing it instead with the three-body problem of orbital motion, one begins to appreciate the simplifications brought by quantum mechanics.

Problem 10.43. Carry out the integrations necessary to derive (10.88).

Problem 10.44. Show that the Lagrangian density of the complex matter field is

$$\mathcal{L} = \frac{i\hbar}{2}(\dot{\psi}^*\psi - \psi^*\dot{\psi}) + \frac{\hbar^2}{2m}|\nabla\psi|^2 + V(r)|\psi|^2 \quad . \tag{10.89}$$

Problem 10.45. Show by calculus of variations that the condition $\delta\langle\hat{H}\rangle = 0$, with ψ normalizable, leads to (10.74).

10.8 Quantized Fields

The quantum analogue of the classical fields discussed up to this point is quite different conceptually, for the field amplitudes are no longer numerical quantities. To allow for the creation and annihilation of field quanta (phonons, photons, etc.) we must allow energy and particle number to jump discontinuously, and the dynamics of the field must involve operators for this purpose.

As an introduction, consider the field describing a noninteracting neutral spinless particle of mass m. These could be neutral pions or, if $m = 0$, pressure waves in a gas, liquid, or solid; the quanta would then be phonons. The field quantity describing such a particle is a real scalar that will be called $u(x, y, z, t)$. In calculations of this kind one usually disposes of unnecessary letters by setting $\hbar = c = 1$, but here everything will be kept. The Lagrangian density for such a field can be written

$$\mathcal{L} = \frac{1}{2}\left[\left(\frac{\partial u}{\partial t}\right)^2 - c^2(\nabla u)^2 - \left(\frac{mc^2}{\hbar}\right)^2 u^2\right] \quad . \tag{10.90}$$

The canonical momentum corresponding to u is

$$\pi = \frac{\partial u}{\partial t} \quad . \tag{10.91}$$

It then follows from variation of $\int \mathcal{L} d^4 x$ (Problem 10.46) that u satisfies

$$\nabla^2 u - \frac{1}{c^2}\frac{\partial^2 u}{\partial t^2} - \left(\frac{mc}{\hbar}\right)^2 u = 0 \quad . \tag{10.92}$$

If $u = \cos(\boldsymbol{k} \cdot \boldsymbol{x} - \omega_{\boldsymbol{k}} t)$, for example, the equation is satisfied provided that

$$(\hbar\omega_{\boldsymbol{k}})^2 = (\hbar c k)^2 + (mc^2)^2$$

and this, with the usual equivalence between dynamical variables and wave quantities, implies the familiar relation

$$E^2 = c^2 p^2 + (mc^2)^2 \quad .$$

There is no way to derive a quantized theory from a classical one; the most one can do is to make the transition plausible. The following remarks are intended for that purpose.

Start with a classical harmonic oscillator defined by

$$H = \frac{1}{2}m(p^2/m^2 + \omega^2 x^2) \quad . \tag{10.93}$$

An arbitrary solution has the form

$$x(t) = x_0 \cos \omega t + \frac{p_0}{m\omega} \sin \omega t$$

and in terms of these constants

$$H = \frac{1}{2}m\left(\frac{p_0^2}{m^2} + \omega^2 x_0^2\right) = \frac{1}{2}m\omega^2\left(x_0 + \frac{ip_0}{m\omega}\right)\left(x_0 - \frac{ip_0}{m\omega}\right) \quad .$$

(Note that H is automatically a constant.) In the Schrödinger representation the dynamical variables are replaced by time-independent operators, and it seems most prudent to use these to replace the constant values p_0 and x_0. Thus H becomes

$$\hat{H} = \frac{1}{2}m\left(\frac{\hat{p}^2}{m^2} + \omega^2 \hat{x}^2\right) \quad , \quad [\hat{p}, \hat{x}] = -i\hbar \tag{10.94}$$

where the commutation relations come from the Poisson bracket. Now proceed as in Sect. 2.6. Noting how H factors, define a pair of operators

$$\hat{a} := \sqrt{\frac{m\omega}{2\hbar}}\left(\hat{x} + \frac{i\hat{p}}{m\omega}\right) \quad , \quad \hat{a}^\dagger := \sqrt{\frac{m\omega}{2\hbar}}\left(\hat{x} - \frac{i\hat{p}}{m\omega}\right) \tag{10.95}$$

which satisfy

$$[\hat{a}, \hat{a}^{\dagger}] = 1 \quad . \tag{10.96}$$

The factorization is then replaced by

$$\hat{H} = \tfrac{1}{2}\hbar\omega(\hat{a}\hat{a}^{\dagger} + \hat{a}^{\dagger}\hat{a}) = \hbar\omega(\hat{a}^{\dagger}\hat{a} + \tfrac{1}{2}) \tag{10.97}$$

and $\hat{a}^{\dagger}$ and $\hat{a}$ take their place in the theory as shift operators that raise and lower the energy eigenvalue by one step. A simple argument shows that the eigenvalues of $\hat{a}^{\dagger}\hat{a}$ are 0, 1, 2, ... The procedure works for the oscillator; let us try to do the same thing for the field.

It is usually most convenient to represent a field by its normal coordinates. Let the field be confined in a cubical box of volume V with periodic boundary conditions. Earlier, in treating a stretched string, we assumed that the displacement vanishes at the boundary. This corresponds to the physical situation but introduces the complication of reflected waves: a wave toward the right is necessarily accompanied by a wave toward the left. With the periodic boundary condition one imagines that the string is a loop in which a wave travels without reflection. The boundary condition is $u(l) = u(0)$, where l is the length of the loop. In three dimensions the situation is harder to visualize but the boundary condition in each dimension is the same. If this is done for the x dimension only, the set of functions

$$1 \quad , \quad \sin\frac{2\pi n x}{l} \quad , \quad \cos\frac{2\pi n x}{l} \quad , \quad n \in \mathbf{Z}_{+}$$

is a basis in which to describe u. In three dimensions, represent u as

$$u = \sum_{k}\left[u_{0k}\cos\left(k\cdot x - \omega_{k}t\right) - \frac{\pi_{0k}}{\omega_{k}}\sin\left(k\cdot x - \omega_{k}t\right)\right]$$

where $\sum_{k}$ means that each component of each k is represented as $n\pi/l$ and the summation is over all the n's. The initial values are taken at the spacetime origin, an arbitrary point.

Consider for a moment a single Fourier component,

$$u_{k} = u_{0k}\cos\left(k\cdot x - \omega_{k}t\right) - \frac{\pi_{0k}}{\omega_{k}}\sin\left(k\cdot x - \omega_{k}t\right) \quad . \tag{10.98}$$

Form the energy density from (10.90), express it in terms of this u_{k}, sum over k, and integrate over the spatial coordinates. The result is

$$H = \frac{1}{2}\sum_{k}(\pi_{0k}^{2} + \omega_{k}^{2}u_{0k}^{2})V \tag{10.99}$$

and so the field is represented in its normal coordinates as a system of independent oscillators. The factor V arises because the Lagrangian is given as a density: the larger the box the more energy it contains. Henceforth set $V = 1$.

The transition to quantum mechanics is made as before. Replace the numerical quantities π_{0} and u_{0} by operators $\hat{\pi}$ and $\hat{u}$ that satisfy

$$[\hat{\pi}_{k'}, \hat{u}_{k'}] = -i\hbar\delta_{kk'} \quad , \quad \text{others zero} \quad . \tag{10.100}$$

The shift operators are

$$\hat{a}_k := \sqrt{\frac{\omega_k}{2\hbar}}\left(\hat{u}_k + \frac{i\hat{\pi}_k}{\omega_k}\right) \quad , \quad \hat{a}_k^\dagger =: \sqrt{\frac{\omega_k}{2\hbar}}\left(\hat{u}_k - \frac{i\hat{\pi}_k}{\omega_k}\right)$$

and they satisfy $[\hat{a}_{k'}, \hat{a}_{k'}^\dagger] = \delta_{kk'}$. In terms of these operators,

$$\hat{H} = \sum_k \hbar\omega_k\left(\hat{N}_k + \frac{1}{2}\right) \quad , \quad \hat{N}_k := \hat{a}_k^\dagger \hat{a}_k \tag{10.101}$$

and one finds in the usual way that the eigenvalues of $\hat{N}_k$ are 0, 1, 2, ... Thus the energy of the entire field is found by summing the energies of the quanta associated with each of its normal modes. The extra $\frac{1}{2}$ is an awkward point, for if the number of degrees of freedom of the field is infinite or very large it adds an unacceptable constant. But except perhaps for gravitational effects nothing in physics depends on where the zero of energy is chosen. The extra energy arises in the transition from classical to quantum description and though one feels uncomfortable, one eliminates it with a stroke of the pen. That being done, the energy of each degree of freedom of the field consists of an integral number of quanta $\hbar\omega_k$ and the shift operators raise or lower this number. Any interaction that creates or destroys quanta can be written in terms of these operators. Note that the field's zero-point motion is not an error in the theory. In matter and in fields, the effects of this motion are both calculable and observable. It is only the extra energy that is by nature unobservable.

Problem 10.46. Show that the equation of motion (10.92) follows from variation of the Lagrangian (10.90).

Indeterminacy Relations. It follows from (10.100) that $\Delta\pi_k\Delta u_k \geq \frac{1}{2}$ and since $\pi_k = \dot{u}_k$ it follows that a field is never at rest. If the field is electromagnetic, for example, the situation in a perfectly dark room is not the same as if there were no such thing as an electromagnetic field. As a single example of this, in thinking about radiative processes one distinguishes between spontaneous and induced emission and treats them separately. When account is taken of field fluctuations it turns out that the radiation called spontaneous is just that induced by the zero-point fluctuations of the field even when no real quanta are present.

How is it possible for people to perform accurate electrodynamic calculations, in antenna theory for example, and take no account of the quantized nature of the field? The conventional answer is that such huge numbers of quanta are involved that their discreteness is lost sight of. There are, however, deeper reasons. At the end of Sect. 7.7 I pointed out that classical dynamics knows nothing of the real dynamics of physical change and replaces it with a sequence of static situations. Exactly the same happens with fields. In concentrating on the field's amplitude one sees nothing of its real dynamics. Consider for example the commutator $[\hat{N}_k, \hat{u}_k]$, where $\hat{u}_k$ is a single Fourier component of the field. This is easily (Problem 10.48) found to be

$$[\hat{N}_k, \hat{u}_k] = -\frac{i}{\omega}\frac{\partial \hat{u}_k}{\partial t}$$

from which follows the indeterminacy relation

$$\Delta n_{\pmb{k}} \Delta u_{\pmb{k}} \geq \frac{1}{2} \left| -\frac{i}{\omega} \frac{\partial}{\partial t} \langle u_{\pmb{k}} \rangle \right| \quad .$$

Suppose now that $\langle u_{\pmb{k}} \rangle = A \sin \phi_{\pmb{k}}$, with $\phi_{\pmb{k}} = \pmb{k} \cdot \pmb{x} - \omega t$. Then

$$\Delta n_{\pmb{k}} \Delta \sin \phi_{\pmb{k}} \geq \tfrac{1}{2} | \cos \phi_{\pmb{k}} | \tag{10.102}$$

and it shows that if one assumes exact knowledge of the phase one can say nothing about the number of quanta and vice versa. In particular, naive remarks about quanta moving along in the manner of a wave are inconsistent with quantum mechanics.[1]

Equation (10.102) clarifies a number of questions in quantum mechanics. An old puzzle refers to the production of interference fringes by light passing through two slits. Put a detector near one of the slits to learn which slit a quantum goes through. Then if it goes through one slit it does not go through the other and so how can there be a two-slit pattern? The counter determines n, with more or less accuracy depending on its operation. If it prepares a state for which Δn is small, then the uncertainty in phase can be large enough to destroy the interference pattern. But one can still talk about the self-interference of quanta. If inspection of a photographic plate reveals that n quanta have passed through the apparatus and been registered, we can mentally assign variables n_1 and n_2 to the two slits as long as we claim to know only their sum. Since one can easily show (Problem 10.49) that

$$\Delta(n_1 + n_2)\Delta \sin (\phi_1 - \phi_2) \geq 0 \tag{10.103}$$

it follows that counting the total number of quanta need not destroy the pattern.

Problem 10.47. Write out the steps leading to (10.99).

Problem 10.48. Derive (10.102).

Problem 10.49. Derive (10.103).

10.9 The Mössbauer Effect

Rather than proceed with the discussion of quantized fields I close with a single example that incorporates the theory of coherent states explained in Sect. 2.6. A useful procedure for investigating the stationary states of nuclei is resonance absorption. A source and a target, two identical samples of material containing a radioactive nucleus and its decay product, are placed side by side. A radioactive process in the source produces an excited nucleus which in turn decays emitting a γ-ray that travels to the target, where it induces excitation in a target nucleus. One would at first expect this experiment to be difficult and require special measures because of energy lost to recoil. Nuclear levels tend to be extremely narrow and so the incident γ must have exactly the energy needed to excite the desired state. Both in emission and absorption, however, the momentum of the γ causes a recoil that subtracts from its energy (Problem 10.50), and so resonance would be missed. But this does not

---

[1] For a detailed discussion of phase as a variable in quantum mechanics see [136] and [137].

necessarily happen, and the reason is an effect discovered by Rudolf Mössbauer in 1958: in a significant fraction of decays in a properly prepared target and source, the emitting and absorbing atoms do not recoil at all. The reason for this can be understood if one thinks about the nature of the coherent states that can be formed from the radiation oscillators just discussed.

I will simplify as much as possible. Suppose the source and target are at absolute zero and the only motion of their atoms is zero-point fluctuations. Let the spinless quantized field (with $m = 0$) represent the motion of these atoms, analyzed in terms of pressure waves. The quanta of the field are phonons. Real sound waves in a solid have three components, one longitudinal and two transverse; we will pretend there is only one. Note that these quanta are not simply related to anything a single atom does, for each quantum involves the motion of every atom in the sample.

When a single atom recoils it excites oscillations in the field. Field oscillators originally in their lowest state are boosted into higher states by a sudden impact. The situation is exactly the one encountered in Sect. 2.6, where the result was a coherent state. The difference is that whereas the earlier coherent states were distributions in space, these are distributions in the amplitude of the quantized field. Again labelling a state of the field by its momentum vector k, define the average excitation of this state in a single process as $\overline{n}_k$. But this is only the average excitation. We saw in Sect. 2.6 that the probability of excitation of any state of a harmonic oscillator is given by a Poisson distribution around the average, and we can infer there is a certain probability that the mode of oscillation labelled k will not be excited at all. By (2.71) this probability is

$$P_{0k} = e^{-\overline{n}_k} \quad . \tag{10.104}$$

If no quanta are excited there is effectively no recoil, for the recoil momentum of the nucleus is absorbed by the sample as a whole. One would at first suppose that even if there is a finite probability that any given mode will not be excited, there is no probability that this will happen in every mode at once. That there is results from the discrete nature of the sample: it consists of a finite number of atoms and therefore it has a finite number N_f of degrees of freedom. In fact, if N atoms have only translational degrees of freedom, $N_f = 3N$, and since the value of N_f is independent of how it is counted, this is also the number of states as labelled by k. No damage is done if we replace sums by integrals. In elementary quantum mechanics one learns that if free-particle states are labelled by k, the number of them in an element dk is $V\,dk/(2\pi)^3$, where V is the volume of the enclosure. The number of states having the magnitude k between k and $k = dk$ is accordingly[2]

$$4\pi V k^2 dk/(2\pi)^3 = V k^2 dk/2\pi^2 \quad .$$

The largest value of k, call it k_m, is given by

$$\frac{V}{2\pi^2} \int_0^{k_m} k^2 dk = 3N \quad , \quad \text{or} \quad k_m = 3\left(\frac{2\pi^2 N}{3V}\right)^{1/3} \quad .$$

---

[2] This quadratic approximation, while valid for a continuous system like an electromagnetic field, is only qualitatively good in solid-state theory. Real distributions must have at least two peaks at which their slope becomes logarithmically infinite (see for example [138]).

This spherical distribution in momentum space is attached to the crystal lattice of the sample, and any movement of the sample moves the distribution along with it. If one atom emits a γ-ray of momentum $\hbar K$, vibrational modes will ordinarily be excited which shift the entire lattice around and it will finally recoil as a whole with infinitesimal speed. But we are interested in the situation in which it does not happen this way. Instead, the entire lattice recoils rigidly at once. If that is the case, the center of the spherical distribution bounded by the sphere k_m becomes displaced; we can handle this most conveniently by representing the state after recoil in the rest system of the recoiling lattice so that momentum is automatically conserved. Energy however, is not automatically conserved, and we must see what happens to it.

The recoiling atom of mass M gains an energy $\hbar^2 K^2/2M$ in addition to the zero-point energy it already had. Before the decay process each vibrational state was in its lowest mode and all the n_k were zero. Afterward, energy is distributed among the modes so that, on the average,

$$\frac{\hbar^2 K^2}{2M} = \sum_k \hbar v k \overline{n}_k$$

where v is the speed of sound in the sample and the energy of a phonon is $\hbar\nu = \hbar v k$. Let us assume that at least for small K^2, $\overline{n}_k$ is proportional to K^2. Write it as AK^2/Nk^2, where the k^2 is to keep $\overline{n}_k$ dimensionless and the reason for the N will be obvious in a moment. Then

$$\frac{\hbar K^2}{2M} = \frac{AK^2 v}{N} \sum_k \frac{k}{k^2} = \frac{AK^2 v}{N} \frac{V}{2\pi^2} \int_0^{k_m} k\, dk = \frac{9AK^2 v}{2k_m}$$

from which $A = \hbar k_m/9Mv$. The reason for the N in the definition of A is clear: it was so that the results will not depend on the size of the sample. Clearly, if the sample is large there will be many photon states close together, and the probability of excitation of any one of them is smaller.

The probability that the mode k is unexcited is $\exp(-\overline{n}_k)$ and since the excitations are statistically independent the probability P_0 that no mode is excited is given by the product of such terms. Its logarithm is

$$\ln P_0 = -\sum_k \overline{n}_k = -\frac{AK^2}{N} \frac{V}{2\pi^2} \int_0^{k_m} dk = -\frac{\hbar K^2}{Mvk_m} \quad .$$

This can be written in terms of more familiar quantities. The frequency corresponding to k_m is $\omega_m = vk_m$. Then

$$\ln P_0 = -\frac{(\hbar c K)^2}{Mc^2 \hbar \omega_m} \quad .$$

Also, in Debye's approximation, the specific heat of a solid depends on its Debye temperature Θ_D, defined by $k_B \Theta_D =: \hbar \omega_m$, where k_B is Boltzmann's constant. Finally, $\hbar c K$ is the gamma energy E_γ, so that

$$P_0 = \exp\left(-\frac{E_\gamma^2}{Mc^2 k_B \Theta_D}\right) \quad . \tag{10.105}$$

Since the kinetic energy E_R of the recoiling nucleus is $E_\gamma^2/2Mc^2$, this can also be written as

$$P_0 = \exp\left(-\frac{2E_R}{k_B \Theta_D}\right) . \tag{10.106}$$

A more exact treatment [139] taking into account the three kinds of phonons and the angular distribution of recoils replaces the 2 in this formula by $\frac{3}{2}$. At finite temperatures small compared with Θ_D the result is not much different from what we have calculated assuming absolute zero.

For experiments one looks for a heavy atom emitting a low-energy gamma in a hard material with high Θ_D. Mössbauer's first experiments were done with the 129-keV decay of Ir^{191}, for which the recoil energy E_R is 0.05 eV, comparable to the Debye temperature ($k_B \Theta_D = 0.025$ eV). The interest of the Mössbauer effect is that many nuclear energy levels are so narrow that in the absence of recoil extremely fine discrimination of frequencies is possible.

Problem 10.50. The beta decay of ^{57}Co produces an excited atom of ^{57}Fe that emits a 14.4-keV gamma ray. With what speed does the atom recoil? The width of the excited level is known to be 4.5×10^{-9} eV. Show that in the absence of the Mössbauer effect resonant absorption between a stationary source and a stationary target would never be seen.

Problem 10.51. A source of ^{57}Co moves toward a target made of ^{57}Fe at speed v. How large can v be before the Doppler effect makes resonance absorption impossible?

Problem 10.52. If none of the field oscillators are excited, something still has to happen to the energy of the recoiling atom. Where does it go? It may be helpful to think about a material oscillator that is hit but not excited.

10.10 Classical and Quantum Descriptions of Nature

As far as anyone knows, quantum mechanics is right, and one might suppose that classical mechanics is therefore wrong, or at best some approximation to the correct theory. Classical physics does not contain h and in the course of deriving it one must somewhere set h equal to zero. But this is not an unambiguous procedure. Take for example de Broglie's relation, as already noted in Chap. 1,

$$\lambda p = h$$

and let h approach zero. If we are not to become involved in trivial changes of units, this can be done in two ways:

p fixed	λ fixed
or	
$\lambda \to 0$	$p \to 0$.

The first leads, as we have shown several times, to classical mechanics; the second, to a classical theory of waves. It was possible for mechanics and wave optics to exist side by side for more than a century without anybody except Hamilton realizing that they might be different limiting cases of the same fundamental theory. The relation between them could be perceived only by a mathematician, since the experiments invoked to support them had little to do with each other and were ordinarily described in quite different languages. To understand the connection of classical physics with quantum physics we must stop worrying about which one is right, and remember Niels Bohr's profound remark: the question physics can answer is not what nature *is*, but rather what can be said about it.

Classical mechanics and quantum mechanics are languages for talking about nature, and classical mechanics provides many of the terms used in quantum mechanics. At the same time, quantum mechanics warns us that the language of classical mechanics is deceptive if we take it too literally, for there are limits to its applicability in microphysics. These limits are embodied in the indeterminacy relations, and how to express physics in ordinary language without transgressing them is the subject of Bohr's principle of complementarity. It is very confusing to explain indeterminacy in terms of "the perturbation of physical quantities by the act of measurement", since the "physical quantities" about to be perturbed can never be directly known to us in any way and hence it is appropriate to think of them as belonging more to our conceptual apparatus than to nature as it "really is".

Much of this book has been concerned with the dynamics of particles. What is a particle? In classical physics the word refers to a massive object, situated at a point or occupying a very small volume, following an exactly definable trajectory in space and time that is determined by external fields. To the student of quantum mechanics the concept of particle may seem weakened and vague, but that is because one must not try to fit a classical concept into another theory in which its role is quite different. "In its essence, it is not a material particle in space and time but, in a way, only a symbol on whose introduction the laws of nature assume an especially simple form." [Ref. 140, p. 62.] If one understands this one can stop worrying about which slit a photon passes through in a diffraction experiment.

The laws of nature themselves are clear and economical descriptions from which one can predict what will happen in experiments that have not yet been performed. By what sometimes seems a miracle, these descriptions are sometimes expressed in such general and universal mathematical language that we can feel as Galileo did when he wrote that in contemplating them we are perceiving the world as God himself perceives it. But these descriptions are made by and for the contemplation of man. They derive from experience, but not in a logical or necessary way, and thus they are not unique; different ones have value for different purposes. This is especially evident when two theories overlap, when they offer a choice between different ways of describing the same set of facts. Such is often the situation with the subjects discussed in this book. Three languages are used here: those of classical optics (*phase, wavelength, ...*); classical mechanics (*position, momentum, mass, ...*); and quantum mechanics, which uses a larger and more mathematical vocabulary. And, of course, in the classical mechanics of elastic media there are also terms of the classical wave theory from which, by analogy, the wave terminology of optics and quantum physics are derived. One does not *derive* one language from another, but one can make accurate translations when the subject is one which can be discussed adequately in both, and it often helps our thoughts to do so.

Classical mechanics talks about a world in which particles are things that can be described in an objective manner. Quantum mechanics typically answers the question: "If I measure a certain physical system using a certain experimental procedure, what is the probability that I will obtain a certain answer?" There is a long and imposing history of debate on how the word "probability" as used here should best be defined [141, 142], but most physicists take it as a primitive concept that can be explained only by giving examples. Probability is involved with any empirically based statement of fact, for one can always be mistaken. I think that outside of purely deductive systems like mathematics (and classical dynamics belongs there) a careful person defining difficult words like fact or truth must speak of probabilities. Probability is also involved, in a somewhat different way, in the arguments that interpret quantum mechanics, but even knowing this, one often says for example that the ground-state energy of a hydrogen atom "is" $-13.6\,\text{eV}$. We know enough about physics to know that we will rarely if ever get into trouble saying such a thing, but the language is not that of quantum mechanics and cannot be derived from it; it is simply borrowed, for convenience, from classical physcis.

It is useful in physics to use words like wave and particle in their classical meanings in order to help explain phenomena which really need quantum mechanics to explain them. Here, in a new situation, words from the old languages are not to be used literally but rather as metaphors. In quantum theory the wave is not a wave in any physical medium but only in a certain mathematical function, and a particle is not a thing. We must always remember that they are metaphors expressed by words which in other contexts have exact and literal meanings. It is by this linguistic process that new theories are formulated and progress is made, and the reader of this book has had an opportunity to perceive not only the close formal relationships between the equations of the old theories and those of the new, but also the vastly different languages in which they describe the world.

References

0   Park, D.: *Classical Dynamics and Its Quantum Analogues*, Lecture Notes in Physics, Vol. 110 (Springer, Berlin, Heidelberg 1979)
1   Park, D.: *Introduction to the Quantum Theory* (McGraw-Hill, N.Y., 2nd ed. 1974)
2   Arnold, V.I.: *Mathematical Methods of Classical Mechanics* (Springer, Berlin, Heidelberg 1978, 1989)
3   Liouville, J.: J. Maths. pures appl. **2**, 16 (1837)
4   Arnaud, J.A.: Am. J. Phys. **42**, 71 (1974)
5   Hamilton, W.R.: *Mathematical Papers*, Vol. 2 (Cambridge University Press 1940)
6   Landau, L.D., E.M. Lifshitz: *Mechanics* (Pergamon, London, 2nd ed. 1969)
7   Shirley, J.W.: Am. J. Phys. **19**, 507 (1951)
8   Goldstine, H.: *History of the Calculus of Variations from the 17th Through the 19th Century* (Springer, Berlin, Heidelberg 1980)
9   Einstein, A.: Ann. Physik **49**, 769 (1916)
10 Einstein, A. et al.: *The Principle of Relativity* (Methuen, London 1923 and many later editions)
11 Robertson, D.S., W.E. Carter: Nature **310**, 572 (1984)
12 Fomalont, E.B., R.A. Sramek: Phys. Rev. Lett. **36**, 1475 (1976)
13 Soldner, J.: Berl. Astron. Jahrb. *1804*, 161 (1901); see also Ann. Physik **65**, 593 (1920)
14 Reasenberg, R.D. et al.: Ap. J. **234**, L219 (1979)
15 Ehrenfest, P.: Z. Physik **45**, 455 (1927)
16 Huygens, C.: *Horologium oscillatorium* (Muguet, Paris 1673; reprod. Culture et Civilization, Bruxelles 1973)
17 Whittaker, E.T., G.N. Watson: *Modern Analysis* (Cambridge University Press, 4th ed. 1927)
18 Stoker, J.J.: *Nonlinear Oscillations* (Interscience, New York 1950)
19 Bogoljubov, N.N. et al.: *Methods of Accelerated Convergence in Nonlinear Mechanics* (Springer, Berlin, Heidelberg 1976)
20 Mathews, J., R.L. Walker: *Mathematical Methods of Physics* (Benjamin, New York, 2nd ed., 1970)
21 Abraham, R.H., C.D. Shaw: *Dynamics – The Geometry of Behavior* (Aerial Press, Santa Cruz 1982–85)
22 Schrödinger, E.: Naturw. **28**, 664 (1926)
23 Schrödinger, E.: *Collected Papers on Wave Mechanics* (Blackie, London 1928)
24 Perelomov, A.: *Generalized Coherent States and Their Applications* (Springer, Berlin, Heidelberg 1986)
25 Klauder, J.R., B.-S. Skagerstam: *Coherent States. Applications in Physics and Mathematical Physics* (World Scientific, Singapore 1985)
26 Minorsky, N. in Margenau, H., G.M. Murphy: *The Mathematics of Physics and Chemistry*, Vol. 2 (Van Nostrand, New York 1964)
27 Lagrange, J.L.: *Méchanique analitique* (Desaint, Paris 1788)
28 De Maupertuis, P.L.M.: *Essaie de cosmologie* (Walter, Dresden 1752; second ed. Bruyset, Lyon 1756)
29 Lanczos, C.: *The Variational Principles of Mechanics* (University of Toronto Press, 4th ed. 1970)
30 Newton, I.: *Sir Isaac Newton's Mathematical Principles of Natural Philosophy and System of the World* (University of California Press, Berkeley 1934)
31 Kepler, J.: *Astronomia nova* (Vögelin, Heidelberg 1609)
32 Kepler, J.: *Harmonices mundi* (Tampach, Linz 1619)
33 Bernoulli, J.: *Opera omnia* (Bousquet, Lausanne 1742)
34 De Laplace, P.S.: *Traité de mécanique céleste* (Duprat, Paris, 1798)
35 Pauli, W.: Z. Physik **36**, 336 (1926)
36 Gerjuoy, E.: Am. J. Phys. **17**, 477 (1949)

37   Bateman Project: *Higher Transcendental Functions,* 3 Vols. (McGraw-Hill, New York 1953)
38   Goldstein, H.: *Classical Mechanics* (Addison-Wesley, Reading, 2nd ed. 1980)
39   ter Haar, D.: *Elements of Hamiltonian Mechanics* (North-Holland, Amsterdam 1961)
40   Corben, H.C., P. Stehle: *Classical Mechanics* (Wiley, New York, 2nd ed. 1960)
41   Clausius, R.: Phil. Mag. **40**, 122 (1870)
42   Jackson, J.D.: *Classical Electrodynamics* (Wiley, New York, 2nd ed. 1975)
43   Lynch, R.: Am. J. Phys. **53**, 176 (1985)
44   Pauli, W.: Nuov. cim. **10**, 1176 (1953)
45   Wigner, E.P.: Phys. Rev. **77**, 711 (1950)
46   Hojman, S., R. Montemayor, Hadronic J. **3**, 1644 (1980)
47   Kennedy, F.J., E.H. Kerner: Am. J. Phys. **33**, 463 (1965)
48   Henneaux, M.: Ann. Phys. (N.Y.) **140**, 45 (1982)
49   Jordan, T.F.: *Linear Operators for Quantum Mechanics* (Wiley, New York 1969)
50   Bloore, F.J.: J. Phys. A: Math., Nucl, Gen. **6**, L7 (1973)
51   Lamb, Jr., W.E.: Phys. Rev. **85**, 259 (1952)
52   Yang, K.-H.: Ann. Phys. (N.Y.) **101**, 62 (1976)
53   Jacobi, C.G.J.: *Vorlesungen über Dynamik* (Reimer, Berlin 1884)
54   Dugas, R.: *A History of Mechanics* (Ed. Griffon, Neuchâtel 1955)
55   Courant, R., D. Hilbert: *Methods of Mathematical Physics,* Vol. 2 (Interscience, New York 1962)
56   Newcomb, S.: Wash. Astron. Papers **6**, 108 (1898)
57   Will, C., in Hawking, Israel, eds.: *Three Hundred Years of Gravitation* (Cambridge University Press 1987)
58   Shapiro, I.I., in Hegy, D., ed.: *Sixth Texas Symposium on Relativistic Astrophysics* (Ann. N.Y. Acad. of Sci. **124**, 1973)
59   Hénon, M., C. Heiles, Astron. J. **69**, 73 (1964)
60   McMillan, E.M., in Brittin and Odabasi, eds.: *Topics in Modern Physics* (Colorado Assoc. Univ. Press, Boulder 1971)
61   Ford, J.: Adv. in Chem. Phys. **24**, 187 (1973)
62   Gustavson, F.G.: Astron. J. **71**, 670 (1966)
63   Chang, Y.F. et al.: J. Math. Phys. **23**, 531 (1982)
64   Contopoulos, G., in N.R. Lebovitz, ed.: *Theoretical Principles in Astrophysics and Relativity* (University of Chicago Press 1978)
64a   Cherry, C.: Proc. Cambridge Phil. Soc. **22**, 287 (1924)
65   Whiteman, K.J.: Rep. Prog. Phys. **40**, 1033 (1977)
66   Arnold, V.I., A. Avez: *Ergodic Problems of Classical Mechanics* (Benjamin, New York 1968)
67   Arnold, V.I.: *Mathematical Methods of Classical Mechanics* (Springer, Berlin, Heidelberg 1978)
68   Chirikov, B.V.: Phys. Repts. **52**, 264 (1979)
69   Hammel, S.M., J.A. Yorke, C. Grebolgi, Bull. Am. Math. Soc. **19**, 465 (1988)
70   Jensen, R.V.: Amer. Scientist **75**, 168 (1987)
71   Percival, I., D. Richards: *Introduction to Dynamics* (Cambridge University Press 1982)
72   Zaslavskii, G.M., B.V. Chirikov: Sov. Phys. Uspekhi **14**, 549 (1972)
73   Ford, J., in E.G.D. Cohen, ed.: *Fundamental Problems in Statistical Mechanics III* (North-Holland, Amsterdam 1975)
74   Jorna, S., ed.: *Topics in Nonlinear Dynamics* (American Institute of Physics, New York 1978)
75   Hénon, M.: Quart. Appl. Math. **27**, 291 (1969)
76   Eisenhart, L.P.: Phys. Rev. **45**, 427 (1934); **74**, 87 (1948)
77   Morse, P.M., H. Feshbach: *Methods of Theoretical Physics* (McGraw-Hill, New York 1953)
78   Wilson, W.: Phil. Mag. **29**, 795 (1915)
79   Einstein, A.: Verh. Deutsch. phys. Ges. **19**, 82 (1917)
80   Percival, I.C.: Adv. Chem. Phys. **36**, 1 (1977)
81   Delos, J.B.: Adv. Chem. Phys. **65**, 161 (1986)
82   Noid, D.W., R.A. Marcus, J. Chem. Phys. **62**, 2119 (1975)
83   Sommerfeld, A.: *Atomic Structure and Spectral Lines* (Dutton, New York 1923)
84   Born, M.: *The Mechanics of the Atom* (Bell, London 1927)
85   Bohr, N., in [86] p. 95
86   Van der Waerden, B.L.: *Sources of Quantum Mechanics* (North-Holland, Amsterdam 1967)
87   Saletan, E.J., A.H. Cromer: *Theoretical Mechanics* (Wiley, New York 1971)
88   Vanderwoort, P.O.: Ann. Phys. (N.Y.) **12**, 436 (1961)
89   Slutskin, A.A.: Sov. Phys. JETP **18**, 676 (1964)
90   Chandrasekhar, S.: *Principles of Stellar Dynamics* (University of Chicago Press 1942)

91   Merzbacher, E.: *Quantum Mechanics* (Wiley, New York, 2nd ed. 1970)
92   Maslov, V.: *Théorie des perturbations* (Dunod, Paris 1972)
93   Leopold, J.G. et al.: J. Phys. **B13**, 1025 (1980)
94   Born, M., W. Heisenberg: Z. Physik **16**, 229 (1923)
95   Pauli, W.: Ann. Phys. **68**, 177 (1922)
96   Heisenberg, W.: Z. Physik **33**, 879 (1925); in [86]
97   Born, M., P. Jordan: Z. Physik **34**, 858 (1925)
98   De Broglie, L.: Comptes rendus **177**, 107, 148 (1923)
99   De Broglie, L.: Ann. Physique **3**, 22 (1925)
100 Schrödinger, E.: Ann. Physik **81**, 109 (1926); in [23]
101 Schrödinger, E.: Ann. Physik **79**, 734 (1926); in [23]
102 Feynman, R.P.: Revs. Mod. Phys. **20**, 367 (1948)
103 Feynman, R.P., A.R. Hibbs: *Quantum Mechanics and Path Integrals* (McGraw-Hill, New York 1965)
104 Dirac, P.A.M.: Phys. Z. Sowjetunion **3**, 64 (1933); in [105]
105 Schwinger, J., ed.: *Quantum Electrodynamics* (Dover, New York 1958)
106 Pauli, W.: *Lectures on Physics*, Vol. 6 (MIT Press, Cambridge 1973)
107 Blokhintsev, D.I., B.M. Barbasov: Sov. Phys. Uspekhi **15**, 193 (1972)
108 DeWitt, C. et al.: Phys. Repts. **50**, 255 (1980)
109 Mannheim, D.: Am. J. Phys. **51**, 328 (1983)
110 Marinov, M.S.: Phys. Repts. **60**, 1 (1980)
111 Faddeev, L.D., A.A. Slavonov: *Gauge Fields* (Benjamin/Cummings, Reading 1980)
112 Cheng, K.-S.: J. Math. Phys. **13**, 1723 (1972); **14**, 980 (1973)
113 Danby, J.M.A.: *Celestial Mechanics* (Macmillan, New York 1962)
114 Berry, M.V.: Proc. R. Soc. Lond. A**392**, 45 (1984)
115 Boersch, H. et al.: Z. Physik **165**, 79 (1961)
116 Konopinski, E.J.: *Classical Descriptions of Motion* (Freeman, San Francisco 1969)
117 Feynman, R.P.: *"Surely You're Joking, Mr. Feynman!"* (Norton, N.Y. 1985)
118 Andrews, F.C.: *Equilibrium Statistical Mechanics* (Wiley, N.Y., 2nd ed. 1975)
119 Bentley, M.A. et al.: Phys. Rev. Lett. **59**, 2141 (1987)
120 Landau, L.D., E.M. Lifschitz: *Quantum Mechanics* (Addison-Wesley, Cambridge 1958)
121 Casimir, H.B.G.: *Rotation of Rigid Body in Quantum Mechanics* (Wolter, Groningen 1931)
122 Arfken, G.: *Mathematical Methods for Physicists* (Academic Press, N.Y., 3rd ed. 1985)
123 Sommerfeld, A.: *Mechanics* (Academic Press, N.Y. 1952)
124 Klein, F.: *Mathematical Theory of the Top* (Scribner, N.Y. 1896)
125 Whittaker, E.T.: *A Treatise on the Analytical Dynamics of Particles and Rigid Bodies* (Cambridge University Press, 1937)
126 Bolker, E.D.: Am. Math. Monthly **80**, 977 (1973)
127 Hartung, R.W.: Am. J. Phys. **47**, 900 (1979)
128 Klein, A.G., G.I. Opat: Phys. Rev. D**11**, 523 (1975); Phys. Rev. Letts. **37**, 238 (1976)
129 van der Waerden, B.L.: *Group Theory and Quantum Mechanics* (Springer, Berlin, Heidelberg 1974)
130 Gottfried, K., V.F. Weisskopf: *Concepts of Particle Physics*, 2 Vols. (Oxford University Press 1984–86)
131 Park, D.: *Introduction to Strong Interactions* (Benjamin, N.Y. 1966)
132 Yourgrau, W., S. Mandelstam: *Variational Principles in Dynamics and Quantum Theory* (Saunders, Philadelphia, 3rd ed. 1968)
133 Abramowitz, M., I.A. Stegun: *Handbook of Mathematical Functions* (Nat. Bur. of Standards, Washington 1964)
134 Sommerfeld, A.: *Partial Differential Equations in Physics* (Academic Press, N.Y. 1949)
135 Schrödinger, E.: Ann. Physik **79**, 361 (1926); in [23]
136 Susskind, L., J. Glogower: Physics (N.Y.) **1**, 49 (1964)
137 Carruthers, P., M.M. Nieto: Am. J. Phys. **33**, 537 (1965)
138 de Launay, J.: *Solid State Physics* **2**, 220 (1956)
139 Frauenfelder, H.: *The Mössbauer Effect* (Benjamin, N.Y. 1962)
140 Heisenberg, W.: *Philosophic Problems of Nuclear Science* (Pantheon, N.Y. 1952, repr. Fawcett 1966)
141 Jammer, M.: *The Conceptual Development of Quantum Mechanics* (McGraw-Hill, N.Y. 1966)
142 Jammer, M.: *The Philosophy of Quantum Mechanics* (Wiley, N.Y. 1974)

Tables of Formulas and Integrals

Dwight, H.B.: *Tables of Integrals and Other Mathematical Data* (Macmillan, New York 1934 and later eds.)

Gradshtein, I.S., I.M. Ryzhik: *Table of Integrals, Series, and Products* (Academic Press, New York 1980)

Peirce, B.O.: *A Short Table of Integrals* (Ginn, New York 1929 and later eds.)

Notation

a: half the major axis of an ellipse.

b: half the minor axis.

$ds^2 = \eta_{ij} dx^i dx^j$ or $g_{ij} dx^i dx^j$ (summed over i and j) = square of an infinitesimal interval of proper time.

E: energy.

g_{ij}: metric tensor in general relativity ($i, j = 0, \ldots 3$).

$\hbar := h/2\pi$.

$H(p, q) := \sum_n p_n \dot{q}_n - L$: Hamiltonian function.

$\mathcal{H}$: Hamiltonian spatial density.

$\mathfrak{H}$: Hamiltonian constructed of four-vectors.

I_{ij}: inertia tensor.

$\hat{J}_n$ ($n = 1, 2, 3$): angular momentum coordinates.

$J_\nu := \oint p_\nu dq_\nu$ integrated over one cycle of a periodic motion: phase integral or action variable.

$k := 1/\lambda$.

$\mathbf{k}$: vector giving wave number and direction of motion.

$\hat{K}_n$ ($n = 1, 2, 3$): angular momentum operators in body-fixed coordinates.

$L(q, \dot{q}, t)$: Lagrangian function.

$\mathcal{L}$: Lagrangian spatial density.

$\mathfrak{L}$: Lagrangian constructed of four-vectors.

L: angular momentum.

m: rest mass.

$p_n := \partial L/\partial \dot{q}_n$: nth generalized momentum.

$\mathfrak{p}_i := m\mathfrak{u}_i$: four-momentum.

q_n: nth generalized coordinate.

s: arc-length in spacetime, see ds^2.

$S(q, t) := \int L(q, \dot{q}, t)dt$: Hamilton's principal function.

T: period of a periodic motion.

$U(q, \beta) := S(q, \alpha) - \sum \alpha_n \beta_n$: orbital function.

u_i (i=1,2): components of the first-rank spinor u.

$u(x, t)$: field variable.

$\hat{u}$: field operator corresponding to u.

$\mathrm{u}_i := dx^i/ds$: four-velocity.

v: an index or subscript not to be summed over.

$W(q) := S(q, t) + Et$: Hamilton's characteristic function for an autonomous system.

W: the virial, Wien's constant.

x^i: ordinary (contravariant) component of the vector x.

$x_i := g_{ij}x^j$: covariant component of the vector x.

$\mathbb{Z}$: the integers; $n \in \mathbb{Z}$ means $n = 0, 1, 2, \ldots$

$\mathbb{Z}_+$: the positive integers.

α: "upward" basis spinor for spin 1/2.

α_n: first integrals of the Hamilton-Jacobi equation, representing constants of the motion.

β: "downward" basis spinor for spin 1/2.

β_n: second integrals of the Hamilton-Jacobi equation, representing initial conditions.

η_{ij}: metric tensor for Minkowski spacetime ($i, j = 0, \ldots 3$), with $\eta_{00} = 1$, $\eta_{11} = \eta_{22} = \eta_{33} = -1/c^2$.

λ: constant Lagrange multiplier; wavelength.

$\lambdabar := \lambda/2\pi$.

$\mu(t)$ or $\mu(s)$: variable Lagrange multiplier.

ϱ: density.

σ_n ($n = 1, 2, 3$): Pauli spin matrices.

$\sum'$ is a summation with one or more terms omitted.

ϕ: a phase or an angle; velocity potential.

φ: a function of x, y, z, such as a time-independent wave function.

χ: an angle.

ψ: an angle; a time-dependent wave function.

$a_{,i} := \partial a/\partial x_i$

$a_t := \partial a/\partial t$, $a_{tt} := \partial^2 a/\partial t^2$, etc.

$V_{i,j} : \partial V_i/\partial x_j$

$\hat{a}$: operator representing a dynamical variable called a (the $\,\hat{}\,$ is often omitted).

$[a, b] := ab - ba$ (commutator of a and b).

a^*: complex conjugate of a (number, function, matrix, or operator).

a^{T}: transpose of matrix a.

$a^\dagger := a^{*\mathrm{T}}$: adjoint of matrix or operator a.

$\hat{a}_{\mathrm{H}}(t)$: operator $\hat{a}$ in the Heisenberg representation.

$\overline{A}$: time average of classical variable A.

$\langle a \rangle$ or $\langle \hat{a} \rangle$: expectation value of quantum variable a or the operator $\hat{a}$ that represents it (the two are numerically the same).

$|A\rangle$: a vector in Hilbert space defined as an eigenvector of the operator $\hat{A}$ (for example): $\hat{A}|A\rangle = a|A\rangle$. Where there is no ambiguity $|A\rangle$ may also be used for the expression that gives the eigenvector in some particular representation.

tr A (trace of the matrix A_{ij}): $= \sum_i A_{ii}$.

$|A|$: determinant of the matrix A_{ij}.

$(A(p,q), B(p,q)) := \sum_n \left(\frac{\partial A}{\partial q_n} \frac{\partial B}{\partial p_n} - \frac{\partial A}{\partial p_n} \frac{\partial B}{\partial q_n} \right)$ (Poisson bracket of A and B).

$\mathbf{1}$: unit matrix or operator.

$:=$ means "is defined as".

$\oint$: integral around a whole cycle.

$\frac{\delta L(q,\dot{q})}{\delta q_n} := \frac{\partial L}{\partial q_n} - \frac{d}{dt} \left(\frac{\partial L}{\partial \dot{q}_n} \right)$ (called a functional derivative).

Summation convention: in expressions such as the definition of H given above, when the summation index occurs in two terms of a product, the summation sign is often omitted: $p_n \dot{q}_n := \sum_n p_n \dot{q}_n$.

$\varheta$: the planet Mercury.

$\oplus$: Earth.

$\mathrm{2\!\!\!\downarrow}$: Jupiter.

Subject Index

Action and angle variables 240
Action integral 155
Action principle 56, 79, 219
— continuous system 292
— general relativity 151, 156
— special relativity 109
Adiabatic theorem
— classical mechanics 192
— quantum mechanics 232
Aharonov-Bohm effect 222
Angular velocity and momentum 253
Arc-sines 32
Atmospheric refraction 180

Bernoulli, Johann 12, 15, 73
Berry's phase 221
Bessel functions 305
Bohr orbits 77
Brachistochrone 15, 18
de Broglie, Louis, matter waves 207

Calculus of variations 14
— auxiliary conditions 19
Center of mass 82
Center of mass coordinates 84, 89
"Chaos" 166
Characteristic function (W) 144, 197
Coherent states 49
Completeness 159
Constraints, motion under 59
Coordinates
— center of mass 84, 89
— curvilinear 173
— cyclic 59, 103
— general relativity 22, 153
— generalized 55
— normal 95
— rotating 257
Correspondance limit 187, 199

Degeneracy 77, 234
Determinism 172
Dido's problem 21
Dynamics, general remarks 87, 211, 320

Ehrenfest's theorems 29
Eikonal equation 14
Einstein, Albert 27
— quantum conditions 189
Electron lens 111
Elliptic functions 35
— points 167
Equinoxes
— precession 154, 250
— calculated 269
Euler, Leonhard 17, 299, 305
— anticipation of Bessel functions 305
— equations for a rigid body 262
— summation formulas 298
Euler-Lagrange equations 17
— for continuous system 293
Expectation values and time averages 197

Fermat's principle 13, 25
— general relativity 155
Feynman, Richard, construction
    of wave function 212
Foucault, Léon
— gyrocompass 267
— pendulum 258

Galaxies, clusters 99
Galileo Galilei 15
Gas, theory 99
Gauge invariance 138
Group and phase velocity 6

Hamilton, William 13, 56
— equations 103
— principle 56, 213, 219
Hamiltonian function 104
Hamilton-Jacobi equation 143
— solution 148
Heisenberg, Werner, version
    of quantum mechanics 200
Heisenberg picture 47
Helium atom, ground state 312
Hénon-Heiles potential 164
Huygens principle 212, 219

Hydrodynamics 100
Hyperbolic points 167

Inertia tensor 255
Integral surfaces 160, 189
Integrals, deduction 132
Invariance 116, 120, 123, 124
— gauge 138

Jacobi, Carl, contribution
   to Hamilton-Jacobi theory 157
— identity 126, 132

KAM theorem 167
Kepler, Johannes, planetary laws 66
Kepler problem, *see* planetary motion
KS entropy 170

Lagrange, Joseph Louis 17, 54
— equations 54
Lagrangian function 55
— continuous system 292, 296
— field 314
— multipliers 19, 60
— special relativity 109
Laplace's integral 72
Larmor's theorem 261
Least time, *see* Fermat's principle
Light, gravitational deflection 25

Magnetic forces 107
Many-particle system 82
Mappings, area-preserving 169
Mathieu's equation 43
Matter field 310
de Maupertuis, Pierre-Louis 57
— principle 79, 157
Momentum, conjugate 59
— generalized 59
Motion, natural, *see* Path
Mössbauer effect 317

Newton, Isaac 15, 17, 67
Newton's laws
— first and second 87
— third 85
Noether's argument 116
Nuclei, *yrast* 278

Old quantum theory 184
— hydrogen atom 190
Optics
— in gravitational field 22, 157
— ray 2, 179
— wave 2

Orbital function (U) 208
Orbital motion
— hyperbolic 75
— planetary 64
— stability 70
— vectorial integral 72
Oscillator
— anharmonic 33, 40, 80, 226, 233
— coupled 93, 242
— driven 38
— quantized 46

Parameters, conjugate 130
Path, natural 116, 117, 126, 213
Pauli matrices 282
Pendulum, simple 33
— double 61, 92
Perturbations
— adiabatic 231
— canonical theory 244
— periodic and secular 226
— quantum mechanics 228
Phase and group velocity 6
Phase shift
— classical mechanics 220
— quantum mechanics 221, 229
Phase space 104, 167
— complex 241
Planetary motion 64, 149
— general relativity 151
— Newtonian precession 246
— special relativity 149
Poincaré tubes 164, 183
Poinsot construction 266
Poisson, Siméon 124
— theorem 127
Poisson brackets 124
— correspondence with commutator 24, 135
— properties 126
Principal function (S) 12, 115, 143
— relation with phase 12, 146, 214
Propagator 214, 218, 238

Quantized fields 313
— indeterminacy relations 316
Quantum conditions 186, 188, 194, 203
— Einstein's form 189
Quantum mechanics, relation
   with classical 31, 320

Refraction 11, 21 *see* Snell's law
Relativity
— general 22, 25, 151
— special 105, 109

Rigid-body motion 253
— quantum mechanics 274
Rotating coordinates 113, 257, 260
Rotator
— asymmetric 263
— symmetric 264
Round-off errors 171

Schrödinger, Erwin, derivation
  of quantum mechanics 210, 310
Schwarzschild metric 24
Separation constants 177
Shell's law 11, 15, 18
Sound waves 100, 307
Spinning particles 286
Spinors 279
— relation with vectors 281, 284
— rotations of 2π and 4π 283
— transformations 281
Stability 42, 70, 169
Stationary phase 8, 13, 208, 219
Stretched string
— motion 290, 297
— four descriptions 291
Sum over histories 214
Surface of section 164
Symmetry
— and conservation 59, 65, 103, 116
— and degeneracy 78
— dynamical 121, 134
— inertia tensor 256

Time
— represented by a function 131
— by an operator 132
Transformations
— canonical 117, 128
— finite by iteration 129
— gauge 138
— infinitesimal 121, 124
— parameter 130
— point 117, 119
— unitary 123

Variation principle
— quantum mechanics 311
— stretched string 300
— two dimensions 302
Velocity potential 101
Vibrating systems 90
Virial theorem 97

Water waves 8, 181
Wave equation
— one dimension 293
— polar coordinates 304
— in space 102, 307
Wave packets 7
— dynamics 10
Waves and rays 2
Wien's law 185
WKB approximation 3, 197, 199
— condition for validity 4

Zeno of Elea, paradox 212

B. N. Zakhariev, A. A. Suzko

Direct and Inverse Problems

Potentials in Quantum Scattering

1990. Approx. 200 pp. 42 figs. Softcover, in prep.
ISBN 3-540-52484-3

This textbook can almost be viewed as a "how-to" manual for solving quantum inverse problems, that is, for deriving the potential from spectra and/or scattering data. The formal exposition of inverse methods is paralleled by a discussion of the direct problem.
In part differential and finite-difference equations are presented side by side. A variety of solution methods is presented. Their common features and (dis)advantages are analyzed.
To foster a better understanding, the physical meaning of the mathematical quantities are discussed in detail. Wave confinement in continuum bound states, resonance and collective tunneling, and the spectral and phase equivalence of various interactions are some of the physical problems covered.

A. G. Sitenko

Scattering Theory

1990. Approx. 320 pp. 32 figs. Hardcover, in prep.
ISBN 3-540-51953-X

This mathematically rigorous introduction to nonrelativistic scattering theory addresses upper level undergraduates in physics. The relationship between the scattering matrix and physical observables is discussed in detail. Among the emphasized topics are the stationary formulation of the scattering problem, the inverse scattering problem, dispersion relations, three-particle bound states and their scattering, collisions of particles with spin and polarization phenomena. The analytical properties of the scattering matrix are discussed. Problems are included to help the reader to gain some experience and more expertise in scattering theory.

Springer-Verlag Berlin
Heidelberg New York
London Paris Tokyo
Hong Kong

Springer

N. G. Chetaev

Theoretical Mechanics

Translated from the Russian by I. Aleksanova

1990. Approx. 500 pp. 188 figs. Hardcover DM 68,–
ISBN 3-540-51379-5

Jointly published by Springer-Verlag
Berlin Heidelberg New York London Paris Tokyo Hong Kong
and MIR Publishers, Moscow, USSR

This university-level textbook reflects the extensive teaching experience of N. G. Chetaev, one of the most influential teachers of theoretical mechanics in the Soviet Union. The mathematically rigorous presentation largely follows the traditional approach, supplemented by material not covered in most other books on the subject. To stimulate active learning numerous carefully selected exercises are provided. Attention is drawn to historical pitfalls and errors that have led to physical misconceptions.

N. R. Sibgatullin

Oscillations and Waves in Strong Gravitational and Electromagnetic Fields

1990. Approx. 380 pp. Hardcover in prep.
ISBN 3-540-19461-4

This book emerged from a course given at Moscow State University and provides an introduction to current research in general relativity, relativistic gas dynamics, and cosmology, touching as well on the different methods used in wave theory. Each chapter begins with an elementary introduction and then proceeds to a more sophisticated discussion including a presentation of the current state of the art.

Springer-Verlag Berlin Heidelberg New York London Paris Tokyo Hong Kong

Springer